# THE THERAPEUTIC PROPERTIES OF MEDICINAL PLANTS

## Health-Rejuvenating Bioactive Compounds of Native Flora

*Innovations in Plant Science for Better Health: From Soil to Fork*

# THE THERAPEUTIC PROPERTIES OF MEDICINAL PLANTS

## Health-Rejuvenating Bioactive Compounds of Native Flora

*Edited by*

**Megh R. Goyal**

**Hafiz Ansar Rasul Suleria**

**Ademola Olabode Ayeleso**

**T. Jesse Joel**

**Sujogya Kumar Panda**

Apple Academic Press Inc.
4164 Lakeshore Road
Burlington ON L7L 1A4, Canada

Apple Academic Press Inc.
1265 Goldenrod Circle NE
Palm Bay, Florida 32905, USA



First issued in paperback 2021

*Exclusive worldwide distribution by CRC Press, a member of Taylor & Francis Group*
No claim to original U.S. Government works

ISBN 13: 978-1-77463-483-7 (pbk)
ISBN 13: 978-1-77188-803-5 (hbk)



**Library and Archives Canada Cataloguing in Publication**

Title: The therapeutic properties of medicinal plants : health-rejuvenating bioactive compounds of native flora/edited by Megh R. Goyal, Hafiz Ansar Rasul Suleria, Ademola Olabode Ayeleso, T. Jesse Joel, Sujogya Kumar Panda.

Names: Goyal, Megh Raj, editor. | Suleria, Hafiz, editor. | Ayeleso, Ademola Olabode, editor. | Joel, T. Jesse, editor. | Panda, Sujogya Kumar, editor.

Series: Innovations in plant science for better health.

Description: Series statement: Innovations in plant science for better health: from soil to fork | Includes bibliographical references and index.

Identifiers: Canadiana (print) 20190169990 | Canadiana (ebook) 20190170018 | ISBN 9781771888035 (hardcover) | ISBN 9780429265204 (ebook)

Subjects: LCSH: Materia medica, Vegetable. | LCSH: Medicinal plants—Research. | LCSH: Botanical chemistry—Research.

Classification: LCC RS164 .T54 2020 | DDC 615.3/21—dc23

**Library of Congress Cataloging-in-Publication Data**

Names: Goyal, Megh Raj, editor. | Suleria, Hafiz, editor. | Ayeleso, Ademola Olabode, editor. | Joel, T. Jesse, editor. | Panda, Sujogya Kumar, editor.

Title: The therapeutic properties of medicinal plants : health-rejuvenating bioactive compounds of native flora/edited by Megh R. Goyal, Hafiz Ansar Rasul Suleria, Ademola Olabode Ayeleso, T. Jesse Joel, Sujogya Kumar Panda.

Other titles: Innovations in plant science for better health

Description: Oakville, ON ; Palm Bay, Florida : Apple Academic Press, [2020] | Series: Innovations in plant science for better health: from soil to fork | Includes bibliographical references and index. | Summary: "This volume, The Therapeutic Properties of Medicinal Plants, provides some informative research on the scientific evidence of the health benefits that can be derived from medicinal plants and how their efficacies can be improved. The volume is divided into three sections covering the phytochemistry of medicinal plants, disease management with medicinal plants, and novel research techniques in medicinal plants. The pharmacological benefits of several specific plants, such as basil, fig, garlic, palm tree, etc., are discussed, addressing health issues including metabolic and mental disorders, acute mountain sickness, polycystic ovarian syndrome, and specific diseases such as Huntington's. It also looks at the role of antioxidants in disease management. Additionally, the book covers recent problems of drug resistance and how medicinal plants can serve as antibiotic, anthelmintic, and antiparasitic drugs that will be helpful for human and animals. Furthermore, it also covers novel approaches for the screening of plant-based medicines, extraction, toxicity and safety issues of essential oils, and nanoparticle-based delivery of plant metabolites. The findings reported in this book will be useful in health policy decisions and will help to motivate the development of new health care products from plants for health benefits. The book will further encourage the preservation of traditional medical knowledge of plants with therapeutic qualities. It will be a valuable reference for researchers, scientists, students, growers, traders, processors, industry professionals, dieticians, medical practitioners, and others"-- Provided by publisher.

Identifiers: LCCN 2019036017 (print) | LCCN 2019036018 (ebook) | ISBN 9781771888035 (hardcover) | ISBN 9780429265204 (ebook)

Subjects: LCSH: Materia medica, Vegetable. | Medicinal plants--Research. | Botanical chemistry--Research.

Classification: LCC RS164 .T522 2019 (print) | LCC RS164 (ebook) | DDC 615.3/21--dc23

LC record available at https://lccn.loc.gov/2019036017

LC ebook record available at https://lccn.loc.gov/2019036018

Apple Academic Press also publishes its books in a variety of electronic formats. Some content that appears in print may not be available in electronic format. For information about Apple Academic Press products, visit our website at **www.appleacademicpress.com** and the CRC Press website at **www.crcpress.com**

# OTHER BOOKS ON PLANT SCIENCE FOR BETTER HEALTH BY APPLE ACADEMIC PRESS, INC.

**Book Series:** *Innovations in Plant Science for Better Health: From Soil to Fork*
Editor-in-Chief: Hafiz Ansar Rasul Suleria, PhD

**Bioactive Compounds of Medicinal Plants: Properties and Potential for Human Health**
Editors: Megh R. Goyal, PhD, and Ademola O. Ayeleso

**Bioactive Compounds from Plant Origins: Extraction, Applications, and Potential Health Claims**
Editors: Hafiz Ansar Rasul Suleria, PhD, and Colin Barrow, PhD

**Human Health Benefits of Plant Bioactive Compounds: Potentials and Prospects**
Editors: Megh R. Goyal, PhD, and Hafiz Ansar Rasul Suleria, PhD

**Plant- and Marine-Based Phytochemicals for Human Health: Attributes, Potential, and Use**
Editors: Megh R. Goyal, PhD, and Durgesh Nandini Chauhan, MPharm

**Plant Secondary Metabolites for Human Health: Extraction of Bioactive Compounds**
Editors: Megh R. Goyal, PhD, P. P. Joy, PhD, and Hafiz Ansar Rasul Suleria, PhD

**Phytochemicals from Medicinal Plants: Scope, Applications, and Potential Health Claims**
Editors: Hafiz Ansar Rasul Suleria, PhD, Megh R. Goyal, PhD, and Masood Sadiq Butt, PhD

**The Therapeutic Properties of Medicinal Plants: Health-Rejuvenating Bioactive Compounds of Native Flora**
Editors: Megh R. Goyal, PhD, PE, Hafiz Ansar Rasul Suleria, PhD, Ademola Olabode Ayeleso, PhD, T. Jesse Joel, and Sujogya Kumar Panda

**The Role of Phytoconstitutents in Health Care: Biocompounds in Medicinal Plants**
Editors: Megh R. Goyal, PhD, Hafiz Ansar Rasul Suleria, PhD, and Ramasamy Harikrishnan, PhD

**Assessment of Medicinal Plants for Human Health: Phytochemistry, Disease Management, and Novel Applications**
Editors: Megh R. Goyal, PhD, and Durgesh Nandini Chauhan, MPharm

# ABOUT THE SENIOR EDITOR-IN-CHIEF

**Megh R. Goyal, PhD**

*Retired Professor in Agricultural and Biomedical Engineering, University of Puerto Rico, Mayaguez Campus; Senior Acquisitions Editor, Biomedical Engineering and Agricultural Science, Apple Academic Press, Inc.*

Megh R. Goyal, PhD, PE, is a Retired Professor in Agricultural and Biomedical Engineering from the General Engineering Department in the College of Engineering at the University of Puerto Rico–Mayaguez Campus; and Senior Acquisitions Editor and Senior Technical Editor-in-Chief in Agriculture and Biomedical Engineering for Apple Academic Press, Inc. He has worked as a Soil Conservation Inspector and as a Research Assistant at Haryana Agricultural University and Ohio State University.

During his professional career of 50 years, Dr. Goyal has received many prestigious awards and honors. He was the first agricultural engineer to receive the professional license in Agricultural Engineering in 1986 from the College of Engineers and Surveyors of Puerto Rico. In 2005, he was proclaimed as "Father of Irrigation Engineering in Puerto Rico for the Twentieth Century" by the American Society of Agricultural and Biological Engineers (ASABE), Puerto Rico Section, for his pioneering work on micro irrigation, evapotranspiration, agroclimatology, and soil and water engineering. The Water Technology Centre of Tamil Nadu Agricultural University in Coimbatore, India, recognized Dr. Goyal as one of the experts "who rendered meritorious service for the development of micro irrigation sector in India" by bestowing the Award of Outstanding Contribution in Micro Irrigation. This award was presented to Dr. Goyal during the inaugural session of the National Congress on "New Challenges and Advances in Sustainable Micro Irrigation" held at Tamil Nadu Agricultural University. Dr. Goyal received the Netafim Award for Advancements in

Microirrigation: 2018 from the American Society of Agricultural Engineers at the ASABE International Meeting in August 2018.

A prolific author and editor, he has written more than 200 journal articles and textbooks and has edited over 70 books. He is the editor of three book series published by Apple Academic Press: Innovations in Agricultural & Biological Engineering, Innovations and Challenges in Micro Irrigation, and Research Advances in Sustainable Micro Irrigation. He is also instrumental in the development of the new book series Innovations in Plant Science for Better Health: From Soil to Fork.

Dr. Goyal received his BSc degree in engineering from Punjab Agricultural University, Ludhiana, India; his MSc and PhD degrees from Ohio State University, Columbus; and his Master of Divinity degree from Puerto Rico Evangelical Seminary, Hato Rey, Puerto Rico, USA.

# ABOUT THE EDITORS

**Hafiz Ansar Rasul Suleria, PhD**

*Alfred Deakin Research Fellow,*
*Deakin University, Melbourne, Australia;*
*Honorary Fellow, Diamantina Institute*
*Faculty of Medicine, The University of*
*Queensland, Australia*

Hafiz Anasr Rasul Suleria, PhD, is currently working as the Alfred Deakin Research Fellow at Deakin University, Melbourne, Australia. He is also an Honorary Fellow at the Diamantina Institute, Faculty of Medicine, The University of Queensland, Australia.

Recently he worked as a postdoc research fellow in the Department of Food, Nutrition, Dietetic and Health at Kansas State University, USA.

Previously, he has been awarded an International Postgraduate Research Scholarship (IPRS) and an Australian Postgraduate Award (APA) for his PhD research at the University of Queens School of Medicine, the Translational Research Institute (TRI), in collaboration with the Commonwealth and Scientific and Industrial Research Organization (CSIRO, Australia).

Before joining the University of Queens, he worked as a lecturer in the Department of Food Sciences, Government College University Faisalabad, Pakistan. He also worked as a research associate in the PAK-US Joint Project funded by the Higher Education Commission, Pakistan, and the Department of State, USA, with the collaboration of the University of Massachusetts, USA, and National Institute of Food Science and Technology, University of Agriculture Faisalabad, Pakistan.

He has a significant research focus on food nutrition, particularly in the screening of bioactive molecules—isolation, purification, and characterization using various cutting-edge techniques from different plant, marine, and animal sources; and *in vitro, in vivo* bioactivities; cell culture; and animal modeling. He has also done a reasonable amount of work on functional foods and nutraceuticals, food and function, and alternative medicine.

Dr. Suleria has published more than 50 peer-reviewed scientific papers in different reputed/impacted journals. He is also in collaboration with more than ten universities where he is working as a co-supervisor/special member for PhD and postgraduate students and is also involved in joint publications, projects, and grants. He is Editor-in-Chief for the book series Innovations in Plant Science for Better Health: From Soil to Fork, published by AAP.

Readers may contact him at: hafiz.suleria@uqconnect.edu.au.

**Ademola Olabode Ayeleso, PhD**

*Senior Lecturer, Department of Biochemistry, Adeleke University, Nigeria*

Ademola O. Ayeleso, PhD, is a Senior Lecturer in the Department of Biochemistry at Adeleke University, Ede, Nigeria. Dr. Ayeleso is working on the use of medicinal plants, plant products, and synthesized compounds in the management of diabetes mellitus that has become a major threat to the lives of people in both urban and rural areas throughout the world. He has presented research papers at different international conferences. He has also worked as a Post-doctoral Research Fellow at three different universities in South Africa. Dr. Ayeleso is a prolific writer with over 20 publications to his credit.

Dr. Ayeleso obtained his BTech degree in Biochemistry at the Federal University of Technology, Akure, Nigeria. He received his MSc degree in Biochemistry at the University of Port Harcourt, Port Harcourt, Nigeria, and his doctoral degree in Biomedical Technology at the Cape Peninsula University of Technology, Bellville, South Africa.

**T. Jesse Joel, PhD**

T. Jesse Joel, PhD, is at present an Assistant Professor in Microbiology at the Biotechnology and Biosciences Department in the School of Agriculture and Biosciences at Karunya University, Coimbatore, India. He is an editorial board member for the International Research Journal of Biological Sciences and is also on the Scientific and Technical Committee on Bioengineering and Life Sciences. He teaches undergraduate and postgraduate courses on microbiology, cell biology, molecular biology, biochemistry, biomaterials and artificial organs, taxonomy and phylogeny, etc.

During his professional career of 13 years, he has received two Awards of Excellence for best paper and best lecture. Currently Dr. Joel is working in the field of nanoscience using green synthesis of nanoparticles from plant extracts, probiotics, and oral bioformulations as well as exploring the ameliorative effects of oral microorganisms. He has published several book chapters and has attended over 36 scientific events where he has chaired sessions and delivered guest lectures and lead talks in conferences and technical symposiums, conducted nationally as well as abroad. He also works as a consultant microbiologist on several projects.

He earned his BSc degree (Triple Major: in Microbiology, Chemistry, and Botany) from Andhra Loyola College, Vijayawada, India; his MSc degree in Applied Microbiology from K.S. Rangasamy College of Arts and Science, Tiruchengode, India, and his PhD degree in Microbiology from the Faculty of Engineering, Sam Higginbottom Institute of Agriculture, Technology and Sciences, Allahabad, India. He also received DMin from the Theological Extension Study Programme (TESPRO), Shillong, Meghalaya, India.

**Sujogya Kumar Panda, PhD**

*Animal Physiology and Neurobiology, KU Leuven, Belgium*

Sujogya Kumar Panda, PhD, was formerly a faculty member in the P.G. Department of Microbiology at M.P.C (Auto) College and in the Department of Biotechnology, North Orissa University, Odisha, India. He has an eclectic background that includes teaching, research, writing, and publishing.

During the last 10 years he has worked in the field of ethno-pharmacology, nanoparticles, infectious diseases, and pharmaco-epidemiology. Dr. Panda is a reviewer and an editorial board member for many professional journals. He has edited three books and published more than 40 research articles in national and international peer-reviewed journals. Dr. Panda has also received a Young Scientist Award from the Science & Technology Department, Government of India; a Erasmus-Mundus postdoc fellowship at KU Leuven, Belgium; and a CICOPS fellowship at the University of Pavia, Italy, to his continue research. At present he is continuing his postdoctorate in Animal Physiology and Neurobiology at KU Leuven, Belgium, and is engrossed in purifying the active compounds from Indian medicinal plants using bioassay guided purification.

# CONTENTS

# CONTRIBUTORS

**Yapo Guillaume Aboua**
Senior Lecturer, Department of Medical Laboratory Science, Faculty of Health and Applied Sciences, Namibia University of Science and Technology, Private Bag 13388 Windhoek Namibia, Mobile: +264-814225104, E-mail: yaboua@nust.na

**Abiola Fatimah Adenowo**
Lecturer, Department of Medical Biochemistry, Lagos State University, College of Medicine, PMB 21266, Ikeja, Lagos, Nigeria; Email: afaatimah@yahoo.com

**Idowu Jesulayomi Adeosun**
PhD Student, Adeleke University, P.M.B 250, Ede, Osun State, Nigeria, Mobile: +2348135209244, E-mail: adeosunidowu6@gmail.com

**Huma Bader-Ul Ain**
Research Associate, Institute of Home and Food Sciences, Government College University, Faisalabad, Pakistan, E-mail: humahums@yahoo.com

**Muhammad Umair Arshad**
Director, Institute of Home and Food Sciences, Government College University, Faisalabad, Pakistan, Mobile: +92-333-6575583, E-mail: umairfood1@gmail.com

**Ademola Olabode Ayeleso**
Senior Lecturer, Department of Biochemistry, Adeleke University, Ede, Nigeria, Mobile: +23408144556529,
E-mails: ademola.ayeleso@gmail.com, ademola.ayeleso@adelekeuniversity.edu.ng

**Olutoyin Omolara Bamigboye**
Associate Professor, Adeleke University, P.M.B 250, Ede, Osun State, Nigeria, Mobile: +234-9039301158, +234-8035724474, E-mail: toyinphd@gmail.com

**Kalpana Kumari Barwal**
Assistant Professor, All India Institute of Medical Sciences, Sijua-751019, Odisha, India, Mobile: +91-9438884026, E-mail: drkalpanabarhwal@hotmail.com

**Satpal Singh Bisht**
Professor, Department of Zoology, Kumaun University, Nainital, Uttarakhand, India

**Nicole Lisa Brooks**
Head, Department of Wellness Sciences, Faculty of Health and Wellness Sciences, Cape Peninsula University of Technology, PO Box 652, Cape Town, 8000, South Africa, Mobile: +278-2939922, E-mail: BrooksN@cput.ac.za

**Ibrahim Chikowe**
Lecturer, Department of Pharmacy, College of Medicine, P/Bag 360, Blantyre, Malawi, Mobile: +265-999328058, E-mail: ichikowe@medcol.mw

**Abhijit Dey**
Ethnopharmacology and Natural Product Research Laboratory, Department of Life Sciences, Presidency University, 86/1 College Street, Kolkata–700073, West Bengal, India, Mobile: +919903214237, E-mail: abhijitbio25@yahoo.com, abhijit.dbs@presiuniv.ac.in

**Sujatha Govindaraj**
Assistant Professor, Department of Botany, Periyar E. V. R. College (Autonomous), Tiruchirappalli-620023, Tamil Nadu, India, Mobile: +91-9894568526, E-mail: sujathagovindaraj@gmail.com

**Megh R. Goyal**
Retired Faculty in Agricultural and Biomedical Engineering from College of Engineering at University of Puerto Rico-Mayaguez Campus, and Senior Technical Editor-in-Chief in Agricultural and Biomedical Engineering for Apple Academic Press Inc., PO Box 86, Rincon–PR–006770086, USA, E-mail: goyalmegh@gmail.com

**Mayeso Gwedela**
Assistant Lecturer, Division of Physiology, College of Medicine, P/Bag 360, Blantyre, Malawi, Mobile: +265-995154446, E-mail: mgwedela@medcol.mw

**Sunil Kumar Hota**
Scientist 'D,' Defense Institute of High Altitude Research (DIHAR), Ministry of Defense-Government of India, C/O 56 APO, Leh-Ladakh-901205, Jammu and Kashmir, India, Mobile: +91-9463998315, E-mail: drsunilhota@hotmail.com

**Muhibah Folashade Ilori**
PhD Candidate, Department of Biochemistry, College of Medicine, University of Lagos, Idi-Araba, Mushin, Lagos, Nigeria; Email: muhibbahalade@yahoo.ca

**T. Jesse Joel**
Assistant Professor, Department of Biotechnology, School of Agriculture and Biosciences, Karunya Institute of Technology and Sciences, Karunya Nagar, Coimbatore - 641114, Tamil Nadu. India; Email: jessejoel@karunya.edu

**Kondwani Katundu**
Lecturer, Division of Physiology, College of Medicine, P/Bag 360, Chichiri, Blantyre 3, Malawi, Mobile: +265-999313978, E-mail: kkatundu@medcol.mw

**Mutiu Idowu Kazeem**
Lecturer, Antidiabetic Drug Discovery Group, Department of Biochemistry, Lagos State University, PMB 001, Ojo, Lagos, Nigeria, Mobile: +234-8030622000, E-mail: mikazeem@gmail.com

**Dadasaheb M. Kokare**
Assistant Professor, Department of Pharmaceutical Sciences, Rashtrasant Tukadoji Maharaj Nagpur University, Nagpur-440033, India, Mobile: +91 9850318502, E-mail: kokaredada@yahoo.com

**Kushal Kumar**
Fellow, Defense Institute of High Altitude Research (DIHAR), Ministry of Defense–Government of India, C/O 56 APO, Leh-Ladakh-901205, Jammu and Kashmir, India, Mobile: +91-9646633864, E-mail: kushal1kumar@gmail.com

**Fanuel Lampiao**
Associate Professor, and Deputy Director, Africa Centre of Excellence in Public Health and Herbal Medicine (ACEPHEM), College of Medicine, P/Bag 360, Blantyre Malawi, Mobile: +265-0995482713, E-mail: flampiao@medcol.mw

**Gail B. Mahady**
Associate Professor Director, Clinical Pharmacognosy Laboratory, University of Illinois at Chicago College of Pharmacy PAHO/WHO Collaborating Centre for Traditional Medicine Department of Pharmacy Practice, Chicago, IL-60612, USA, Mobile: +1-630931669, E-mail: gail.mahady@gmail.com

**Suman Kalyan Mandal**
Research Scholar, Department of Botany (UGC-DRS-SAP & DST-FIST Sponsored), Visva Bharati, Santiniketan-731235, West Bengal, India, Mobile: +91-9851248512, E-mail: skmandal.vb@gmail.com

**Simeon A. Materechera**
Professor, Indigenous Knowledge Systems, North-West University, Mafikeng Campus, Private Bag X2046, Mmabatho-2745, South Africa, E-mail: albert.materechera@nwu.ac.za

**Rojita Mishra**
Lecturer, Department of Botany, Polosara Science College, Polosara, Ganjam, Odisha–761105, E-mail: rojitamishra@gmail.com

**Boitumelo Rosemary Mosito**
PhD Candidate, Department of Wellness Sciences, Faculty of Health and Wellness Sciences, Cape Peninsula University of Technology, PO Box 1906 Bellville 7535, South Africa, E-mail: mositoboitumelo@yahoo.com

**Anuradha Mukherjee**
Anuradha Mukherjee, M.Sc., PhD Research Scholar, Assistant Teacher of Biological Sciences, Moriswar Motilal High School (MMHS), Joynagar 7433337, District South 24 Parganas, West Bengal, India; Email: anuradhamukherjee1980@gmail.com

**Emmanuel Mukwevho**
Professor, Diabetes and Obesity Therapeutics Research Group, Department of Biochemistry, North-West University, Mmabatho, South Africa, Mobile: +27713839197, E-mail:emmanuel.mukwevho@nwu.ac.za

**Banadipa Nanda**
Ethnopharmacology and Natural Product Research Laboratory, Department of Life Sciences, Presidency University, 86/1 College Street, Kolkata–700073, West Bengal, India, Mobile: +918240976279, E-mail: banadipa@gmail.com

**Samapika Nandy**
Ethnopharmacology and Natural Product Research Laboratory, Department of Life Sciences, Presidency University, 86/1 College Street, Kolkata–700073, West Bengal, India, Tel.: +9800203311, E-mail: samapika.nandy25@gmail.com

**Francis N. Nkede**
Postgraduate Student, Department of Botany and Plant Physiology. University of Buea, P.O Box 63, Buea, Cameroon, E-mail: nkedefrancis1@gmail.com

**Olugbenga Kayode Ogidan**
Senior Lecturer, Department of Electrical & Computer Engineering, Elizade University, P.M.B 002 Ilara Mokin, Ondo State, Nigeria, Mobile: +234-8035825897, E-mail: olugbengaogidan@gmail.com

**Ganiyat Abiola Oladunmoye**
Undergraduate Student, Department of Animal Production and Health, Federal University of Agriculture, PMB 2240, Abeokuta. Nigeria, Mobile: +234-7033277064, E-mail: ganiyatoladunmoye@gmail.com

**Abiodun Emmanuel Onile**
Assistant Lecturer, Department of Electrical & Computer Engineering, Elizade University, P.M.B 002 Ilara Mokin, Ondo State, Nigeria, Mobile: +234-8168305311, E-mail: abiodun.onile@elizadeuniversity.edu.ng

**Wilfred Otang-Mbeng**
Senior Lecturer of Botany, School of Biology and Environmental Sciences,
Faculty of Natural Sciences and Agriculture, University of Mpumalanga,
Mbombela Campus, P/bag X11283, Nelspruit, 1200, South Africa,
Mobile: +271-30020235, E-mail: wilfred.mbeng@ump.ac.za

**Shesan John Owonubi**
Post-Doctoral Researcher, Department of Chemistry, University of Zululand, KwaDlangezwa,
KwaZulu-Natal, South Africa, Mobile: +27849718848, E-mail: oshesan@gmail.com

**Chowdhury Habibur Rahaman**
Associate Professor, Department of Botany (UGC-DRS-SAP & DST-FIST Sponsored),
Visva Bharati, Santiniketan-731235, West Bengal, India, Mobile: +919434210136,
E-mail: habibur_cr@yahoo.co.in, habibur_cr@rediffmail.com

**Sujogya Kumar Panda**
Postdoctoral Researcher, Animal Physiology and Neurobiology Section, Department of Biology,
University of Leuven, Naamsestraat 59 - box 2465, 3000 Leuven, KU, Belgium;
Email: sujogyapanda@gmail.com; sujogya.panda@kuleuven.be

**Mahendra Rana**
Assistant Professor, Department of Pharmaceutical Sciences, Kumaun University Campus,
Bhimtal, Uttarakhand, India

**Nishikant A. Raut**
Assistant Professor, Department of Pharmaceutical Sciences,
Rashtrasant Tukadoji Maharaj Nagpur University, Nagpur-440033, India,
Mobile: +91 9422803768, E-mail: nishikantraut29@gmail.com

**Anindya Sundar Ray**
Research Scholar, Department of Botany (UGC-DRS-SAP & DST-FIST Sponsored),
Visva Bharati, Santiniketan-731235, West Bengal, India, Mobile: +919474311389,
E-mail: argharay90@gmail.com

**Neerish Revaprasadu**
Professor, Department of Chemistry, University of Zululand, KwaDlangezwa,
KwaZulu-Natal, South Africa, Tel: +27(0)359026137, E-mail: revaprasadun@unizulu.ac.za

**Farhan Saeed**
Assistant Professor, Institute of Home and Food Sciences, Government College University,
Faisalabad, Pakistan, E-mail: f.saeed@gcuf.edu.pk

**Hafiz Ansar Rasul Suleria**
Alfred Deakin Postdoctoral Research Fellow, Centre for Chemistry and Biotechnology,
School of Life and Environmental Sciences, Faculty of Science, Engineering and Built Environment,
Deakin University, 75 Pigdons Road, Geelong, VIC 3216, Australia,
Mobile: +61-470439670, E-mail: hafiz.suleria@uqconnect.edu.au

**Azeez Olanrewaju Yusuf**
Lecturer II, Department of Animal Production and Health, Federal University of Agriculture,
PMB 2240, Abeokuta, Nigeria, Mobile: +234-8036250504, E-mail: yusufao@funaab.edu.ng

# ABBREVIATIONS

| | |
|---|---|
| 2-AAF | 2-acetylaminofluorene |
| ABC | ATP binding cassette |
| ABPP | activity-based probe profiling |
| ABTS | 2,2'-azino-bis (3-ethylbenzothiazoline-6-sulphonic acid) |
| ACE | angiotensin converting enzymes |
| AD | Alzheimer's disease |
| ADH | antidiuretic hormone |
| AFB1 | aflatoxin B1 |
| AGEs | advanced glycation end-products |
| AIDS | acquired immune deficiency syndrome |
| ALAD | δ-aminolevulinic acid dehydratase |
| $AlCl_3$ | aluminum chloride |
| ALP | alkaline phosphatase |
| ALT | alanine aminotransferase |
| ALT | aspartate aminotransferase |
| AMI | age-induced memory impairment |
| AMP | adenosine monophosphate |
| AMS | acute mountain sickness |
| ANN | artificial neural network |
| ANP | atrial natriuretic peptide |
| ARE | antioxidant response element |
| ASOs | antisense oligonucleotides |
| ATBC | alpha-tocopherol beta-carotene |
| ATP | adenosine triphosphate |
| AVP | average precision |
| BaP | benzo[a]pyrene |
| BBB | blood-brain-barrier |
| BCA | bicinchoninic acid |
| BDNF | brain-derived neurotrophic factor |
| BHT | butylated hydroxytoluene |
| bZIP | basic-leucine zippe |
| C2C12 | immortalized mouse myoblast cell line |
| CAGE | cap analysis of gene expression |
| CAM | complementary and alternative medicines |

| | |
|---|---|
| CAMD | compacted air microwave distillation |
| CAMs | cell adhesion molecules |
| CAT | catalase |
| CCCP | compound centric chemical proteomics |
| CDRI | Central Drug Research Institute |
| CLMT | Chaihu-Jia-Longgu-Muli Tan |
| CNC | Cap'n'Collar |
| CNS | central nervous system |
| COMT | catechol-O-methyl transferase |
| COX-2 | cyclooxygenase 2 |
| CP | crude protein |
| CUS | chronic unpredictable stress |
| CV | computer vision |
| CVD | cardiovascular disease |
| CVS | chronic variable stress |
| CYP1A2 | cytochrome P450 1A2 enzyme |
| DDS | drug delivery system |
| DETAPAC | diethylenetriaminepentaacetic acid |
| DMAD | dimethylallyl diphosphate |
| DMBA | 7,12-dimethylbenz(a)anthracene |
| DMSO | dimethyl sulfoxide |
| DNA | deoxyribonucleic acid |
| DPPH | diphenyl picryhydrazyl |
| DSS | dextran sodium sulfate |
| DUFARMS | directorate of university farms |
| EDTA | ethylenediaminetetraacetic acid |
| EGCG | epigallocatechin gallate |
| EOs | essential oils |
| EPG | egg count per gram |
| ERE | electrophile response element |
| ERK | extracellular regulated kinases |
| ESBL | extended spectrum beta lactamase |
| ESC | embryonic stem cell |
| EST | embryonic stem cell test |
| FAB-MS | fast atom bombardment mass spectrometry |
| FAO | Food and Agriculture Organization |
| $FB_1$ | fumonisin B1 |
| FDA | Food and Drug Administration |
| FEC | faecal egg count |
| $FeSO_4$ | ferrous sulfate |

| | |
|---|---|
| FLBP | fuzzy local binary patterns |
| FRAP | ferric reducing antioxidant power |
| FSH | follicle stimulating hormone |
| FST | forced swim test |
| G1 phase | growth 1 phase |
| GABA | gamma amino butyric acid |
| GAD | generalized anxiety disorders |
| GAE | gallic acid equivalent |
| GC | gas chromatography |
| GCLC | glutamate cysteine ligase catalytic subunit |
| GDNF | glial cell derived neurotrophic factor |
| GGT | γ-glutamyl transferase |
| GIT | gastrointestinal tract |
| GLCM | gray level co-occurrence matrix |
| GLUT | glucose transporter |
| GPCRs | G-protein coupled receptors |
| GPx | glutathione peroxidase |
| GR | glutathione reductase |
| GUI | graphical user interface |
| $H_2O_2$ | hydrogen peroxide |
| HbAA | hemoglobin A |
| HbSS | hemoglobin S |
| HBV | hepatitis B virus |
| HCl | hydrochloric acid |
| HD | Huntington's disease |
| HD | hydrodistillation |
| HDAC | histone deacetylases |
| HDL-C | high-density lipoprotein cholesterol |
| HDL | high-density lipoprotein |
| HHA axis | hypothalamus-adrenocortical axis |
| HHT | homoharringtonine |
| HIV | human immunodeficiency virus |
| HMF | hydroxymethylfurfurals |
| HPA | hypothalamus-pituitary-adrenocortical |
| HPBMDM | human peripheral blood monocyte derived macrophages |
| HPLC | high performance liquid chromatography |
| HPV | human papilloma virus |
| HS-CRP | high sensitive C reactive protein |
| HUVEC | human umblical vein endothelial cells |
| IC | inhibition concentration |

| | |
|---|---|
| $IC_{50}$ | 50% median inhibition concentration |
| ICT | information communication technologies |
| IgM | immunoglobulin M |
| IL | interleukin |
| IL-10 | interleukin 10 |
| IL-6 | interleukin 6 |
| iNOS | inducible nitric oxide synthase |
| IPD | isopentenyl diphosphate |
| IPP | isopentenyl pyrophospahte |
| ISO | International Standard Organization |
| IVF | *in vitro* fertilization |
| kDa | Kilo Dalton |
| $KH_2PO_4$ | potassium phosphate |
| KV | kolaviron |
| LBP | local binary pattern |
| LBPV | local binary pattern variance |
| LC | liquid chromatography |
| LDH | lactate dehydrogenase |
| LDL-C | low density lipoprotein cholesterol |
| LH | luteinizing hormone |
| LLS | Lake Louise Score |
| LPO | lipid peoxidation |
| LPS | lipopolysaccharide |
| MAE | microwave-assisted extraction |
| MATLAB | matrix laboratory |
| MCV | mean cell volume |
| MDA | malondialdehyde |
| MDG | microwave dry-diffusion and gravity |
| MDR | multidrug resistant |
| MDR1 | multi drug resistance gene-1 |
| MedLeaf | medicinal leaf |
| mg/KgBW | milligram per kilogram body weight |
| MGDG | monogalactosyldiacyl glycerol |
| MHG | microwave hydrodiffusion and gravity |
| MIC | minimum inhibitory concentration |
| MILDA | microplate luminescence automated digital analyzer |
| MMC | mitomycin C |
| MNRET | micronucleated reticulocyte |
| MPA | metaphosphoric acid |
| MTT | 3-(4,5-dimethylthiazol-2-yl)-2,5-diphenyltetrazolium bromide |

| | |
|---|---|
| MudPIT | multidimensional protein identification technology |
| MUFAs | mono-unsaturated fatty acids |
| MySQL | my structured query language |
| NAC | N-acetylcysteine |
| NAD | nicotinamide adenine dinucleotide |
| NAD(P)H | nicotinamide adenine dinucleotide phosphate |
| NADH | nicotinamide adenine dehydrogenase |
| NADP(H) | nicotinamide adenine dinucleotide phosphate |
| NaOH | sodium hydroxide |
| *NB* | Naivy Bayes |
| NCCIH | National Center for Complementary and Integrated Health |
| NCI | National Cancer Institute |
| NEMPS | nuclear-encoded mitochondrial proteins |
| NEUT | neutrophils |
| NF-κβ | nuclear factor-κβ |
| NIRS | near infrared spectroscopy |
| NMR | nuclear magnetic resonance |
| NO | nitric oxide |
| NPs | nanoparticles |
| NRF | nuclear receptor factor |
| O-PA | orthophosphoric acid |
| OPG | oocyst per gram |
| ORAC | oxygen radical absorbance capacity |
| OSRC | Oxidative Stress Research Centre |
| PCA | principal component analysis |
| PCOS | polycystic ovarian syndrome |
| PD | Parkinson Disease |
| PDR | pan drug resistant |
| PDR | product decision rule |
| PGC-1 | peroxisome proliferator-activated receptor gamma coactivator-1 |
| PKA | protein kinase A |
| PLE | pressurized liquid extraction |
| PLGA | poly(lactic-co-glycolic) acid |
| PNN | probabilistic neural network |
| PNPs | polymeric nanoparticles |
| PPI | prepulse inhibition |
| PQQ | pyroloquinoline quinone |
| Prx-1 | peroxiredoxin-1 |
| RAAS | renin-angiotensin-aldosterone system |

| | |
|---|---|
| RAGE | receptor the advanced glycation endproducts |
| *RF* | random forest |
| RIN-5F | rat islet tumor cell line |
| RNA | ribonucleic acid |
| RNAi | RNA interference |
| ROS | reactive oxygen species |
| RPMI | Roswell Park Memorial Institute |
| S phase | synthesis phase |
| SAGE | serial analysis of gene expression |
| SCA | sickle cell anemia |
| SCD | sickle cell disease |
| SCF | supercritical fluid technology |
| SCT | sickle cell trait |
| SD | sprague dawley |
| SDH | succinate dehydrogenase |
| SFE | supercritical fluid extraction |
| SFME | solvent-free microwave extraction |
| *SL* | simple *logistic* |
| SLNs | solid lipid nanoparticles |
| SMEDDS | self-microemulsifying drug delivery system |
| SOD | superoxide dismutase |
| SRC | standard rat chow |
| SSI | Sense of Smell Institute |
| SSRIs | selective serotonin reuptake inhibitors |
| STAT3 | signal transduction as well as transcription 3 |
| STZ | streptozotocin |
| SVMRFE | support vector machine recursive feature elimination |
| TBA | thiobarbituric acid |
| TBAR | thiobarbituric acid reactive |
| TBARS | thiobarbituric acid reactive substances |
| t-BHP | tertiary-butyl hydroperoxide |
| TC | total cholesterol |
| TFA | trifluoroacetic acid |
| Tfam | mitochondrial transcription factor A |
| TG | triglycerides |
| THPs | traditional health practitioners |
| TLC | thin layer chromatography |
| TNF-α | tumor necrosis factor-α |
| TPA | 12-O-tetradecanoyl phorbol-13-acetate |
| TRAP | free radical trapping abilities |

| | |
|---|---|
| Txn-1 | thioredoxin-1 |
| UAE | ultrasound-assisted extraction |
| URF | University Research Fund |
| VEGF | vascular endothelial growth factor |
| VMHD | vacuum microwave hydro-distillation |
| VOCs | volatile organic compounds |
| WAD | West African Dwarf |
| WBC | white blood cells |
| WHCO5 | human oesophageal cancer cells |
| WHO | World Health Organization |
| XDR | extensive drug resistant |
| YGS | Yi-gan san |

# SYMBOLS

| | |
|---|---|
| μM | micromole |
| α-T | alpha-tocopherol |
| β-T | beta-tocopherol |
| γ-GT | γ-glutamyl transferase |
| γ-T | gamma-tocopherol |
| δ-T | delta-tocopherol |

# PREFACE

*To be healthy is our moral responsibility;*
*Towards Almighty God, ourselves, our family, and our society;*
*Eating fruits and vegetables makes us healthy;*
*Believe and have faith;*
*Reduction of food waste can reduce the world hunger*
*and can make our planet ecofriendly.*

**—Megh R. Goyal, PhD**
*Senior Editor-in-Chief*

Medicinal plants contain certain chemicals in their organs, such as leaves, stem, root, and fruits, which can provide therapeutic benefits against different kinds of diseases. These chemicals are often referred to as "phytochemicals." The word "phyto" is a Greek word, that means "plant." Phytochemicals are natural non-essential bioactive compounds found in plants/plant foods. Thousands of phytochemicals have already been identified, and more are still being discovered year by year. Plants that are used for medicinal purposes are often considered to be less toxic and induce fewer side effects than synthetic medicine. In our world today, many commercially available drugs have plant-based origins, with more than 30% of modern medicines directly or indirectly derived from medicinal plants. Indeed, plants can be a major source of pharmaceutical agents in the treatment of many life-threatening diseases.

The use of medicinal plants has largely increased because they are locally accessible, economical, as well as vital in promoting health. However, scientific data and information regarding the safety and efficacy of these medicinal plants are inadequate.

We introduce this book volume under the book series Innovations in Plant Science for Better Health: From Soil to Fork. This book mainly covers the current scenario of the research and case studies and contains scientific evidence on the health benefits that can be derived from medicinal plants and how their efficacies can be improved. The findings reported in this book can be useful in health policy decisions. It will also motivate the development of health care products from plants for health benefits. The book will further encourage the preservation of traditional medical knowledge of medicinal plants. Therefore, these plant products are drawing the attention of researchers and policymakers because of their demonstrated beneficial effects against diseases with high global burdens such as diabetes, hypertension, cancer, and neurodegenerative diseases.

This book volume is a treasure house of information and an excellent reference for researchers, scientists, students, growers, traders, processors, industries, dieticians, medical practitioners, and others. We hope that this compendium will be useful for the students and researchers as well as those working in the food, nutraceutical, and herbal industries.

The contributions by the cooperating authors to this book volume have been most valuable in the compilation. Their names are mentioned in each chapter and in the list of contributors. We appreciate you all for having patience with our editorial skills. This book would not have been written without the valuable cooperation of these investigators, many of whom are renowned scientists who have worked in the field of plant science and food science throughout their professional career.

The goal of this book volume is to guide the world science community on how plant-based secondary metabolites can alleviate us from various conditions and diseases.

We will like to thank editorial and production staff, and Ashish Kumar, Publisher and President at Apple Academic Press, Inc., for making every effort to publish this book when all are concerned with health issues.

We request the reader to offer your constructive suggestions that may help to improve the next edition.

We express our admiration to our families and colleagues for their understanding and collaboration during the preparation of this book volume. As an educator, We give a piece of advice to one and all in the world: "*Permit that our almighty God, our Creator, provider of all and excellent Teacher, feed our life with Healthy Food Products and His Grace—; and Get married to your profession.*"

—Editors

# PART I
# Phytochemistry of Medicinal Plants

# CHAPTER 1

# ANTIOXIDANT AGENTS FROM GREEN LEAFY VEGETABLES: A REVIEW

ABIOLA FATIMAH ADENOWO, MUHIBAH FOLASHADE ILORI, and MUTIU IDOWU KAZEEM

## ABSTRACT

Despite various reports on the biochemical importance of several green leafy vegetables abundant in Nigeria and other countries of sub-Saharan Africa, not enough research has been done on the utilization of these vegetables as nutraceuticals and drugs. It is highly recommended that further studies be carried out for:

1. Identification and isolation of the bioactive compounds responsible for the observed antioxidant properties of green leafy vegetables;
2. Utilization of isolated bioactive compound in the development of nutraceuticals and drugs to alleviate oxidative stress-related diseases and complications;
3. Utilization of bioactive component by the food manufacturing industry to retard oxidative degradation in foods in order to improve food quality and improve shelf-life.

## 1.1 INTRODUCTION

Green vegetables are herbaceous plant species, whose parts are eaten as auxiliary food or core dishes. The consumption of green leafy vegetables is a major aspect of cultural heritage that plays vital functions in the customs, traditions, and food culture of traditional households. Nigeria is bestowed with a diversity of traditional vegetables; and different ethnic groups consume different vegetables for various beneficial reasons [58].

Vegetables are reaped at different phases of growth and consumed either in fresh, processed, or semi-processed state by humans, whereas they are generally given in the fresh form to livestock. The nutritional constituents of vegetables differ substantially, but generally, they are not main sources of carbohydrates in comparison with starchy foods. Nevertheless, vegetables are packed with ample quantities of crude fiber, vitamins, minerals, carotene, and essential amino acids [38, 79].

Green leafy vegetables are useful for the preservation of health and prevention of diseases, due to treasured food nutrients necessary for body build-up as well as repair [1, 35]. Aside from rich nutritional values of vegetables, they are the cheapest and most abundant source of proteins due to their ability to produce amino acids from simple materials like water, carbon dioxide as well as atmospheric nitrogen [13, 20]. Aside from their low methionine content, most species of green leafy vegetables have amino acid profiles comparable with those of egg, fish, meat, and soybean; and their amino acid profile exceeds the Food and Agriculture Organization (FAO) stipulated pattern of essential amino acids [13].

Vegetables have low calories as well as insignificant amounts of utilizable energy; and therefore, these are ideal for obese individuals who can gratify their appetite without fear of accumulating calories. Additionally, vegetables are valuable in conserving alkaline reserve of the human body by acting as buffering mediators for acidic substances in the GI tract [4].

This chapter presents an overview of the antioxidant potential of green leafy vegetables and their role in the prevention and/or mitigation of oxidative stress-related diseases.

## 1.2 GREEN LEAFY VEGETABLES VERSUS OXIDATIVE STRESS-RELATED DISEASES

Numerous human degenerative conditions (such as cancer, atherosclerosis, *diabetes mellitus*, heart disease, stroke, ulcers, rheumatoid arthritis, osteoporosis, cataract, sunburn, and aging) have been documented as the outcome of damage by free radicals and reactive oxygen species (ROS) [3]. Several studies have also been undertaken on how to prevent or avoid the onset of such diseases. Nevertheless, the most possible and practical approach for combating degenerative ailments is by increasing the body antioxidant status, which can be realized by more intake of fruits and vegetables. Green leafy vegetables typically contain high amounts of natural antioxidants, which are able to scavenge free radicals [24, 25, 84].

The antioxidant action of leafy vegetables may be due to the presence of biocompounds like flavones, isoflavone, flavonoids, catechin, isocatechin, and anthocyanin rather than only vitamins C, E, and β-carotene [65, 70]. Dietary antioxidants are essential to manage ROS, which cause damage to the DNA, RNA, alter proteins, and affect lipid peroxidation (LPO) in the cells. Antioxidants can inhibit the commencement or proliferation of oxidative stress in organisms [64]. Researchers have shown a keen interest in the study of antioxidants for combating the deleterious consequence of free radicals; and natural products such as fruits and vegetables are in the limelight of such studies. Green leafy vegetables with reported antioxidant properties are discussed in this section.

### 1.2.1 TALINUM TRIANGULARE (WATER LEAF)

*Talinum triangulare* belongs to the family *Portulacaceae,* and originated from the tropical region of Africa, but it is extensively cultivated in Asia, South America and West Africa (especially in Nigeria) as a food crop [55]. It is greatly distributed in many ecological zones of Nigeria, where it is called '*Gbure*' in Yoruba, '*Nte-oka*' in Igbo and '*Alenyruwa*' in Hausa [64]. *T. triangulare* is a perennial plant and is popularly called 'waterleaf' due to its high moisture content of approximately 90.8mg/100mg of edible leaf [86]. It is an herbaceous plant with succulent stem and pink flowers, which is used as a sauce, flavoring as well as condiment in foods. Furthermore, it is utilized in folk medicine to alleviate diuretic ailments, gastrointestinal disorder, and edema [9, 19, 58].

This plant is rich in protein, essential oils (EOs), total lipids, cardiac glycosides, flavonoids as well as polyphenols. Phytochemical studies have revealed the occurrence of omega-3-fatty acids and copious amounts of essential minerals such as calcium, potassium as well as magnesium. It also contains soluble fibers like pectin as well as vitamins C; α-tocopherol, β-tocopherols, and β-carotene that are required for growth and development [54, 68].

Anyasor et al., [19] investigated the *in-vitro* antioxidant activity of aqueous and methanol extracts of *T. triangulare*. They showed that both extracts tested positive to rapid thin layer chromatography (TLC) screening for the presence of antioxidant activity. The color change upon spraying with diphenyl picrylhydrazyl (DPPH) (deep violet to yellow spots) suggested the presence of free radical scavengers. Notably, the free radical scavenging ability of the extracts was increased with increase in the concentration of the extract. Additionally, thiobarbituric acid reactive (TBAR) substances assays indicated that the extract prevented LPO at a concentration of 100 μg/ml. A similar study by Amorim et al., [16] showed that the stems of this vegetable have phenolic compounds possessing high antioxidant activity based on the results of 1,1-diphenyl-2-picrylhydrazyl (DPPH) scavenging activity.

Liang et al., [55] evaluated the antioxidant activity of polysaccharides from *T. triangulare*. The polysaccharides were extracted with boiling water and deproteinized using the Savage method. The polysaccharides demonstrated varied degrees of antioxidant activities in a dose-dependent fashion, which necessitated further studies towards its utilization for the management of oxidative stress-induced diseases. Methanol and hydro-ethanol extracts of *T. triangulare* were also studied for their $Fe^{3+}$ reducing ability and free radical scavenging activity. Results showed that both extracts demonstrated antioxidant activity, though the hydro-ethanol extract had stronger antioxidant properties [71].

### *1.2.2 TELFAIRIA OCCIDENTALIS (FLUTED PUMPKIN)*

*Telfairia occidentalis* (fluted pumpkin) belongs to the family *Cucurbitaceae*. Once harvested, the leaves are cautiously detached from the stems because the stems are regarded as poisonous and are thus thrown away as waste [31]. Fluted pumpkins occur chiefly in the forest zones of Central and West Africa, mostly in Cameroun, Republic of Benin and Nigeria. The leaves of this plant are widely consumed due to its acknowledged nutritional and medicinal benefits. In Nigeria, it is called by various traditional names such as 'Ugu' in Igbo, 'Aporoko/Iroko' in Yoruba, 'Ubong' in Efik and 'Umeke' in Edo language [51, 67].

The seeds are cooked and used as a protein supplement in a wide range of local foods. The seeds are also utilized as composite flours for the manufacture of bakery produce like bread and cookies [39]. Mineral elements in this plant include calcium, magnesium, phosphorus, sodium, and iron. Interestingly, the high iron content (approximately 700 ppm) scientifically authenticate the folk tradition of administering aqueous extract of the leaves as a blood tonic to convalescing individuals [51, 67]. *T. occidentalis* also contain phenols, flavonoids, alkaloids, oxalates, saponins, resins, and glycosides [67, 81]. Additionally, it is enriched with amino acids, including alanine, aspartate, cysteine, methionine, and phenylalanine [20, 51].

Oboh et al., [67] investigated the antioxidant activity of aqueous and ethanol extracts of *T. occidentalis*. Both extracts showed antioxidant activity, but the aqueous extract exhibited significantly higher reducing power (1.9 $O.D_{700}$) and free radical scavenging power (92%) than the ethanol extract. Several other researchers have equally established, with both *in vivo* and *in vitro* experiments, the ability of different extracts of *T. occidentalis* to prevent production of free radicals, scavenge free radicals, lower LPO as well as raised levels of antioxidant enzymes like catalase (CAT) and superoxide dismutase (SOD) [2, 46, 50, 63].

The recorded antioxidant potential of *T. occidentalis* might be attributed to the presence of bioactive constituents such as polyphenols, flavonoids, and vitamin C. An *in vivo* study showed the ability of aqueous leaf extract to enhance the oxidative status of the reproductive system of male Sprague-Dawley rats. The experimental rats were orally administered with 200, 400, and 800 mg/kg/day of freshly prepared aqueous extracts for 56 days (for spermatogenesis to take place). Thereafter, the animals were sacrificed, and the testicular oxidative status was assessed by measuring CAT activity, SOD activity, and glutathione peroxidase (GPx) activity. Results showed a dose-dependent (200 mg/kg/day gave best results) testiculo-protective ability of the aqueous extract of *T. occidentalis* [81].

### 1.2.3 GONGRONEMA LATIFOLIUM (UTAZI)

*Gongronema latifolium* (called 'Utazi' and 'Arokeke' in Igbo and Yoruba languages respectively in Nigeria) is a tropical rain forest plant predominantly used as a vegetable and spice. It is a member of the *Asclepiadaceae* family [32, 34]. This bitter-tasting green leafy vegetable is used traditionally in the management of anorexia, malaria, and nausea. Furthermore, the liquor of *G. latifolium* obtained after slicing the plant and boiling with lime extract or steeping in water for about three days is commonly taken as a purgative for colic, stomach upset and to treat signs associated with worm infections [35, 87].

Reports showed that this plant contains appreciable amounts of protein, fiber, saponins, EOs, flavonoids, and minerals (calcium, potassium, sodium, phosphorus, and cobalt). The amino acid profile revealed the plant is made up of both essential and non-essential amino acids and its pattern of amino acids is comparable with the World Health Organization (WHO) standards. Noteworthy is also the occurrence of high amounts of aspartic acid, glycine, and glutamic acid (13.8%, 10.3%, and 11.9%, respectively) [32, 33].

Fasakin et al., [37] studied the antioxidant activity of polyphenol extracts of *Gongronema latifolium in vitro*. Various assays (DPPH radical scavenging, chelation of metal ion, hydroxyl radical scavenging assay, superoxide scavenging assay, and ferric reducing activity) confirmed the antioxidant property of various extracts of *G. latifolium*. The authors suggested that the leaf extract can function as a prospective source of natural antioxidants, while further research is needed for identification and isolation of the bioactive compounds for utilization in the food industry [37]. Similarly, other studies also confirmed the antioxidant activity of aqueous and ethanol extracts of *G. latifolium* [14, 87].

### *1.2.4 VERNONIA AMYGDALINA (BITTER LEAF)*

*Vernonia amygdalina* belongs to the *Asteraceae* family, which grows in many African countries, including Cameroun, Zimbabwe, and Nigeria. It is commonly referred to as bitter leaf due to its bitter taste, which can be eliminated by continuous soaking in water and cooking. It is consumed as a vegetable food and in herbal preparations. Traditionally, it is used as a tonic, for tick control and in the treatment of constipation, fever, dysentery, hypertension, cough, and sexually transmitted diseases [36, 44]. Organic extracts of *V. amygdalina* have been established to possess cytotoxic activity against human carcinoma cells of nasopharynx as well as having antiparasitic and antimicrobial activities [36]. Numerous compounds with potent biological activities have been isolated from *V. amygdalina*, such as:

- Flavonoids;
- Anthraquinone;
- Steroids;
- Alkaloids;
- Glycosides;
- Luteolin;
- Luteolin 7-0-glucuronide;
- Luteolin 7-0-glucoside;
- Vernonioside [36, 47].

Johnson et al., [48] investigated the *in vitro* antioxidant activity of methanolic extract of *V. amygdalina*. The extract was established to possess DPPH scavenging activity of 96.65 μg/ml and Ferric reducing antioxidant power (FRAP) of 0.708. The results indicated that the extract possesses strong antioxidant activity [44]. Atangwho et al., [21] also reported a dose-dependent *in vitro* antioxidant activity of aqueous, methanol, chloroform as well as petroleum ester extracts of *V. amygdalina*.

Erasto et al., [34] also investigated the antioxidant effects (using reducing power assay and DPPH radical scavenging activity) of ethanol extract and two formerly isolated sesquiterpene lactones (vernodalol and vernolide) were reported. They showed that vernolide exhibited a higher reducing power than ethanol extract and vernodalol, while the ethanol extract showed higher radical scavenging activity. However, all three samples from *V. amygdalina* demonstrated appreciable antioxidant properties. Similarly, the antioxidant activity of boiled, cold, and methanolic extracts of *V. amygdalina* was estimated using DPPH free radical assay. Results confirmed the antioxidant action of this green leafy vegetable at various concentrations studied [43].

### 1.2.5 *OCIMUM GRATISSIMUM* (CLOVE BASIL OR SCENT LEAF)

*Ocimum gratissimum* is a perennial plant that is broadly distributed in the tropics of Africa as well as warm temperature regions. *O. gratissimum* (popularly called 'Scent leaf') is a traditional vegetable condiment used in Nigeria and some other countries to increase food flavor, as well as oral care products. This aromatic plant is widely used as expectorant and carminative, and traditionally employed in the treatment of epilepsy, rheumatism, paralysis, gonorrhea, diarrhea, influenza, and mental illness in India and Africa [22, 78]. It is used in south-eastern Nigeria in the management of umbilical cord of neonates to sustain sterility of the wound surfaces. Its roots are also used by Brazilian tropical forest inhabitants as part of the constituents of a decoction as a sedative for children [77]. Volatile aromatic oil from the leaves comprises principally of thymol and eugenol as well as xanthones, terpenes, and lactones, which possess antiseptics, antibacterial, and antifungal activities as well as an insect repellent [10].

The free radical scavenging ability of methanol extract of the leaves of *O. gratissimum* was evaluated by assessing its ability to scavenge 2,2-diphenyl-1-picrylhydrazyl, superoxide anion, hydroxyl, nitric oxide (NO) radicals, and its capability to inhibit LPO. It was concluded that the plant is a potential source of natural antioxidants that would be valuable in food manufacturing to retard oxidative degradation of lipids and thus increase food quality [22].

Akinmoladun et al., [10] showed that methanol extract of *O. gratissimum* had a DPPH scavenging activity of 84.6% at 250 μg/ml and a reductive potential of 0.77 at 100 μg/ml. These values were comparable with values obtained for standards, gallic acid, and ascorbic acid. These results affirm to the wide traditional use of the plant extract in the management of various human maladies. The essential oil of *O. gratissimum* has also been confirmed to possess good antioxidant activity through the DPPH radical scavenging activity and β-carotene-linoleic acid bleaching assay in a dose-dependent fashion [78]. This is an indication that the essential oil will be valuable in food processing and preservation.

### 1.2.6 CORCORUS OLITORIUS (JUTE)

*C. olitorius* (Jute) belongs to the *Tiliaceae* family. This annual herb is native to Africa. It is also cultivated in several parts of India and Bangladesh [40, 65]. Jute plant is a common traditional vegetable, which is prepared into a slimy soup or sauce in many of West African culinary traditions. In Western Nigeria, it is known as 'ewedu' while the Songhay people of Mali refer to it as 'fakohoy.' Furthermore, it is a common meal ingredient in the Northern province of the Philippines, where it is known as 'saluyot' and in Taiwan where it is boiled with sweet potatoes into a nourishing meal. The plant is rich in flavonoids, α-tocopherol, carotenoids, polyphenols, as well as vitamin C, iron, and calcium [65, 91]. It is popularly used in folk medicine for treating ailments like dysentery, bone pains, fever, gastroenteritis, and diabetes. The seeds are also used traditionally as a contraceptive [26, 40, 92].

Azuma et al., [23] identified six phenolic compounds from the leaves of *C. olitorus* using nuclear magnetic resonance (NMR) and Fast atom bombardment mass spectrometry (FAB-MS). The compounds are 5-caffeoylquinic acid (chlorogenic acid), 3,5-dicaffeoylquinic acid, quercetin 3-galactoside, quercetin 3-glucoside, and quercetin 3-(6-malonylglucoside). Thereafter, the content of these phenolic acids, α-tocopherol, and ascorbic acid was assessed, and their antioxidant potential was measured with radical generator-initiated peroxidation of linoleic acid. The results revealed that 5-caffeoylquinic acid was the major phenolic antioxidant in the leaves of *C. olitorus* [23].

Similarly, Oboh et al., [69] assessed the antioxidant activity of hydrophilic (water) and lipophilic (hexane) extracts of *C. olitorusin vitro*. Their study showed that both extracts exhibited significant antioxidant properties. However, the hydrophilic extract showed higher free radical scavenging ability, reducing power and trolox equivalent antioxidant capacity than the lipophilic extract, due to the higher content of total phenol, total flavonoid, and ascorbic acid compared with the lipophilic extract [69].

### 1.2.7 GNETUM AFRICANUM (UKAZI)

*Gnetum africanum* belongs to the family *Gnetaceae*, which is extensively distributed in Cameroun, Equatorial Guinea, Central Africa Republic, Gabon, and Nigeria. It is widely consumed in the South East region of Nigeria, where it is known as 'Afang' or 'Ukazi.' It is often cooked with water-leaves and also consumed as vegetable salad [15, 30].

Traditionally, *G. africanum* is used in the management of several illnesses including nausea, boils, enlarged spleen, and neutralization of some poisons and to reduce the pain during childbirth. It is also used for the treatment of hemorrhoids, diabetes, and as worm expeller [46, 76]. The leaves have been reported to have great nutritional value constituting a significant source of protein, minerals, and essential amino acids. Numerous molecular compounds associated to the families of stilbenes, flavonostilbenes, and glycosylflavones have been isolated and identified in the leaf extract of this plant. These compounds may be responsible for the interesting properties and biological activities of the plant [15, 29].

The antioxidant potential of both raw and cooked leaf extracts of *G. africanum* was investigated by Ogbonnaya and Chinedum [72]. The results for vitamin assay showed that both the raw and cooked leaf extracts had a significant content of vitamin C and E, and their concentrations were not affected by cooking. Analysis of the scavenging activities of both extracts using 2,2-diphenyl-1-picrylhydrazyl radical showed that both extracts had significant scavenging activity. The results obtained can be attributed to their polyphenolic contents, which possess varying levels of antioxidant activity and the hydrogen donating capacity of the OH groups of the phenolic compounds. Similar results were also reported by other researchers [6, 11].

### 1.2.8 PIPER GUINEENSE (AFRICAN BLACK PEPPER)

*Piper guineense* belongs to the family *Piperaceae*, and there are about 700 species spread across tropical and subtropical regions of the world. It is generally called African black pepper or hot leaf in various parts of Africa such as Ghana, Cameroon, and Nigeria, where it is consumed as a spice due to its nutritional and therapeutic properties. In Nigeria, it is known by various local names such as 'Uziza' in Igbo and 'Iyere' in Yoruba [18, 60]. In Nigeria, the seeds are used by women after childbirth to increase uterine contraction in order to ease the ejection of the placenta as well as other residues from the womb. The leaves are also used in the management of cough, male infertility, respiratory, and intestinal diseases. They also have carminative, appetitive, and eupeptic properties [7]. The antimicrobial, antiparasitic, antifungal, and insecticidal activities of the leaves and seeds have also been reported. Its phytoconstituents include:

- Alkaloids;
- Sterols;
- Flavonoids;
- Saponins;
- Glycosides [18, 60].

The hydro-ethanolic and ethanolic extracts of leaves and stems of the *P. guineense* were investigated for their free radical scavenging activities as well as antioxidant potential. The results showed that the extracts significantly inhibited the DPPH, $ABTS^{+}$, NO, and $OH^{-}$ radicals in a concentration-dependent manner. A significant ferrous ion chelating ability was also observed through FRAP and phosphomolybdenum antioxidant potential assays. The study also revealed that the polyphenol content of this plant differs subject to the form of extracts and the solvent used. However, the hydro-ethanolic and ethanolic extracts of leaves presented a higher level of phenolic compounds [60]. Hence further studies need to be done to isolate and characterize the bioactive compounds responsible for the observed action. Similarly, Agbor et al., [7] investigated the *in vitro* antioxidant potential of three Piper species including *P. guineense*, and their results showed that methanolic leaf extract of all species studied possess antioxidant activity in a dose-dependent manner [7].

### *1.2.9 CELOSIA ARGENTIA (PLUMED COCKSCOMB)*

*Celosia argentea* is a tropical herbaceous plant from the family *Amaranthaceae*, known for its attractive and distinctive brightly-colored flowers. It is commonly referred to as plumed cockscomb or the feathery amaranth. It is widely distributed in Southern Asia, China, and some parts of Africa. It is popularly consumed as a vegetable in the Western part of Nigeria, where it is known as 'red soko' due to red pigmentation on its leaves [49, 57, 90]. The seeds are very small and are used traditionally for the management of jaundice, fever, gonorrhea, and wounds; and for the treatment of itching, sores, ulcers, and fever. The plant has been studied for antibacterial, anti-inflammatory, diuretic, antipyretic, antidiarrheal and antidiabetic properties. Amino acids, carbohydrates, flavonoids, tannins, saponins, phytosterols, and glycosides are some of the reported phytochemicals present in the plant [54, 82].

Mahadik et al., [56] investigated the antioxidant activity of ethanolic extract of *C. argentea in vitro* via the reducing power assessment. They showed that the ethanolic extracts of *C. argentea* compared significantly with that of the standard antioxidant agent ascorbic acid. Combination of the extract and ascorbic acid also produced a synergistic effect. These results indicated that ethanolic extract of the plant can be utilized as natural antioxidant either alone or in combination with ascorbic acid. A similar study with methanolic leaf extracts of *Celosia argentea* using DPPH, NO, and hydrogen peroxide ($H_2O_2$) radical showed a concentration-dependent free radical scavenging property [89]. *In vivo* and *in vitro* studies revealed the antioxidant ability of various extracts of *C. argentea*, and the observed activity was attributed to the presence of phenolic compounds in the plant [41, 80].

### 1.2.10 SOLANUM MACROCARPON (EGGPLANT)

*Solanum macrocarpon* is a flowering herbaceous plant usually used for its nutritional content and medicinal properties. It is commonly cultivated in Western Africa countries, where it functions as a source of fruit and leafy vegetable [27, 52]. It is found in the southwestern region of Nigeria, where it is called igbagba or 'Igbo' [62, 66]. The leaves are packed with fat, protein, calcium, zinc, crude fiber, as well as appreciable amounts of the amino acid and methionine [52].

Besides the use of the leaves for culinary purpose, different parts of the plants are used traditionally for diverse purposes. In Sierra Leone, the leaves are heated and masticated to treat throat complications; while in Kenya, the liquor from boiled roots is drinking to purge out hookworms, whereas the crushed leaves are taken to alleviate stomach disorders [27, 83]. The fruits are eaten in Nigeria as purgatives and are also used in the management of heart disease, while the flowers are chewed to maintain oral hygiene. The roots are also used for treating wounds, body aches, bronchitis, itch, and asthma while the seeds are used to cure toothache. Several reports on aqueous extract of its fruits have shown its hemolytic, hepatoprotective, and hypolipidemic activities [27, 52, 62, 66].

Study on the free radical scavenging activity of the ethanolic leaves extract using the DPPH assay showed a dose-dependent activity, which may be attributed to its high phenolic content [52]. Similarly, Olajire and Azeez [74] investigated the total antioxidant activity of various commonly consumed vegetables in Nigeria. Their findings showed that *S. macrocarpon* has the best antioxidant properties, compared to other vegetables and standard antioxidant agent tocopherol. This was revealed by its lowest $IC_{50}$ value (6.21 mg $ml^{-1}$), compared to α-tocopherol ($IC_{50}$: 13.20 mg $mL^{-1}$). It also possesses the highest total phenolic, flavonoid as well as ascorbic acid contents compared with the other vegetables studied. Similar works also confirmed the ability of various extracts of *S. macrocarpon* leaves to alleviate oxidative stress due to its antioxidant activity [52, 59, 73].

### 1.2.11 *BASELLA ALBA* (VINE SPINACH)

*Basella alba* is a perennial plant, which belongs to the *Basellaceae* family. It is commonly called Malabar spinach, Red vine spinach or creeping spinach. It is generally found in the tropical countries of the world, including Nigeria, where it is common in the south-west region. It is composed of fat, proteins, vitamins A, B (folic acid), C, E, and K, as well as riboflavin, niacin, thiamine. It also has minerals, including iron, calcium, and magnesium. Studies showed that it has some unique constituents like betalain, basellasaponins, and kaempferol [5, 42, 75].

Various parts of the plant are used for several purposes in traditional medicines. In Nepal, the leaf extract is used in the treatment of cold and cough while the pastes are used externally in treating boils. The boiled leaves and stems are administered as laxatives, and the flowers are useful as an antidote for poisons [88]. The leaves and stem are also used in the management of diarrhea, dysentery, anemia, cough, cold, headaches, and ringworm. It also aids in the removal of placenta after childbirth as well as increase the flow of milk in nursing mother [5, 85].

Sridevi et al., [85] investigated the antioxidant potential of ethanolic leaf extract of *Basella alba* using *in vitro* assays and using 2,2-diphenyl-1-picryl hydrazyl (DPPH), reducing power assay, and phosphomolybdenum assay. The results showed that the extract exhibited significant free radical scavenging activity compared with the standard, gallic acid.

In an effort to study the antioxidant potential of ethyl acetate and chloroform extracts of *Basella alba* leaves, the phenol and flavonoid content of the extracts were measured by aluminum chloride assays and Folin Ciocalteu assay, respectively. Results showed that the total phenolic contents of the ethyl acetate and chloroform extracts were 0.029 mg/g and 0.030 mg/g, respectively, whereas the total flavonoids content were 0.045 mg/g and 0.085 mg/g, respectively.

Furthermore, both extracts showed radical scavenging activity in a concentration-dependent manner. Hence, *Basella alba* contains ample antioxidant agents and may be valuable for the formulation of drugs/nutraceuticals to combat oxidative stress-related disorders [75].

## 1.3 SUMMARY

This review summarizes some reported findings on the antioxidant potential of some commonly consumed green leafy vegetables in Nigeria. It focuses on the reported ability of various extracts of the studied green leafy vegetables to alleviate conditions associated with oxidative stress both *in vivo* and *in vitro*. Hence, these green vegetables are recommended for further study towards the development of nutraceutical and drugs for combating oxidative stress associated conditions.


## KEYWORDS

- **antioxidants**
- **bitter leaf**
- **clove basil**
- **epilepsy**
- **fluted pumpkin**
- **free radicals**
- **kaempferol**
- **omega-3 fatty acid**
- **oxidative stress**
- **plumed cockscomb**
- **polyphenols**
- **saponins**
- **superoxide dismutase**
- **nuclear magnetic resonance**
- **vernodalol**
- **waterleaf**

## REFERENCES

1. Acho, C. F., Zoue, L. T., Akpa, E. E., Yapo, V. G., & Niame, S. L., (2014). Leafy vegetables consumed in Southern Côte d'Ivoire: A source of high value nutrients. *Journal of Animal and Plant Science*, *20*(3), 3159–3170.
2. Adaramoye, O. A., Achem, J., Akintayo, O. O., & Fafunso, M. A., (2007). Hypolipidemic effect of *Telfairia occidentalis* (fluted pumpkin) in rats fed a cholesterol-rich diet. *Journal of Medicinal Foods*, *10*(2), 330–336.
3. Adefegha, S. A., & Oboh, G., (2011). Enhancement of total phenolics and antioxidant properties of some tropical green leafy vegetables by steam cooking. *Journal of Food Processing and Preservation*, *35*(5), 615–622.
4. Adeniyi, S. A., Ehiagbonare, J. E., & Nwangwu, S. C. O., (2012). Nutritional evaluation of some staple leafy vegetables in Southern Nigeria. *International Journal of Agriculture and Food Science*, *2*(2), 37–43.
5. Adhikari, R., Naveen, K. H. N., & Shruthi, S. D., (2012). A review on medicinal importance of *Basella alba. L. International Journal of Pharmaceutical Science and Drug Research*, *4*(2), 110–114.
6. Agbor, G. A., Moumbegna, P., Oluwasola, E. O., Nwosu, L. U., Njoku, R. C., Kanu, S., Emekabasi, E., Akin, F., Obasi, A. P., & Abudei, F. A., (2011). Antioxidant capacity of some plants foods and beverages consumed in the eastern region of Nigeria. *African Journal of Traditional Complementary and Alternative Medicine, 8*(4), 362–369.
7. Agbor, G. A., Vinson, J. A., Oben, J. E., & Ngogang, J. Y., (2008). *In vitro* antioxidant activity of three piper species. *Journal of Herbal Pharmacotherapy*, *7*(2), 49–64.
8. Agbor, G. A., Vinson, J. A., Sortino, J., & Johnson, R., (2012). Antioxidant and anti-atherogenic activities of three Piper species on atherogenic diet fed hamsters. *Experimental and Toxicologic Pathology*, *64*(4), 387–391.
9. Aja, P. M., Okaka, A. N. C., Ibiam, U. A., Uraku, A. J., & Onu, P. N., (2010). Proximate analysis of *Talinum triangulare* (Water Leaf) leaves and its softening principle. *Pakistan Journal of Nutrition*, *9*(6), 524–526.
10. Akinmoladun, A. C., Ibukun, E. O., Afor, E., Obuotor, E. M., & Farombi, E. O., (2007). Phytochemical constituent and antioxidant activity of extract from the leaves of *Ocimum gratissimum*. *Science Research and Essay*, *2*(5), 163–166.
11. Akintola, A. O., Ayoola, P. B., & Ibikunle, G. J., (2012). Antioxidant activity of two Nigerian green leafy vegetables. *Journal of Pharmaceutical and Biomedical Science*, *14*(14), 110–119.
12. Akuodor, G. C., Idris-Usman, M. S., Mbah, C. C., Megwas, U. A., Akpan, J. L., Ugwu, T. C., Okoroafor, D. O., & Osunkwo, U. A., (2010). Studies on anti-ulcer, analgesic and antipyretic properties of the ethanolic leaf extract of *Gongronema latifolium* in rodents. *African Journal of Biotechnology*, *9*(15), 2316–2321.
13. Aletor, O., Oshodi, A. A., & Ipinmoroti, K., (2002). Chemical composition of common leafy vegetables and functional properties of their leaf protein concentrates. *Food Chemistry*, *78*(1), 63–68.
14. Alhaji, U. I., Samuel, N. U., Aminu, M., Chidi, A. V., Umar, Z. U., Umar, U. A., & Adewale, B. M., (2014). *In vitro* antitrypanosomal activity, antioxidant property and phytochemical constituents of aqueous extracts of nine Nigerian medicinal plants. *Asian Pacific Journal of Tropical Diseases*, *4*(5), 348–355.

15. Ali, F., Assanta, M. A., & Robert, C., (2011). *Gnetum africanum*: A wild food plant from the African forest with many nutritional and medicinal properties. *Journal of Medicinal Foods*, *14*(11), 1289–1297.
16. Amorim, A. P. O., De Oliveira, M. C. C., De Azevedo Amorim, T., & Echevarria, A., (2013). Antioxidant, iron chelating and tyrosinase inhibitory activities of extracts from *Talinum triangulare* leach stem. *Antioxidants*, *2*, 90–99.
17. Andarwulan, N., Batari, R., Sandrasari, D. A., Bolling, B., & Wijaya, H., (2010). Flavonoid content and antioxidant activity of vegetables from Indonesia. *Food Chemistry*, *121*(4), 1231–1235.
18. Anyanwu, C. U., & Nwosu, G. C., (2014). Assessment of the antimicrobial activity of aqueous and ethanolic extracts of *Piper guineense* leaves. *Journal of Medicinal Plant Research*, *8*(10), 436–440.
19. Anyasor G, N., Ogunwenmo, K. O., Ogunnowo, A. A., & Olabisi, A. S., (2010). Comparative antioxidant, phytochemical and proximate analysis of aqueous and methanolic extracts of *Vernonia amygdalina* and *Talinum triangulare*. *Pakistan Journal of Nutrition*, *9*(3), 259–264.
20. Asaolu, S. S., Adefemi, O. S., Oyakilome, I. G., Ajibulu, K. E., & Asaolu, M. F., (2012). Proximate and mineral composition of Nigerian leafy vegetables. *Journal of Food Research*, *1*(3), 214–218.
21. Atangwho, I. J., Egbung, G. E., Ahmad, M., Yam, M. F., & Asmawi, M. Z., (2013). Antioxidant versus anti-diabetic properties of leaves from *Vernonia amygdalina Del.* growing in Malaysia. *Food Chemistry*, *141*(4), 3428–3434.
22. Awah, F. M., & Verla, A. W., (2010). Antioxidant activity, nitric oxide scavenging activity and phenolic contents of *Ocimum gratissimum* leaf extract. *Journal of Medicinal Plants Research*, *4*(24), 2479–2487.
23. Azuma, K., Nakayama, M., Koshioka, M., Ippoushi, K., Yamaguchi, Y., Kohata, K., Yamauch, Y., Ito, H., & Higashio, H., (1999). Phenolic antioxidants from the leaves of *Corchorus olitorius* L. *Journal of Agricultural and Food Chemistry*, *47*(10), 3963–3966.
24. Boivin, D., Lamy, S., Lord-Dufour, S., Jackson, J., Beaulieu, E., Côté, M., Moghrabi, A., Barrette, S., Gingras, D., & Béliveau, R., (2009). Antiproliferative and antioxidant activities of common vegetables: A comparative study. *Food Chemistry*, *112*(2), 374–380.
25. Chu, Y. F., Sun, J. I. E., Wu, X., & Liu, R. H., (2002). Antioxidant and antiproliferative activities of common vegetables. *Journal of Agricultural and Food Chemistry*, *50*(23), 6910–6916.
26. Dewanjee, S., Sahu, R., Karmakar, S., & Gangopadhyay, M., (2013). Toxic effects of lead exposure in Wistar rats: Involvement of oxidative stress and the beneficial role of edible jute (*Corchorus olitorius*) leaves. *Food and Chemical Toxicology*, *55*, 78–91.
27. Dougnon, T. V., Bankolé, H. S., Johnson, R. C., Klotoé, J. R., Dougnon, G., Gbaguidi, F., Assogba, F., Gbenou, J., Sahidou, S., Ategbo, J., & Rhin, B., (2012). Phytochemical screening, nutritional and toxicological analyses of leaves and fruits of *Solanum macrocarponLinn (Solanaceae)* in Cotonou (Benin). *Food and Nutritional Science*, *3*(11), 1595–1608.
28. Eddy, N. O., & Ebenso, E. E., (2010). Corrosion inhibition and adsorption properties of ethanol extract of *Gongronema latifolium* on mild steel in $H_2SO_4$. *Pigment and Resin Technology*, *39*(2), 77–83.
29. Edet, U. U., Ehiabhi, O. S., Ogunwande, I. A., Walker, T. M., Schmidt, J. M., Setzer, W. N., & Ekundayo, O., (2005). Analyses of the volatile constituents and antimicrobial

activities of *Gongronema latifolium (Benth.)* and *Gnetum africanum L. Journal of Essential Oil Bearing Plants*, *8*(3), 324–329.
30. Ekop, A. S., (2007). Determination of chemical composition of *Gnetum africanum* (Afang) seeds. *Pakistan Journal of Nutrition*, *6*(1), 40–43.
31. Ekpete, O. A., & Horsfall, M. J. N. R., (2011). Preparation and characterization of activated carbon derived from fluted pumpkin stem waste (*Telfairia occidentalis* Hook F). *Research Journal of Chemical Science*, *1*(3), 10–17.
32. Eleyinmi, A. F., (2007). Chemical composition and antibacterial activity of *Gongronema latifolium. Journal of Zhejiang University of Science B*, *8*(5), 352–358.
33. Eleyinmi, A. F., Sporns, P., & Bressler, D. C., (2008). Nutritional composition of *Gongronema latifolium* and *Vernonia amygdalina*. *Nutritional and Food Science*, *38*(2), 99–109.
34. Erasto, P., Grierson, D. S., & Afolayan, A. J., (2007). Antioxidant constituents in *Vernonia amygdalina* leaves. *Pharmaceutical Biology*, *45*(3), 195–199.
35. Falade, O. S., Sowunmi, O. R., Oladipo, A., Tubosun, A., & Adewusi, S. R., (2003). The level of organic acids in some Nigerian fruits and their effect on mineral availability in composite diets. *Pakistan Journal of Nutrition*, *2*(2), 82–88.
36. Farombi, E. O., & Owoeye, O., (2011). Antioxidative and chemopreventive properties of *Vernonia amygdalina* and *Garcinia* biflavonoids. *International Journal of Environmental Research and Public Health*, *8*(6), 2533–2555.
37. Fasakin, C. F., Udenigwe, C. C., & Aluko, R. E., (2011). Antioxidant properties of chlorophyll-enriched and chlorophyll-depleted polyphenolic fractions from leaves of *Vernonia amygdalina* and *Gongronema latifolium. Food Research International*, *44*(8), 2435–2441.
38. Fasuyi, A. O., (2006). Nutritional potentials of some tropical vegetable leaf meals: Chemical characterization and functional properties. *African Journal of Biotechnology*, *5*(1), 49–53.
39. Giami, S. Y., Achinewhu, S. C., & Ibaakee, C., (2005). The quality and sensory attributes of cookies supplemented with fluted pumpkin (*Telfairia occidentalis* Hook) seed flour. *International Journal of Food Science and Technology*, *40*(6), 613–620.
40. Gupta, M., Mazumder, U. K., Pal, D. K., & Bhattacharya, S., (2003). Anti-steroidogenic activity of methanolic extract of *Cuscuta reflexa Roxb.* stem and *Corchorus olitorius Linn.* seed in mouse ovary. *Indian Journal of Experimental Biology*, *41*(6), 641–644.
41. Hamzah, R. U., Jigam, A. A., Makun, H. A., & Egwim, E. C., (2014). Phytochemical screening and *in-vitro* antioxidant activity of methanolic extract of selected Nigerian vegetables. *Asian Journal of Basic and Applied Sciences*, *1*(1), 1–14.
42. Haskell, M. J., Jamil, K. M., Hassan, F., Peerson, J. M., Hossain, M. I., Fuchs, G. J., & Brown, K. H., (2004). Daily consumption of Indian spinach *(Basella alba)* or sweet potatoes has a positive effect on total-body vitamin A stores in Bangladeshi men. *American Journal of Clinical Nutrition*, *80*(3), 705–714.
43. Iwalewa, E. O., Adewunmi, C. O., Omisore, N. O. A., Adebanji, O. A., Azike, C. K., Adigun, A. O., Adesina, O. A., & Olowoyo, O. G., (2005). Pro- and antioxidant effects and cytoprotective potentials of nine edible vegetables in southwest Nigeria. *Journal of Medicinal Foods*, *8*(4), 539–544.
44. Iwalokun, B. A., Efedede, B. U., Alabi-Sofunde, J. A., Oduala, T., Magbagbeola, O. A., & Akinwande, A. I., (2006). Hepatoprotective and antioxidant activities of *Vernonia amygdalina* on acetaminophen-induced hepatic damage in mice. *Journal of Medicinal Foods*, *9*(4), 524–530.

45. Iweala, E. E. J., & Obidoa, O., (2009). Some biochemical, haematological and histological responses to a long term consumption of *Telfairia occidentalis* supplemented diet in rats. *Pakistan Journal of Nutrition*, *8*(8), 1199–1203.
46. Iweala, E. E., Uhuegbu, F. O., & Obidoa, O., (2009). Biochemical and histological changes associated with long term consumption of *Gnetum africanum Welw*. leaves in Rats. *Asian Journal of Biochemistry*, *4*(4), 125–132.
47. Izevbigie, E. B., (2003). Discovery of water-soluble anticancer agents (edotides) from a vegetable found in Benin City, Nigeria. *Experimental Biology and Medicine*, *228*(3), 293–298.
48. Johnson, M., Kolawole, O. S., & Olufunmilayo, L. A., (2015). Phytochemical analysis, *in vitro* evaluation of antioxidant and antimicrobial activity of methanolic leaf extract of *Vernonia amygdalina* (bitter leaf) against Staphylococcus aureus and *Pseudomonas aeruginosa. International Journal of Current Microbiology and Applied Science*, *4*(5), 411–426.
49. Kachchhi, N. R., Parmar, R. K., Tirgar, P. R., Desai, T. R., & Bhalodia, P. N., (2012). Evaluation of the antiurolithiatic activity of methanolic extract of *Celosia argentea* roots in rats. *International Journal of Phytopharmacy*, *3*(3), 249–255.
50. Katerere, D. R., Graziani, G., Thembo, K. M., Nyazema, N. Z., & Ritieni, A., (2012). Antioxidant activity of some African medicinal and dietary leafy African vegetables. *African Journal of Biotechnology*, *11*(17), 4103–4108.
51. Kayode, A. A. A., & Kayode, O. T., (2011). Some medicinal values of *Telfairia occidentalis*: A review. *American Journal of Biochemistry and Molecular Biology*, *1*, 30–38.
52. Komlaga, G., Sam, G. H., Dickson, R. A., Mensah, M. L. K., & Fleischer, T. C., (2014). Pharmacognostic studies and antioxidant properties of the leaves of *Solanum macrocarpon*. *Journal of Pharmaceutical Science Research*, *6*(1), 1–4.
53. Kumar, A., Prasad, M. N. V., & Sytar, O., (2012). Lead toxicity, defense strategies and associated indicative biomarkers in *Talinum triangulare* grown hydroponically. *Chemosphere*, *89*(9), 1056–1065.
54. Kumar, C. P., & Mohana, K. N., (2014). Phytochemical screening and corrosion inhibitive behaviour of *Pterolobium hexapetalum* and *Celosia argentea* plant extracts on mild steel in industrial water medium. *Egyptian Journal of Petroleum*, *23*(2), 201–211.
55. Liang, D., Zhou, Q., Gong, W., Wang, Y., Nie, Z., He, H., Li, J., Wu, J., & Zhang, J., (2011). Studies on the antioxidant and hepatoprotective activities of polysaccharides from *Talinum triangulare*. *Journal of Ethnopharmacology*, *136*(2), 316–321.
56. Mahadik, S. U., Jagtap, V. N., Oswal, H., Kumawat, N., & Kothari, R., (2011). *In vitro* Antioxidant activity of ethanolic extracts of *Celosia argentea* aerial parts, fresh fruits of *Fragaria vesca, Tamarindus indica, Psidium guajava, Zizyphus mauritiana*. *Research Journal of Pharmaceutical Technology*, *4*(11), 1782–1784.
57. Malomo, S. O., Ore, A., & Yakubu, M. T., (2011). *In vitro* and *in vivo* antioxidant activities of the aqueous extract of *Celosia argentea* leaves. *Indian Journal of Pharmacology*, *43*(3), 278–285.
58. Mensah, J. K., Okoli, R. I., Ohaju-Obodo, J. O., & Eifediyi, K., (2008). Phytochemical, nutritional and medical properties of some leafy vegetables consumed by Edo people of Nigeria. *African Journal of Biotechnology*, *7*(14), 2304–2309.
59. Morrison, J. F., & Twumasi, S. K., (2013). Comparative studies on the *in vitro* antioxidant properties of methanolic and hydro-ethanolic leafy extracts from eight edible leafy vegetables of Ghana. *African Journal of Biotechnology*, *9*(32), 5177–5184.

60. Moukette, B. M., Anatole, P. C., Biapa, C. P. N., Njimou, J. R., & Ngogang, J. Y., (2015). Free radicals quenching potential, protective properties against oxidative mediated ion toxicity and HPLC phenolic profile of a Cameroonian spice: *Piper guineensis. Toxicology Reports*, *2*, 792–805.
61. Nantia, E. A., Manfo, F. P., Beboy, N. S., & Moundipa, P. F., (2013). *In vitro* antioxidant activity of the methanol extract of *Basella alba L (Basellaceae)* in rat testicular homogenate. *Oxid. Antioxid. Med. Sci.*, *2*(2), 131–136.
62. Nwanna, E. E., Ibukun, E. O., & Oboh, G., (2013). Inhibitory effects of methanolic extracts of two eggplant species from South-western Nigeria on starch hydrolysing enzymes linked to type-2 diabetes. *African Journal of Pharmacy and Pharmacology*, *7*(23), 1575–1584.
63. Nwanna, E. E., & Oboh, G., (2007). Antioxidant and hepatoprotective properties of polyphenol extracts from *Telfairia occidentalis* (Fluted Pumpkin) leaves on acetaminophen induced liver damage. *Pakistan Journal of Biological Sciences*, *10*(16), 2682–2687.
64. Nwauzoma, A. B., & Dappa, M. S., (2013). Ethnobotanical studies of Port Harcourt metropolis, Nigeria. *ISRN Botany*, 1–11.
65. Oboh, G., Ademiluyi, A. O., Akinyemi, A. J., Henle, T., Saliu, J. A., & Schwarzenbolz, U., (2012). Inhibitory effect of polyphenol-rich extracts of jute leaf (*Corchorus olitorius*) on key enzyme linked to type 2 diabetes (α-amylase and α-glucosidase) and hypertension (angiotensin I converting) *in vitro*. *Journal of Functional Foods*, *4*(2), 450–458.
66. Oboh, G., Ekperigin, M. M., & Kazeem, M. I., (2005). Nutritional and haemolytic properties of eggplant (*Solanum macrocarpon*) leaves. *Journal Food Composition and Analysis*, *18*(2), 153–160.
67. Oboh, G., Nwanna, E. E., & Elusiyan, C. A., (2006). Antioxidant and antimicrobial properties of *Telfairia occidentalis* (fluted pumpkin) leaf extracts. *Journal of Pharmacology and Toxicology*, *5*(8), 539–547.
68. Oboh, G., Raddatz, H., & Henle, T., (2008). Antioxidant properties of polar and non-polar extracts of some tropical green leafy vegetables. *Journal of the Science of Food and Agriculture*, *88*(14), 2486–2492.
69. Oboh, G., Raddatz, H., & Henle, T., (2009). Characterization of the antioxidant properties of hydrophilic and lipophilic extracts of jute *(Corchorus olitorius)* leaf. *International Journal of Food Science and Nutrition*, *60*, 124–134.
70. Oboh, G., & Rocha, J. B. T., (2007). Antioxidant in foods: A new challenge for food processors. *Leading Edge Antioxidants Research*, 35–64.
71. Ofusori, D. A., Adelakun, A. E., Ayoka, A. O., Oluwayinka, O. P., Omotoso, E. O., Odukoya, S. A., & Adeyemi, D. O., (2008). Waterleaf (*Talinum triangulare*) enhances cerebral functions in Swiss albino mice. *Journal of Neurological Science*, *25*(4), 239–246.
72. Ogbonnaya, E. C., & Chinedum, E. K., (2013). Health promoting compounds and *in vitro* antioxidant activity of raw and decoctions of *Gnetum africanum* Welw. *Asian Pacific Journal of Tropical Diseases*, *3*(6), 472–479.
73. Ogunlade, I., Tucker, G., Fisk, I., & Ogunlade, A., (2009). Evaluation of antioxidant activity and vitamin E profile of some selected indigenous vegetables in Nigerian diet. *Journal of Food Agriculture and Environment*, *7*(2), 143–145.
74. Olajire, A. A., & Azeez, L., (2011). Total antioxidant activity, phenolic, flavonoid and ascorbic acid contents of Nigerian vegetables. *African Journal of Food Science and Technology*, *2*(2), 22–29.

75. Olaniyi, T. A., Adeniran, S. A., Adebayo, L. A., Olusegun, K. A., & Oyedeji, T. A., (2013). Antioxidant and anti-lipid peroxidation potentials of the ethylacetate and chloroform extracts of *Basella alba* leaves. *Asian Journal of Natural and Applied Science*, *2*(2), 81–88.
76. Omoregie, E. S., & Osagie, A. U., (2012). Antioxidant properties of methanolic extracts of some Nigerian plants on nutritionally-stressed rats. *Nigerian Journal of Basic and Applied Science*, *20*(1), 7–20.
77. Prabhu, K. S., Lobo, R., Shirwaikar, A. A., & Shirwaikar, A., (2009). *Ocimum gratissimum*: A review of its chemical, pharmacological and ethnomedicinal properties. *Open Complementary Medicine Journal*, *1*, 1–15.
78. Prakash, B., Shukla, R., Singh, P., Mishra, P. K., Dubey, N. K., & Kharwar, R. N., (2011). Efficacy of chemically characterized *Ocimum gratissimum L.* essential oil as an antioxidant and a safe plant based antimicrobial against fungal and aflatoxin $B_1$ contamination of spices. *Food Research International*, *44*(1), 385–390.
79. Prasad, K. N., Shivamurthy, G. R., & Aradhya, S. M., (2008). *Ipomoea aquatica*, an underutilized green leafy vegetable: A review. *International Journal of Botany*, *4*(1), 123–129.
80. Rub, R. A., Patil, M. J., Siddiqui, A. A., Ghorpade, P. R., & Moghe, A. S., (2015). Free radical scavenging and cytotoxic potential of *Celosia argentea*. *Pharmacognosy Journal*, *7*(3), 191–197.
81. Saalu, L. C., Kpela, T., Benebo, A. S., Oyewopo, A. O., Anifowope, E. O., & Oguntola, J. A., (2010). The dose-dependent testiculoprotective and testiculotoxic potentials of *Telfairiaoccidentalis* hook f. leaves extract in rat. *International Journal of Applied Research and Natural Products*, *3*(3), 27–38.
82. Sharma, P., Vidyasagar, G., Singh, S., Ghule, S., & Kumar, B., (2010). Antidiarrhoeal activity of leaf extract of *Celosia Argentea* in experimentally induced diarrhoea in rats. *Journal of Advanced Pharmaceutical and Technology Research*, *1*(1), 41–47.
83. Sodipo, O. A., Abdulrahman, F. I., Akan, J. C., & Akinniyi, J. A., (2008). Phytochemical screening and elemental constituents of the fruit of *Solanum macrocarpumLinn*. *Continental Journal of Applied Science*, *3*, 88–97.
84. Song, W., Derito, C. M., Liu, M. K., He, X., Dong, M., & Liu, R. H., (2010). Cellular antioxidant activity of common vegetables. *Journal of Agricultural and Food Chemistry*, *58*(11), 6621–6629.
85. Sridevi, K., Ravishankar, K., & Bhandhavi, P. P., (2013). *In vitro* antioxidant activities of ethanolic seed extract of *Millettia pinnata* and leaf extract of *Basella alba*. *International Journal of Pharmaceutical Chemical and Biological Science*, *3*(2), 269–274.
86. Swarna, J., & Ravindhran, R., (2013). Pharmacognostical and phytochemical evaluation of *Talinum triangulare* (Jacq.) wild. *International Journal of Pharmacy and Pharmaceutical Science, 5*, 249–256.
87. Ugochukwu, N. H., & Babady, N. E., (2003). Antihyperglycemic effect of aqueous and ethanolic extracts of *Gongronema latifolium* leaves on glucose and glycogen metabolism in livers of normal and streptozotocin-induced diabetic rats. *Life Science*, *73*(15), 1925–1938.
88. Ugochukwu, N. H., & Babady, N. E., (2002). Antioxidant effects of *Gongronema latifolium* in hepatocytes of rat models of non-insulin *diabetesmellitus*. *Fitoterapia*, *73*(7), 612–618.
89. Urmila, G. H., Ganga, R. B., & Satyanarayana, T., (2013). Phytochemical and *in-vitro* antioxidant activity of methanolic extract of *Lactuca scariola* and *Celosia argentea* leaves. *Journal of Drug Delivery and Therapy*, *3*(4), 114–117.

90. Wu, Q. B., Wang, Y., Liang, L., Jiang, Q., Guo, M. L., & Zhang, J. J., (2013). Novel triterpenoid saponins from the seeds of *Celosia argentea* L. *Natural Products Research, 27*(15), 1353–1360.
91. Yan, Y. Y., Wang, Y. W., Chen, S. L., Zhuang, S. R., & Wang, C. K., (2013). Anti-inflammatory effects of phenolic crude extracts from five fractions of *Corchorus Olitorius* L. *Food Chemistry, 138*(2), 1008–1014.
92. Zakaria, Z. A., Sulaiman, M. R., & Gopalan, H. K., (2007). Antinociceptive and anti-inflammatory properties of Corchorus capsularis leaves chloroform extract in experimental animal models. *Yakugaku Zasshi, 127*(2), 359–365.

# CHAPTER 2

# *CARPOBROTUS EDULIS* L. (SOUR FIG): PHYTOCHEMISTRY, PHARMACOLOGY, AND TOXICOLOGY

FRANCIS N. NKEDE, SIMEON A. MATERECHERA, and
WILFRED OTANG-MBENG

## ABSTRACT

*Carpobrotus edulis* (Sour fig) is traditionally used for the management of diabetes and other diseases in South Africa. Despite its frequent usage by the local population, no comprehensive review has been furnished. Scientific literature on *C. edulis* was sourced from the databases, such as Scopus, PubMed, Science Direct, and Google Scholar. Several phytochemical studies on *C. edulis* revealed the presence of β-amyrin, isoterpinolene, oleanolic acid, flavonoids with great antimicrobial potentials. The literature revealed that certain pharmacological areas of *C. edulis* (such as its antiplasmodial, anticancer, antibacterial, anti-oxidant, and antifungal properties) had already been investigated. Published findings on *C. edulis*' isolated phytochemicals and crude extract bioactivity have further confirmed its traditional usage in the treatment of respiratory diseases. Notwithstanding to fully comprehend its medicinal value, more antiplasmodial, anti-inflammatory, and toxicological investigations on main active compounds are necessary.

## 2.1 INTRODUCTION

Traditional medical practitioners in developing countries are known to use medicinal plants to alleviate many diseases [1, 16]. About 80% of South Africans use traditional medicines to meet their primary health care needs [24]. *C. edulis* belongs to the family Aizoaceae and is mostly succulent herbs with simple leaves, and the fruit is a capsule [20, 22]. It is commonly known in South Africa as sour fig and in other countries as pigface, Hottentot-fig or ice plant. There is no previous review available with regard to the botany, phytochemistry, and pharmacological properties of *C. edulis*.

In sub-Saharan Africa, leaf gel of *C. edulis* has been used in traditional medicine to treat different disorders and infections including sinusitis, tuberculosis, vaginal thrust burns, and toothache [10, 17, 21]. Although evidence on traditional use of sour fig has been examined, yet there is a need for future investigation on this plant for certain pharmacological properties such as:

- Anti-malaria;
- Anti-inflammatory;
- Antiplasmodial activities.

This chapter provides a comprehensive review on the phytochemistry, botany, toxicology, and pharmacology of *C. edulis*.

## 2.2 LITERATURE REVIEW

### *2.2.1 TAXONOMY, VERNACULAR NAMES, AND DISTRIBUTION*

The Aizoaceae is a dicotyledonous flowering plant that consists of 135 genera and about 1,900 species and is largely endemic to South Africa, while a few species are found in Australia and the Pacific regions. The plant has central branched or unbranched solitary inflorescences, and the flowers are usually yellow fading to pale pink [23]. The plants are robust with rooting at the nodes forming dense mats. The genus *Carpobrotus* comprises of 13 species, among which seven of these species exist in South Africa [23].

*C. edulis*, also known as ghaukum (Khoi)/sour fig (English)/suurvy (Afrikaans), has the synonyms, namely [12] *Carpobrotus acinaciformis, Carpobrotus aequilaterus, Carpobrotus edulis var.* chrysophthalmus and *Mesembryanthemum edule.* These plants are indigenous to the Flora of Southern African and are widely distributed in the sandy flats and slopes of Namaqualand and the Eastern Cape.

### *2.2.2 BOTANY AND TRADITIONAL USES*

The stem of the plant spreads out, and roots sometimes emerge at nodes. The green leaves are waxy, flowers are purple or pink and measure 4–9 cm in diameter [23]. They are unique in having fleshy, indehiscent fruits (Figure 2.1). These plants thrive well in sandy soils.

Traditional medical practitioners in South Africa have used *C. edulis* to treat various diseases. The Eastern Cape communities use the leaves as traditional medicine for vaginal infections, dysentery, tuberculosis, and constipation [2, 14].

**FIGURE 2.1 (See color insert.)** *C. edulis* (sour fig).
*Source:* Top: Llez. https://en.wikipedia.org/wiki/Carpobrotus_edulis#/media/File:Carpobrotus_edulis_004.jpg
Bottom: Winfried Bruenken (Amrum). https://en.wikipedia.org/wiki/Carpobrotus_edulis#/media/File:Carpobrutus_edulis-P9200072.jpg

### *2.2.3 CHEMICAL CONSTITUENTS*

Phytochemical studies on the extract of leaves of *C. edulis* resulted in the isolation of biocompounds such as tetradecamethylcycloheptasiloxane, 1-octadecane, naphthalene, n- octanol andeicosane [2, 3, 10, 13, 17]. Martins et al., [9] isolated phytochemicals, such as epicatechin, procyanidin B5, β-amyrin, uvaol, catechin, oleanolic acid, and monogalactosyldiacylglycerol (acyls=linolenoyl group; MGDG). Phytochemical investigations conducted on the extracts from stems and flowers of *C. edulis* indicated the presence of flavonoids, saponins, alkaloids as well as anthraquinones [2, 3].

## 2.3 MATERIALS AND METHODS

Scientific literature on *C. edulis* was obtained from Google-Scholar, Science Direct, Scopus, the Plant List databases. These were comprehensively and systematically reviewed for this study.

## 2.4 RESULTS AND DISCUSSION

### *2.4.1 PHYTOCHEMICALS ISOLATED FROM C. EDULIS*

Table 2.1 and Figure 2.2 shows some key isolated phytochemicals from *C. edulis*.

**TABLE 2.1** Isolated Phytochemicals of *C. Edulis*

| Phytochemicals Isolated | Parts Observed | References |
|---|---|---|
| Isoterpinolene | Leaf | [14] |
| Naphthalene, 1,2-dihydro-2,5,8-tri | Leaf | |
| Bistrimethylsilyl N-acetyl EICOSAS | Leaf | |
| β-amyrin, oleanolic acid | Leaf | [8] |
| uvaol, monogalactosyldiacylglycerol | Leaf | |
| (acyls=linolenoyl group; MGDG), | Leaf | |
| epicatechin | Leaf | |
| N-octanol | Leaf | [8] |
| Nonylaldehyde | Leaf | |
| Trans-β-demascenone | Leaf | |

**TABLE 2.1** *(Continued)*

| Phytochemicals Isolated | Parts Observed | References |
|---|---|---|
| Trans-2-tridecenal | Leaf | |
| Tetradecamethylcycloheptasiloxane | Leaf | |
| Tetradecamethylcycloheptasiloxane | Leaf | |
| Tetradecamethylcycloheptasiloxane | Leaf | |
| Octadecane | Leaf | |
| Octadecane | Leaf | [14] |
| 1-octadecene | Leaf | |
| Nonadecane | Leaf | |
| 2-pentadecanone,6,10,14-trimethyl | Leaf | |
| Eicosane | Leaf | |
| Eicosane | Leaf | |
| Phytol (2-Hexadecen-1-o1, 3,7,11, 15-tetramethyl) | Leaf | |
| Tetrasiloxane,1,1,1,5,7,7,7- heptamethyl-3bis (trimethylsilyl)oxy | Leaf | |
| Tetracosamethylcyclododecasiloxane | Leaf | |
| Saponins | Stem, leaf, flower | |
| Alkaloids | Stem, leaf, flower | |
| Flavonoids | Stem, leaf, flower | |
| Anthraquinones | Stem, leaf, flower | |
| Sterols | Stem, leaf, flower | |

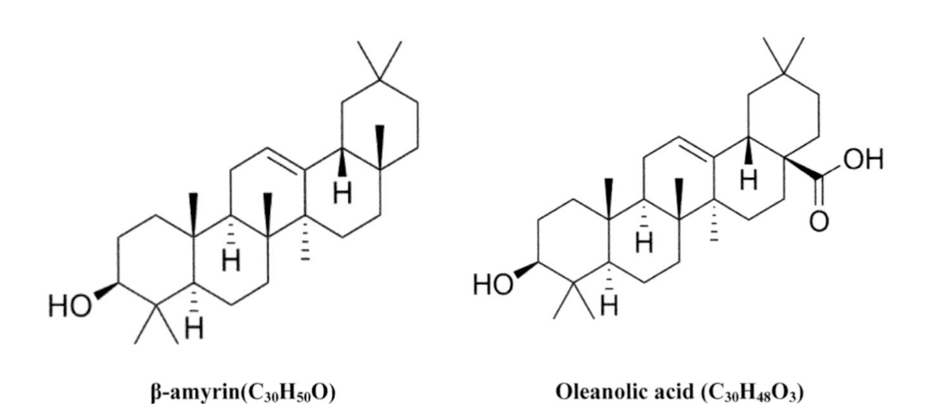


**β-amyrin($C_{30}H_{50}O$)** **Oleanolic acid ($C_{30}H_{48}O_3$)**

**FIGURE 2.2** Chemical structure of some key biocompounds isolated from *C. edulis*.

### 2.4.2 REPORTS ON PHARMACOLOGICAL STUDIES

Literature search has shown that *C. edulis* has been investigated for five pharmacological activities (Table 2.2), such as:

- Antimicrobial;
- Antifungal;
- Anticancer;
- Anti-oxidant; and
- Antiplasmodial.

### 2.4.3 ANTICANCER ACTIVITY

Martins et al., [8] screened two cell lines for anticancer activity using methanol extracts of *C. edulis*. The proliferation effects observed in the cell lines were significant. They concluded that the proliferative activity of uvaol and oleanolic acid on the parental cells were higher than those of epicatechin and MGDG, which showed effects that were significant against MDR1-transfected cells.

### 2.4.4 ANTIOXIDANT ACTIVITY

Ibtissem et al., assessed the antioxidant potential of *C. edulis* using water extract and the synthetic antioxidant (BHT) was used as standard. High antioxidant activity was observed up to 1mg/ml dose, and DPPH inhibition showed 94.64% ± 0.45% at a dose of 2 mg/ml [6].

### 2.4.5 TOXICITY

Ibtissem et al., [6] reported no appreciable cytotoxicity effects in ethanol extracts of *C. edulis* in the treatment of Periodontitis. Ordway et al., [15] reported that HPBMDM and THP-1 cell line showed toxicity to methanol extract of *C. edulis* leaf after 5 days. Active compounds isolated from *C. edulis* leaf [15] did not show any toxicity at doses that were inhibitory to efflux pumps of L5178 mouse cell line. Within one day of exposure at doses associated with Phyto-haemagglutinin in the extract, induced proliferation of THP-1 cells was observed. This affirms that the plant may have potential for cancer chemotherapy.

**TABLE 2.2** Some Pharmacological Studies Conducted on *C. Edulis*

| Bioactivity Investigated | Assay Used | Test Material | Extract Used | Dose | Control | Findings | References |
|---|---|---|---|---|---|---|---|
| **Antibacterial** | Nutrient agar and *Staphylococcus aureus Bacillus cereus, K. pneumoniae, Mycobacterium aurum* | Leaf | Water, ethanol, and dichloromethane | Not determined | Negative control: water, ethanol, and dichloromethane, streptomycin was the Positive control | All extracts were bactericidal against bacteria which were Gram-positive at 0.163–0.098 mg/ml. | [3, 18] |
| **Antibacterial** | Broth dilution assay on *P. aeruginosa. E. coli,* and *S. aureus* and disc diffusion assay | Flavonoids and phenols. | Ethanol and water | 100–600 mg | Positive controls: vancomycin and ceftazidime. Negative control: None | No inhibition against *E. coli* and *Pseudomonas aeruginosa*. *Staphylococcus aureus* was strongly inhibited. | [5, 6] |
| **Antibacterial** | Agar diffusion assay on *Porphyromonas gingivalis, Tannerella forsythensis*, *Actinobacillus actinomycetemcomitans*. | Shoot | Ethanol | Not mentioned | Negative control: broth indicator solution and bacterial suspension. Positive control: resazurin indicator solution. | Extracts strongly inhibited *Porphyro*monas *gingivalis, Tannerella forsythensis* and *Actinobacillus actinomycetemcomitans*. | [6] |
| **Anticancer** | 96 well microplates were used to assay Extracts of *C. edulis* in cell lines of L5178, mouse T-cell. | Leaf | Methanol | 4–10 mg/L | Negative control: DMSO Positive control: Verapamil | Strong anticancer property was observed against the parental cells. | [9] |
| **Anticancer** | *C. edulis* leaf extracts were assayed in L5178 mouse T cell lymphoma in 24 well microplates. | Leaf | Methanol | | Negative control: DMSO. Positive control: Verapamil | THP-1 Cells were proliferated within 24 hours of exposure to *C. edulis* leaf extract. | [15] |

**TABLE 2.2** *(Continued)*

| Bioactivity Investigated | Assay Used | Test Material | Extract Used | Dose | Control | Findings | References |
|---|---|---|---|---|---|---|---|
| **Antifungal** | Micro-dilution assay against five fungal species. | Leaf | Hexane, Acetone, ethanol and water | 0.005–5 mg/ml | Positive control: Nystatin. Negative control: Amphotericin B | Growth of *C. albican* and *C. krusei* was inhibited at higher concentration of Ethanol extract (1.25 mg/ml respectively). Zero inhibition was observed at all concentrations of the water extract | [4, 14] |
| **Antioxidant** | ABTS | Leaf | Hexane, acetone, ethanol, and distilled water | 0.025–0.5 mg/ml. | Negative and positive controls were the corresponding solvent and ascorbic acid, respectively. | Ethanol leaf extract showed high ABTS scavenging activity than BHT at 0.025mg/ml. | [13] |
| **Antioxidant** | FRAP | Leaf | Hexane, acetone, ethanol, and water | 0.025–0.5 mg/ml. | Ascorbic acid and BHT were positive controls, and the respective solvents were used as negative controls. | The extracts significantly recorded lower FRAP values than those of BHT and ascorbic acid. | [13] |
| **Antioxidant** | DPPH | Leaf | Hexane, acetone, ethanol, and distilled water | 0.025–0.5 mg/ml. | Ascorbic acid and BHT were positive controls, and the respective solvents were used as negative controls. | High DPPH radical scavenging activity high for the water and ethanol extracts at 0.025 mg/ml, higher than that of BHT. | [13] |

**TABLE 2.2** *(Continued)*

| Bioactivity Investigated | Assay Used | Test Material | Extract Used | Dose | Control | Findings | References |
|---|---|---|---|---|---|---|---|
| **Antioxidant** | DPPH | Leaf | Ethanol and water | 0.024–0.1 mg/ml. | Negative and positive controls were the corresponding solvent and BHT, respectively. | Extract of *C. edulis* showed High DPPH activity than BHT at 1 mg/ml. | [6] |
| **Antiplasmodial** | *In vitro* antiplasmodial activity was done against *Plasmodium falciparum* strains. | Root | Chloroform, methanol and water. | | Negative control: DMSO+RPMI Positive control: solvent | Methanol extract was mildly active while no activity was observed for chloroform and water extracts. | [7, 11] |
| **Toxicity** | MTT | Leaf | Ethanol | Not indicated. | Not indicated | *C. edulis* was the least cytotoxic medicinal plant when compared with other plants that were tested. | [6] |
| **Toxicity** | Toxicity of *Carpobrotus edulis* methanol extract using MTT. | Leaf | Methanol | 0.01–0.09ml | Not mentioned. | In the first three days of culture, Methanol leaf extract of *C. edulis* showed no toxicity. Toxicity was only evident after day 5. | [15] |

### 2.4.6 ANTIBACTERIAL ACTIVITY

The water extract of *C. edulis* inhibited *Pseudomonas aeruginosa, Staphylococcus aureus,* and *Escherichia coli* [6] and the occurrence of polyphenolics compounds found in the extract accounted for this observation. Other *Carpobrotus* species have been reported to be active against pathogenic bacteria. Ethyl acetate and water extracts obtained from *C. quadrifidus* and *C. muirii* were reported to be inhibitory against *Mycobacterium smegmatis* and *Staphylococcus aureus*, but failed to inhibit *Pseudomonas aeruginosa* and *Candida albicans* [6].

## 2.5 SUMMARY

*C. edulis* is well known by traditional healers in South Africa for its medicinal value and has been frequently utilized to treat various diseases, infections, and ailments. Among the Xhosa communities, it is frequently prescribed as a traditional remedy. Many studies on the phytochemical content and bioactivity of its crude extracts have justified usage of this medicinal plant as a traditional remedy. Therapeutic potential of *C. edulis* has been demonstrated for treatment of mouth ulcers, burns, throat infections, stomach ailments, dysentery, and diarrhea. This is attributed to the presence of various phytochemicals like Nephthalene, Eicosane, Bistrimethulesilyl N-acetyl, Isoterpinolene, flavonoids, saponins, and other compounds present.

## KEYWORDS

- **antimicrobial**
- **antioxidant**
- *Carpobrotus edulis*
- **cytotoxicity**
- **herbs**
- **pharmacology**
- **phytochemistry**
- **sour fig**
- **toxicity**
- **toxicology**
- **Xhosa**

## REFERENCES

1. Abbasi, A. M., Khan, M. A., Ahmad, M., Zafar, M., Jahan, S., & Sultana, S., (2010). Ethnopharmacological application of medicinal plants to cure skin diseases and in folk cosmetics among the tribal communities of North-West Frontier Province, Pakistan. *Journal of Ethnopharmacology, 41*, 22–25.
2. Alam, E. A., (2011). Phytochemical screening on different plant parts of some succulent plants of Egypt. *New York Science Journal, 4*, 5–18.
3. Buwa, L. V., & Afolayan, A. J., (2009). Antimicrobial activity of some medicinal plants used for the treatment of tuberculosis in the Eastern Cape Province, South Africa. *African Journal of Biotechnology, 8*, 6683–6687.
4. Debruyne, D., (1997). Clinical pharmacokinetics of fluconazole in superficial and systemic mycoses. *Clinical Pharmacology, 33*, 52–57.
5. Green, E., Samie, A., Obi, C. L., Bessong, P. O., & Ndip, R. N., (2010). Inhibitory properties of selected South African plants against *Mycobacterium tuberculosis*. *Journal of Ethnopharmacology, 130*, 151–157.
6. Ibtissem, B., Chedly, A., & Souad, S., (2012). Antioxidant and antibacterial properties of *Mesembryanthemum crystallinum* and *Carpobrotus edulis* extracts. *Advances in Chemical Engineering and Science, 2*, 359–365.
7. Jeruto, P., Nyangacha, R. M., & Mutai, C., (2015). *In vitro* and *in vivo* antiplasmodial activity of extracts of selected Kenyan medicinal plants. *African Journal of Pharmacy and Pharmacology, 9*, 500–505.
8. Martins, A., Vasas, Z., Schelz, M., Viveiros, J., Molnár, J., Hohmann, J., & Amaral, L., (2010). Constituents of *Carpobrotus edulis* inhibit P-Glycoprotein of MDR1-transfected mouse lymphoma cells. *Anticancer Research, 30*, 829–836.
9. Martins, A., Vasas, Z., Schelz, M., Viveiros, J., Molnár, J., Hohmann, J., & Amaral, L., (2011). Antibacterial properties of compounds isolated from *Carpobrotus edulis*. *International Journal of Antimicrobial Agents, 37*(5), 438.
10. Mathabe, M. C., Nikolova, R. V., Lall, N., & Nyazema, N. Z., (2006). Antibacterial activities of medicinal plants used for treatment of diarrhea in Limpopo Province, South Africa. *Journal of Ethnopharmacology, 105*, 286–293.
11. Muregi, F. W., Chhabra, S. C., Njagi, E. N., Langat-Thoruwa, C. C., Njue, W. M., Orago, A. S., Omar, S. A., & Ndiege, I. O., (1985). *In vitro* antiplasmodial activity of some plants used in Kisii, Kenya against malaria and their chloroquine potentiation effects. *Journal of Ethnopharmacology, 84*, 235–239.
12. Omoruyi, B. E., Afolayan, A. J., & Graeme, B., (2014). The inhibitory effect of *Mesembryanthemum edule* (L.) bolus essential oil on some pathogenic fungal isolates. *Complementary and Alternative Medicine, 14*, 168–176.
13. Omoruyi, B. E., Graeme, B., & Afolayan, A. J., (2012). Antioxidant and phytochemical properties of *Carpobrotus edulis* (L.) bolus leaf used for the management of common infections in HIV/AIDS. *BMC Complementary and Alternative Medicine, 12*, 215–219.
14. Omoruyi. B. E., Afolayan, A. J., & Graeme, B., (2014). Chemical composition profiling and antifungal activity of the essential oil and plant extracts of *Mesembryanthemum edule* (L.) bolus leaves. *African Journal of Traditional Complementary and Alternative Medicine, 11*(4), 19–30.
15. Ordway, D., Judit, H., Miguel, V., Antonio, V., Joseph, M., Clara, L., Maria, J. A., Maria, A. G., & Leonard, A., (2003). *Carpobrotus edulis* methanol extract inhibits the

MDR efflux pumps, enhances killing of phagocytosed *S. aureus* and promotes immune modulation. *Phytotherapy Research, 17*, 512–519.

16. Rybicki, E. P., Chikwamba, R., Koch, M., Rhodes, J. I., & Groenewald, J. H., (2012). Plant-made therapeutics: Emerging platform in South Africa. *Biotechnology Advances, 30*, 449–459.
17. Scott, G., & Hewett, M. L., (2008). Pioneers in ethnopharmacology: The Dutch East India Company (VOC) at the Cape from 1650 to 1800. *Journal of Ethnopharmacology, 115*, 339–360.
18. Shai, L. J., McGaw, L. J., Aderogba, M. A., Mdee, L. K., & Eloff, J. N., (2008). Triterpenoids with antifungal and antibacterial activity from *Curtisia dentate* (Burm) leaves. *Journal of Ethnopharmacology, 119*, 238–241.
19. Van Der watt, E., & Pretorius, J. C., (2001). Purification and identification of active antibacterial components in *Carpobrotus edulis* L. *Journal of Ethnopharmacology, 76*, 87–91.
20. Van Wyk, B. E., Van Oudtshoorn, B., & Gericke, N., (1997). *Medicinal Plants of Southern Africa* (p. 88). Briza Publications, Pretoria, South Africa.
21. Watt, J. M., & Breyer-Brandwijk, M. G., (1962). *The Medicinal and Poisonous Plants of Southern and Eastern Africa* (p. 162). Livingstone, London.
22. Weber, E., D'Antonio, C., (1999). Germination and growth responses of hybridizing *Carpobrotus* species (Aizoaceae) from coastal California to soil salinity. *American Journal of Botany, 86*, 1257–1263.
23. Wisura, W., & Glen, H. F., (1993). The South African species of *Carpobrotus* (*Mesembryanthema*, Aizoaceae) contribution. *Boletim Herbage, 15*, 76–107.
24. World Health Organization (2008). *World Malaria Report,* http://whqlibdoc.who.int/publications/2008/9789241563697_eng.pdf?ua=1 (Accessed on 29 July 2019).

# CHAPTER 3

# PHYTOCHEMISTRY, PHARMACOLOGY, AND SAFETY ISSUES OF ESSENTIAL OILS: APPLICATIONS IN AROMATHERAPY

ANINDYA SUNDAR RAY, SUMAN KALYAN MANDAL, and CHOWDHURY HABIBUR RAHAMAN

## ABSTRACT

In aromatherapy (an alternative and complementary therapy), essential oils (EOs) are used to cure a wide range of health conditions that are not fully controlled by the conventional therapy. Out of 3,000-reported EOs, nearly 300 compounds have been found important in pharmaceutical, food, and cosmetic industries, and household use. Chemical studies have revealed that terpenoids, aromatic, and aliphatic compounds are major components of EOs, and they are characterized as low-molecular-weight aroma chemicals. Pharmacological and clinical studies have proved that EOs have a wide range of therapeutic potentials in combating the microbial infections, inflammatory, and cancerous diseases, and many more. Review in this chapter embodies the historical perspective of Aromatherapy, its categories, and various applications of the EOs. It also highlights the toxicological aspects of EOs for its safer use in human health care. All the information was reviewed from the printed books as well as electronic databases like Google Scholar, PubMed, Science Direct, Web of Science, library search, etc. using different search keywords. Article selection was restricted by using the keywords "aromatherapy" or "essential oil" and sometimes with either of these words; the connector 'AND' was used for the following phrases: traditional healing, medicinal value of EOs, its chemistry, pharmacology, toxicity, safety, etc. Finally, the review has been concluded with a brief discussion on the present status and future prospect of essential oil research and its implications on human health care.

## 3.1 INTRODUCTION

From the time immemorial, volatile aromas have been used in various cultures of human beings. About 6000 years ago, many ancient civilizations of India, China, and Egypt had developed the practices on the use of volatile substances for treating many common diseases [89, 112]. This age-old tradition of healing with aromatic compounds resulted in "aromatherapy" (a combination of two words: "aroma" means fragrance and "therapy" means treatment). In 400 B.C., Hippocrates mentioned the benefit of regular uses of aromas and fragrance in maintaining good health. According to him, daily massage and bath with the scented substances is the proper way to health (https://en.wikipedia.org/wiki/Hippocrates).

Aromatherapy is defined as the practice of healing health complications through psycho-somatic approaches using the essential oils (EOs), which first act on the mind by boosting the patient's mood and then subsequently by physiochemical changes in the body. Aromatherapy has been specified as an effective healing process for curing a wide range of health complications, and this branch of therapy has been recognized as holistic medicine [12]. At the end of the 20th century, this therapy gained attention for its healing benefits in many of the diseases. In the 21st century, it has been very popular among both the patients and practitioners of different systems of medicine. Aromatherapy is now recognized as aroma science due to its popularity, therapeutic importance, and widespread acceptance [47].

The EOs, used as therapeutic agents in aromatherapy, are highly concentrated substances extracted from different parts of the aromatic plants [44, 147]. About 2,000 plant species have been investigated, and nearly 3,000 volatile oil compounds have so far been identified, most of which have diversified commercial value. In recent years, annual production of commercially valued aromatic oils is estimated at 60,000 tons with an estimated market value of US$700 million [43]. EOs contain a complex mixture of bioactive compounds from different chemical groups like mono-, sesqui-, and sometimes di-terpenoids, phenylpropanoids, fatty acids and their fragments, benzenoids, etc. [37]. They are colorless liquids with the amusing smell and high refractive index.

EOs are administered in small quantities through inhalation, massage, or sometimes by simple application on the skin surface to get relief from different kinds of stress by stimulating olfactory nerves. In the aromatherapy, the healing principle of EOs is basically to stabilize [https://en.wikipedia.org/wiki/Hippocrates] the mind-body equilibrium of an individual.

Aromatic compounds boost up the mood, alter the alertness, and relieve the mental stress of individuals. Investigations are underway to understand the effects of EOs on the workability of a person and its actions on the human brain [19]. Apart from its stress-relieving potentials, scientists have already established broad-spectrum biological efficacy of the volatile compounds against a number of health disorders, such as Alzheimer's disease (AD), cardiovascular problems, burn wound, labor pain in pregnancy, sleep-related disorders and cancer [70, 77, 96, 113, 135, 163].

The review in this chapter discusses the present status and future prospects of essential oil research and its implications on human healthcare.

### 3.1.1 BRIEF HISTORY OF AROMATHERAPY

Uses of aromatic plants for curing ailments, offering to Gods and as sex-stimulator were very common practices in ancient civilizations of India, Rome, and China [6, 113]. Galbanum, frankincense, and many other aromatic herbs were mixed with vegetable oil or animal fat, and it was given as after-bath massage onto the bodies of pyramid building workers. In ancient Egypt, the rich people were also used to take such aromatic massages after their bath [113]. Sometimes incense smoke was developed by burning resinous plants to create a sacrosanct atmosphere in the holy places during worship for making sacrifices. Sometimes, narcotics like cannabis were added to the aromatic plant materials to anesthetize sacrificial animals and humans [40].

In the Indian epic Ramayana, the wounded Laxman who was in a deep coma, gained consciousness after ingestion of the *Sanjeevani*, a specific aromatic Himalayan herb; and however, the identity of this herb has still not been confirmed. Scientific discussion on the authenticity of the Sanjeevani proposed several plant species, which include *Selaginella bryopteris*, *Dendrobium plicatile* (synonym *Desmotrichum fimbriatum*), *Cressa cretica* and others. In Vedic literature, there are references of more than 700 aromatic substances mostly of plant origin that were used as healing agents. Among these, cinnamon, ginger, spikenard, coriander, myrrh, and sandalwood were considered as effective agents. Use of EOs as healing option is evidenced in different Traditional Indian systems of medicine like Ayurveda, Siddha, Unani, etc. The aromatic herb Sweet Basil (*tulsi*) has traditionally been related with Indian mythology, various aspects of religion and household remedies and this herb is known as one of the sacred plants of Indian culture.

The term 'Aromatherapy' is originated from the French word '*Aromatherapie*.' Dr. Rene-Maurice Gottefosse, a French chemist, was one of the first, who used the term 'Aromatherapy' in 1927. Later in 1937, Gottefosse published a book on EOs in French, where he used the word 'aromatherapy' in its title, "*Aromathérapie: Les Huiles essentielles hormones végétales.*" Afterward, the book was translated in English entitled "*Gottefosse's Aromatherapy*" [170; < http://www.wikiphyto.org/wiki/Huile_essentielle>]. Magical power of the lavender oil in wound healing had been observed by Dr. Rene Maurice Gottefosse, when he badly burned his hand after a laboratory accident. This observation encouraged him to study the healing properties of the EOs. Dr. Gottefosse then published a good number of scientific works on aromatherapy. Though, scientific discussion on EOs had been initiated long back in 1880 with an article published by Wood and Reichut entitled "*Note on the action upon the circulation of certain Volatile oils*." Later, various books were published on the application of aromatherapy in various health conditions [44].

### 3.1.2 MODES OF ADMINISTRATION OF ESSENTIAL OILS (EOS)

Mostly EOs are used to boost up the mood of the patients by acting upon the body and mind of a person. Many times, their use is restricted to regulate or improve the physical conditions of the patient. In aromatherapy, EOs are administered through various modes, such as:

a. Inhalation for infections in the respiratory system;
b. Topical application on burns, muscle, and skin;
c. Compress for swelling;
d. Ingestion for intestinal complaints and other internal problems.

Addition of a few drops of EOs in water turns it into an enchanted bath. Another unique mode of application is the use of 1–3% of EOs in carrier oils (cold-pressed vegetative oils) during massage therapy. They are applied in various combinations also that work synergistically to get better results.

### 3.1.3 THE TERMINOLOGIES USED IN AROMATHERAPY AND ITS TYPES

Several terminologies are used in different parts of the world in aroma science to define different patterns of usages of EOs, their various modes of action and purpose of uses. Terminologies most commonly used in aroma science are aromachology, aromatherapy, and aromatology, which are precisely defined here according to the concept given by the Sense of Smell Institute (SSI), USA in 1982.

Aromachology is the interrelationship between the psychology of a person (relaxation, exhilaration, sensuality, happiness, and achievement) and the sweet odor of the EOs. Aromatherapy is defined as the therapy of both physical (menstrual disorder, digestive problem, and ache) and psychological conditions only through the topical application of aromatic compounds. On the other hand, Aromatology is associated with the internal application of EOs via different body-openings (oral, anus, vagina, or any other possible openings) by the qualified experts like medical doctors or herbalists. In most of the European countries, except the United Kingdom, the term aromatology is synonymously used with the term aromatherapy. Due to diversified therapeutic approaches, aromatherapy is categorized into several types, such as:

- **Cosmetic Aromatherapy:** (EOs are used in different cosmetic products, which act as a cleanser, moisturizer, toner, etc.).
- **Massage Aromatherapy:** (along with the vegetable oils, EOs provide a healing touch during massage).
- **Medical Aromatherapy:** (aromas are used as sedative and reliever of anxiety and fear).
- **Olfactory Aromatherapy:** (direct inhalation of aromatic compounds stimulates the human olfactory system, which ultimately helps to improve the mood, enhances the calmness of mind or increases the body function).
- **Psycho-Aromatherapy:** (psychological stability can be achieved through creating certain states of moods and emotions by the application of EOs.

### *3.1.4 MOLECULAR BASIS OF AROMATHERAPY*

For thousands of years, the EOs are being used as the source of fragrance molecules that which have a wide range of therapeutic properties [11, 22]. Scientists from all over the world have documented versatile bioactivities of EOs, such as: antibacterial, antiviral, anti-inflammatory, and immunomodolatory, etc. It has been documented that the aromatic biocompounds exhibit these bioactivities by controlling the emotion, enhancing the alertness and providing the calm effect to the mind, by regulating the secretions of hormones, glands, and circulatory systems of the body [44, 168]. Many pilot projects have also been carried out on human subjects to identify the nature and activity of the volatile oils in controlling the diseases and health disorders [105].

Stimulation of the mammalian olfactory system depends on the accuracy of recognizing and discriminating different chemically distinct odorant molecules that are present in EOs. Detection of these odorants probably results from the association of odorous ligands with specific receptors present on the neurons of the olfactory system. It is estimated that humans can identify more than 10,000 distinct odorous ligands [20]. However, it does not imply that humans possess an equally large number of receptors for all structurally distinct odors. Each individual receptor can only sense structurally related (stereochemically) odor molecules. The genes, which code for about 1000 different types of odorant receptors, form the largest gene family so far described in mammals: larger than the immunoglobulin and T-cell receptor gene families combined. The amino acid sequences of odorant receptors (Olfactory receptors) are very diverse, but all odorant receptors are coupled to heterotrimeric G-proteins.

Transduction of olfactory stimulus is probably initiated after binding of odorant molecules to the specific receptor proteins present in the cilia of olfactory receptor cells. Recently, it has been discovered that the specialized olfactory receptor neurons in the lining of the nose use specific G-protein coupled receptors (GPCRs, a seven-pass transmembrane protein) to recognize odors. These receptors are exhibited themselves on the extended ciliary surface. When odorant molecules bind to these olfactory receptors, they activate an olfactory-specific G protein (known as $G_{olf}$), which makes adenylyl cyclase active. For effective functioning, all these receptors use cyclic AMP, and thus an increase in cyclic AMP concentration occurs to help open-up cyclic-AMP-gated cation channels. An influx of $Na^{+}$ cations occurs through these ion channels, which in turn depolarizes the olfactory receptor neuron and produces a nerve impulse that journeys through its axon to the brain. The probable olfactory signaling mechanism is outlined in Figure 3.1.

Mucous membrane in the nose contains specialized cells, which respond to different aroma chemicals. The receptor organ for smell is the olfactory bulb, which upon detection of aroma signals the central cortex of the brain to send a message to the limbic system (the emotional center of the brain). The aroma-molecules stimulate the limbic system causing changes in biochemistry of the brain and affecting the emotions, desires, and memories. An increase in the cerebral blood flow was observed in the cortex of brain in human after inhalation of cineole of eucalyptus oil [74]. Molecular biology in relation to smelling of different EOs suggests that there is a family of about 1000 genes in humans that give rise to a huge variety of receptor proteins, which are expressed only in olfactory epithelium cells in nose to communicate with the brain to translate and perceive different smells.

## 3.2 METHODS

For this review, a literature search was conducted thoroughly, and relevant information was assembled from printed books and electronic databases like Science Direct, Web of Science, Google Scholar, PubMed, library search, etc. using different keywords. Article selection was restricted by using the keywords "aromatherapy" or "essential oil" and sometimes with either of these words the connector 'AND' was used for the following phrases:

- Traditional healing;
- Chemical nature of essential oil;
- Pharmacological activity;
- Medicinal properties;
- Toxicity;
- Safety issue and others.

## 3.3 RESULTS AND DISCUSSIONS

In total 172 published research articles including book chapters and 10 books on relevant topics were reviewed thoroughly and published data were analyzed with great care to give an insight on 'EOs' and its use in 'aromatherapy.' Different aspects of the present topic were further investigated different aspects of EOs.

### *3.3.1 CHEMISTRY OF ESSENTIAL OILS (EOS)*

EOs are mainly of complex mixtures of volatile, odorous chemical compounds. EOs are also called volatile oils, because they evaporate at room temperature in contrast to fixed oils. Molecular weight of the EOs is generally below 300 Daltons. They are usually hydrophobic [70]. In fresh form, they are colorless liquid, but chamomile oil is violet in color. Very few EOs are found in a crystalline or nebulous form. EOs readily disseminate in ether, alcohol, and other organic solvents. They are also slightly soluble in water. In most of the cases, the density of essential oil is less than water, though clove or cinnamon oils are heavier than water. Due to its high refractive index, most of the EOs can rotate the plane of polarized light. There are two chemical groups of plant-based EOs, namely:

- Terpenoids; and
- Phenylpropanoids.

1. **Terpenoids:** These are one of the largest and most diverse chemical classes produced by the plants. They have a wide range of varieties in their carbon skeleton, oxygenated derivatives and side chains including alcohols, esters, aldehydes, ketones, ethers, peroxides, and phenols. Terpenes and their oxygenated products are closely similar regarding their molecular weight, and this similarity between many of these structures reflects the difficulty in their chemical characterization. Many isomers of terpenoids present in nature are optically active. Some common terpenoids present in essential oil are hydrocarbon and oxygenated derivatives of monoterpenes, sesquiterpenes, and phenylpropanoids. Sometimes, diterpenes, sulfur-, and nitrogen-containing compounds and lactones are also associated with EOs.
2. **Phenylpropanoids:** These are the most versatile class of organic molecules that act as parent molecules for the synthesis of almost all plant polyphenols. The polyphenols synthesized in plants provide them protection from ultraviolet light, herbivores, and pathogens. Polyphenolic compounds are also responsible for various colors of the flowers and its sweet aroma, which attracts pollinator and thus helps in plant-pollinator interactions. Amino acids like phenylalanine and tyrosine act as the mother compounds, from which most of the polyphenols are derived through the shikimic acid pathway. More than 4,000 compounds of Phenylpropanoid family have been described, which are structurally and functionally very diverse. Plant-derived phenylpropanoids and their derivatives share considerable space among the most common biologically active constituents present in food, spices, aromas, fragrances, and medicinal plants. Most of the members of the phenylpropanoid family, including its aromatic compounds, have shown various biological activities, such as:

- Antioxidant;
- UV screening;
- Anticancerous;
- Wound healing;
- Anti-viral;
- Antibacterial;
- Anti-inflammatory.

Recently, cosmetic and perfume industries have shown much interest in phenylpropanoids of both natural and synthetic origin.

Generally, 20–60 biochemical components are present in variable quantities in the EOs. Among them, only two or three components are found in fairly high concentrations (20–70%), which are basically responsible for the effective bioactivity of the EOs. For example, essential oil of *Origanum compactum* consists of carvacrol (30%) and thymol (27%); in *Coriandrum sativum* oil, 68% is of linalool; α- and β-thuyone (57%) and camphor (24%) are the major components of *Artemisia herba-alba* essential oil. *Mentha piperita* consists of menthol (59%) and menthone (19%) as major components.

Chemical structures of some important biochemical components in EOs are given in Figure 3.1.

### 3.3.2 PHARMACOLOGICAL ACTIVITIES OF THE ESSENTIAL OILS (EOS)

Many EOs were screened for their wide range of pharmacological activities. Important pharmacological actions of number of EOs have been summarized in Tables 3.1–3.3. Some of the pharmacological activities of the EOs have been discussed in this section.

#### 3.3.2.1 ANTIMICROBIAL ACTIVITY

Simultaneous increase of microbial infections and microbial resistance to many antibiotics becomes a challenge towards the wellbeing of humans. To overcome these critical issues, search for new antibiotics is going on. Plant secondary metabolites are being studied by scientists to develop potent antimicrobial agents to surpass or reduce this problem [29, 50]. In this point of view, EOs have proved their potency in several scientific studies as promising antimicrobial agents.

Several studies have so far been executed by the scientists to establish the antimicrobial properties of EOs. In some pioneering studies, mode of action of few volatile molecules has been explained in detail, but a large number of such aromatic compounds remains unexplored concerning their mechanism of action. Knowledge regarding the mechanism of action of the volatile antimicrobial compounds is very crucial to understand the magnitude of effectiveness on various kinds of microorganisms, their working principle when combined with other antimicrobial compounds, and their interaction with the components of food matrix. Many antimicrobial activities have been exhibited by the EOs (Figure 3.2).

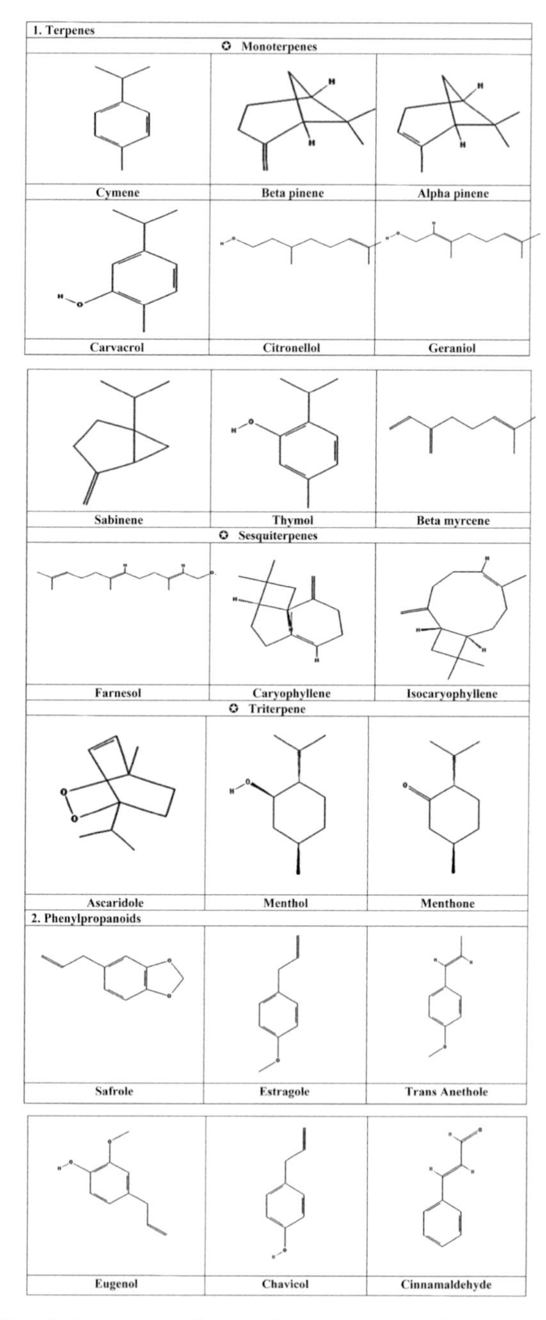


**FIGURE 3.1** Chemical structures of selected phytochemicals in essential oils.

Lipophilic nature of the volatile compounds offers an advantage to be accumulated in the cellular membrane [158]. Therefore, they can cause distortion to the biological membranes by degrading its membrane component like lipopolysaccharide, outer membrane-associated material, increasing its permeability to protons and various cells contents and by inactivating the enzymes implanted in the cell membrane [33, 51, 57, 65, 173, 174].

The growth and viability of fungi largely depends on the structural integrity of its cell wall. The three major structural elements of the fungal cell wall are glucan, chitin, and mannan, which are generally considered as therapeutic targets. Different EOs from plant sources exhibited intense antifungal activity. Trans-anethole present in *Pimpinella anisum* (anise oil) has shown the ability to distort the hyphal morphology of filamentous fungi (*Mucor mucedo*) [183]. Likewise, the growth of *Aspergillus niger* can be inhibited by applying EOs of *Citrus sinensis* [58]. On the other hand, EOs of *Anethum graveolens* seriously affect the TCA cycle and stop mitochondrial ATP synthesis in *Candida albicans* cell [28].

The essential oil components like lupeol and tetraterpenoid present in *Litsea cubeba* can damage cell wall and cell membrane to the various extent and exhibited antifungal potentiality against a number of microorganisms like *Alternaria alternate, Alternaria niger*, *Fusarium moniliforme* and *Fusarium solani* [101]. Sometimes it also causes cytoplasm leakage and partially inhibition of DNA, RNA, protein, and peptidoglycan biosynthesis [64].

Some other components of EOs like cinnamaldehyde, piperidine, citral, furfuraldehyde, and indole were found potent inhibitors of growth and viability of *Candida albicans* by distorting their ergosterol biosynthesis pathway [142].

#### 3.3.2.2 ANTICANCER ACTIVITY

Various types of malignancies like, glioma, colon cancer, gastric cancer, human liver tumor, pulmonary tumor, breast cancer, and leukemia are reported to be lowered after treatment with biocomponents of EOs extracted from plant sources [19]. Therefore, aroma compounds can be opted as preventive and therapeutic agents in cancer therapy [45, 63, 79].

Loutrari et al., [108] have shown that the mastic EOs (*Pistacia lentiscus*) inhibits the growth and metastasis of undifferentiated proliferative tumor cell lines (K562, B16) [108]. Geraniol, a monoterpenoid alcohol from *Cymbopogon martini* was found to have the ability to restrict the functions of the membrane, ion homeostasis, and cell signaling pathways of cancer cell lines. It also inhibits the DNA synthesis, which in turn reduces of the size of colon

**TABLE 3.1** Major Chemical Groups of Essential Oils, its Source Plants and Bioactivities

| Chemical Nature and Major Constituents in Essential Oils | | Plant Source | Bioactivities | References |
|---|---|---|---|---|
| **Monoterpenes (Monoterpenoid phenol)** | Thymol | *Ocimum gratissimum* L., *Trachyspermum ammi* (L.) *Thymus vulgaris* L. | Antimicrobial, anticancerous, significant radical scavenging activities | [17, 153, 157] |
| | Carvacrol | *Origanum vulgare* L., *Thymus vulgaris* L. | Antimicrobial, strong antiplatelet aggregation activity, reduced plasma glucose concentrations, strong antioxidant, anticancerous, and tumor suppression activities | [1, 7, 46, 80, 88, 152, 157] |
| | Cymene | *Cuminum cyminum* L., *Citrus aurantiifolia* (Christm.) *Origanum vulgare* L., | Strong antioxidant, vasorelaxant, and antinociceptive activities | [36, 160] |
| | β-pinene | *Cuminum cyminum* L., *Humulus lupulus* L., *Pinus pinaster Cannabis sativa* L. | Anti-depressant, antibacterial, antimicrobial, and cytotoxic activities | [9, 66] |
| | α-pinene | *Pinus pinaster Origanum vulgare* L., *Rosmarinus officinalis* L., *Cannabis sativa* L., *Satureja myrtifolia* | Anti-inflammatory, bronchodilator, hypoglycemic, sedative, antioxidant, anti-inflammatory, anticatabolic, broad-spectrum antibiotic activities | [126, 136] |
| | Citronellol | *Citrus hystrix* DC., *Cymbopogon nardus* (L.) | Marginal antitumor & antifungal activities | [48, 134] |
| | Geraniol | *Cymbopogon nardus* (L.) *Cymbopogon martini* (Roxb.) *Rosa damascene* Mill., *Rosa centifolia* L. | Insecticidal, antifeedant, hepatoprotective, anti-ulcer, neuroprotective, anthelmintic antioxidant, antimicrobial, anti-inflammatory, and anticancerous activities | [27, 92, 184] |
| | Sabinene | *Zanthoxylum limonella* (Dennst.) *Myristica fragrans Juniperus Sabina* L. | Antifungal, larvicidal, and antioxidant activities | [122, 133] |

**TABLE 3.1** *(Continued)*

| Chemical Nature and Major Constituents in Essential Oils | | Plant Source | Bioactivities | References |
|---|---|---|---|---|
| | β-myrcene | *Coleonema album* (Thunb.) *Apium graveolens* L., *Cannabis sativa* L., *Rosmarinus officinalis* L., *Humulus lupulus* L., *Thymus vulgaris* L., *Cymbopogon citrates* (DC.) *Verbena officinalis* L. | Anti-nociceptive, anti-inflammatory, antiulcerous, and anticancer activities | [15, 26, 143, 166] |
| **Sesquiterpenes** | Farnesol | *Citrus aurantium* L., *Cyclamen persicum* Mill., *Cymbopogon citratus* (DC.) *Polianthes tuberose* L., *Myroxylon balsamum* (L.) | Antimicrobial, anticancerous, antioxidant, anti-inflammatory, and antiallergic | [38, 73, 75, 90, 96, 97] |
| | Caryophyllene | *Syzygium aromaticum* (L.) *Cannabis sativa* L., *Rosmarinus officinalis* L., *Humulus lupulus* L., *Origanum vulgare* L., *Lavandula angustifolia* Mill., *Cinnamomum tamala* | Anticancerous, local anaesthetic activity, anti-arthritic, and anti - inflammatory activities | [34, 55, 98, 175] |
| **Triterpene** | Ascaridole | *Dysphania ambrosioides* (L.) *Peumus boldus* | Anthelmintic, antimalarial, analgesic, anti-fungal, and antirheumatic activities | [128, 138] |
| | Menthol | *Mentha piperita* L., *Mentha arvensis* L., *Ocimum citriodorum* L. | Anaesthetic, anti-inflammatory, antitussive and antitumor activities | [54, 78, 94, 104] |
| | Menthone | *Mentha pulegium* L., *Mentha arvensis* L., *Mentha piperita* L., *Pelargonium cucullatum* (L.) | Antioxidant, antimicrobial, antispasmodic, antiseptic, and anti-HSV | [39, 66, 72] |
| **Phenyl propanoids** | Safrole | *Pimpinella anisum* L., *Sassafras albidum* (Nutt.) *Myristica fragrans Cinnamomum verum* J. *Piper nigrum* L. | Antimicrobial and antioxidant activities | [83, 109] |

**TABLE 3.1** *(Continued)*

| Chemical Nature and Major Constituents in Essential Oils | Plant Source | Bioactivities | References |
|---|---|---|---|
| Estragole | *Artemisia dracunculus* L., *Pinus sylvestris* L., *Clausena anisata* (Willd.) *Syzygium anisatum Pistacia terebinthus* L., *Pinus pinaster Foeniculum vulgare* Mill., *Pimpinella anisum* L. | Anti-inflammatory, antimicrobial, and anti-toxoplasma | [5, 129, 179] |
| Trans Anethole | *Pimpinella anisum* L., *Foeniculum vulgare* Mill., *Syzygium anisatum Glycyrrhiza glabra* L., *Cinnamomum camphora* (L.) *Illicium verum* | Antifertility, fungicidal and antidiabetic | [41, 53, 68, 154] |
| Eugenol | *Syzygium aromaticum Cinnamomum tamala Myristica fragrans Ocimum basilicum* L., *Ocimum tenuiflorum* L., *Artemisia cina* | Anaesthetic, anti-inflammatory, antioxidant, antimicrobial, anticarcinogenic, neuroprotective, hypolipidemic, and antidiabetic. | [82] |
| Chavicol | *Piper beetle* L. | Anti-cancerous, antiplatelet effect. | [3, 25] |
| Cinnamaldehyde | Different species of the genera: *Cinnamomum* and *Cassia* | Antimicrobial, anticancerous, anti-inflammatory and antidiabetic. | [8, 185] |

**TABLE 3.2** Modes of Action of Certain Essential Oils against Different Microorganisms

| Mode of Action | Source Plants with Families | Model Micro-Organisms (Bacteria and Fungi) | References |
|---|---|---|---|
| Induce membrane leakage; Imbalances in pH homeostasis and equilibrium of inorganic ions within the cytoplasm | *Allium sativum* L. (Amaryllidaceae) | *Escherichia coli* | [71] |
| | *Origanum vulgare* L. (Lamiaceae) | *Pseudomonas aeruginosa*, *Staphylococcus aureus* | [94] |
| $K^+$ ion and ATP leakage | *Ocimum gratissimum* L. (Lamiaceae) | *Escherichia coli, Klebsiella* sp., *Pseudomonas aeruginosa, Salmonella enteritidis, Staphylococcus aureus, Proteus mirabilis* | [30, 123, 125] |
| Inhibition of chitin synthase activity | *Pimpinella anisum* L. (Apiaceae) | *Mucor mucedo* | [183] |
| Irreversible change of cell morphology, loss of cellular integrity | *Citrus sinensis* (L.) Osbeck (Rutaceae) | *Aspergillus niger* | [58] |
| Disturbance of the citric acid cycle and inhibition of ATP synthesis in the mitochondria | *Anethum graveolens* L. (Apiaceae) | *Candida albicans* | [28] |
| Inhibition of Histidine decarboxylase activity | *Cinnamomum verum* (Lauraceae) | *Campylobacter jejuni* | [178] |
| Depolarization, permineralization, leakage of the plasma membrane and inhibition of the respiratory activity of the cell | *Cinnamomum verum* (Lauraceae) | *Campylobacter jejuni*, *Salmonella enteritidis*, *Escherichia coli*, *Staphylococcus aureus, Listeria monocytogenes* | [164] |
| Depletion of intracellular ATP pool; Changing the plasma membrane permeability for cations: $K^+$ and $H^+$ leakage | *Origanum vulgare* L. (Lamiaceae) | *Bacillus cereus* | [173, 174] |
| Dissipation of ion gradients; Inhibition of respiration without compromising with ATP production | *Vanilla planifolia* (Orchidaceae) | *Escherichia coli, Lactobacillus plantarum, Listeria innocua* | [51] |

**TABLE 3.2** *(Continued)*

| Mode of Action | Source Plants with Families | Model Micro-Organisms (Bacteria and Fungi) | References |
|---|---|---|---|
| Disruption of the permeability barrier of cell membrane structures and the accompanying loss of chemiosmotic control; Inhibition of respiration | *Melaleuca alternifolia* (Myrtaceae) | *Escherichia coli*, *Staphylococcus aureus*, *Candida albicans* | [31] |
| Destruction of outer and inner membrane of Mitochondria | *Litsea cubeba* (Lour.) (Lauraceae) | *Escherichia coli* | [101] |
| Cell wall damage | *Mentha longifolia* (L.) (Lamiaceae) | *Micrococcus luteus*, *Salmonella typhimurium* | [71] |
| Inhibition of ergosterol biosynthesis and disruption of membrane integrity | *Coriaria nepalensis* (Coriariaceae) | *Candida* sp. | [2] |
| | *Curcuma longa* L. (Zingiberaceae) | *Aspergillus flavus* | [67] |
| Decreased membrane melting temperature and transition enthalpy; Altered membrane potential; Perturbed membrane structure | *Melaleuca alternifolia* (Myrtaceae) | *Candida albicans*, *Escherichia coli*, *Staphylococcus aureus*, *Pseudomonas aeruginosa* | [23, 31, 33] |
| Inhibition of cell wall synthesizing enzymes and cytokinesis of bacteria | *Ocimum sanctum* L. (Lamiaceae) | *Enterobacter aerogenes* | [167] |
| Cell lysis | *Pelargonium graveolens* (Geraniaceae) | *Escherichia coli*, *Staphylococcus aureus*, *Bacillus subtilis* | [147] |

**TABLE 3.3** Activity Against Cancer Cell Lines Shown by Different Components of Essential Oils

| Types of Cancer | Cell Line Used | Source Plant with Families | Mode of Actions | References |
|---|---|---|---|---|
| Leukemia | HL-60 | *Guatteria friesiana* (Annonaceae) | Cytotoxicity | [18] |
| | | *Cymbopogon flexuosus* (Poaceae) | Antiproliferative activity | [91] |
| | | *Casearia sylvestris* Sw. (Salicaceae) | Cytotoxicity | [16] |
| | | *Croton regelianus* (Euphorbiaceae) | Antiproliferative activity | [13] |
| | THP-1 | *Artemisia indica* Willd. (Asteraceae) | Cytotoxicity | [114] |
| | | *Malus domestica* (Rosaceae) | Cytotoxicity | [176] |
| | K562, B16 | *Pistacia lentiscus* L. (Anacardiaceae) | Growth inhibition, apoptosis, and angiogenesis | [108] |
| Liver | HepG2 | *Schefflera heptaphylla* L. (Araliaceae) | Cytotoxicity | [102] |
| | | *Curcuma aromatica* (Zingiberaceae) | Cytotoxicity | [180] |
| | | *Azadirachta indica* (Meliaceae) | Arrest the G0/G1 phase of cell cycle and cell death via ROS induction; Cytochrome C release in mitochondria | [140] |
| | Bel- 7402 | *Aristolochia mollissima* (Aristolochiaceae) | Cytotoxicity | [182] |
| | HepG2, H1299 | *Salvia pisidica* (Lamiaceae) | Protective effect against $H_2O_2$ induced toxicity | [132] |
| Breast | MCF-7, HU02 | *Bassia scoparia* (L.) (Amaranthaceae) and *Trifolium pratense* L. (Leguminosae) | Antiproliferative activity | [84] |
| | MDA-MB 231 | *Afrostyrax lepidophyllus* (Huaceae) and *Scorodophloeus zenkeri* (Leguminosae) | Cytotoxicity | [52] |
| | MCF-7 | *Annona muricata* L. (Annonaceae) | Cytotoxicity | [131] |
| | MCF-7 | *Abies balsamea* (L.) (Pinaceae) | Antitumor activity induced by ROS | [98, 99] |
| Melanoma | MDA-MB435 | *Guatteria friesiana* (Annonaceae) | Antiproliferative activity | [18] |

**TABLE 3.3** *(Continued)*

| Types of Cancer | Cell Line Used | Source Plant with Families | Mode of Actions | References |
|---|---|---|---|---|
| Brain | SF- 295 | *Guatteria friesiana* (Annonaceae) | Cytotoxicity | [18] |
| | *T98G* | *Afrostyrax lepidophyllus* (Huaceae) | Antiproliferative activity | [52] |
| | *T98G* | *Scorodophloeus zenkeri* (Leguminosae) | Cytotoxicity | [52] |
| Colon | HCT-8 | *Guatteria friesiana* (Annonaceae) | Cytotoxicity | [18] |
| | TC-118 human tumors trans-planted in Swiss nu/nu mice | *Cymbopogon martini* (Roxb.) (Poaceae) | Restrict the functions of the membrane, ion homeostasis and cell signaling procedures of cancer cells | [22] |
| | SW480 | *Citrus aurantiifolia* (Rutaceae) | Apoptosis via caspase-3 activation and inhibition of some inflammatory proteins like cox-2 and IL-6 | [76] |
| | HCT116, LN-CaP, PPC1 and MDA-MB231 | *Azadirachta indica* (Meliaceae) | p53 independent apoptosis | [61] |
| Cervix | HeLa cells | *Croton gratissimus* (Euphorbiaceae) | Interference with cell division processes | [127] |
| | HeLa cells, Hep2 | *Ocimum basilicum* L. (Lamiaceae) | Antiproliferative activity | [81] |
| | HeLa cells | *Aristolochia mollissima* (Aristolochiaceae) | Cytotoxic | [182] |
| Kidney | ACHN | *Aristolochia mollissima* (Aristolochiaceae) | Cytotoxic | [182] |

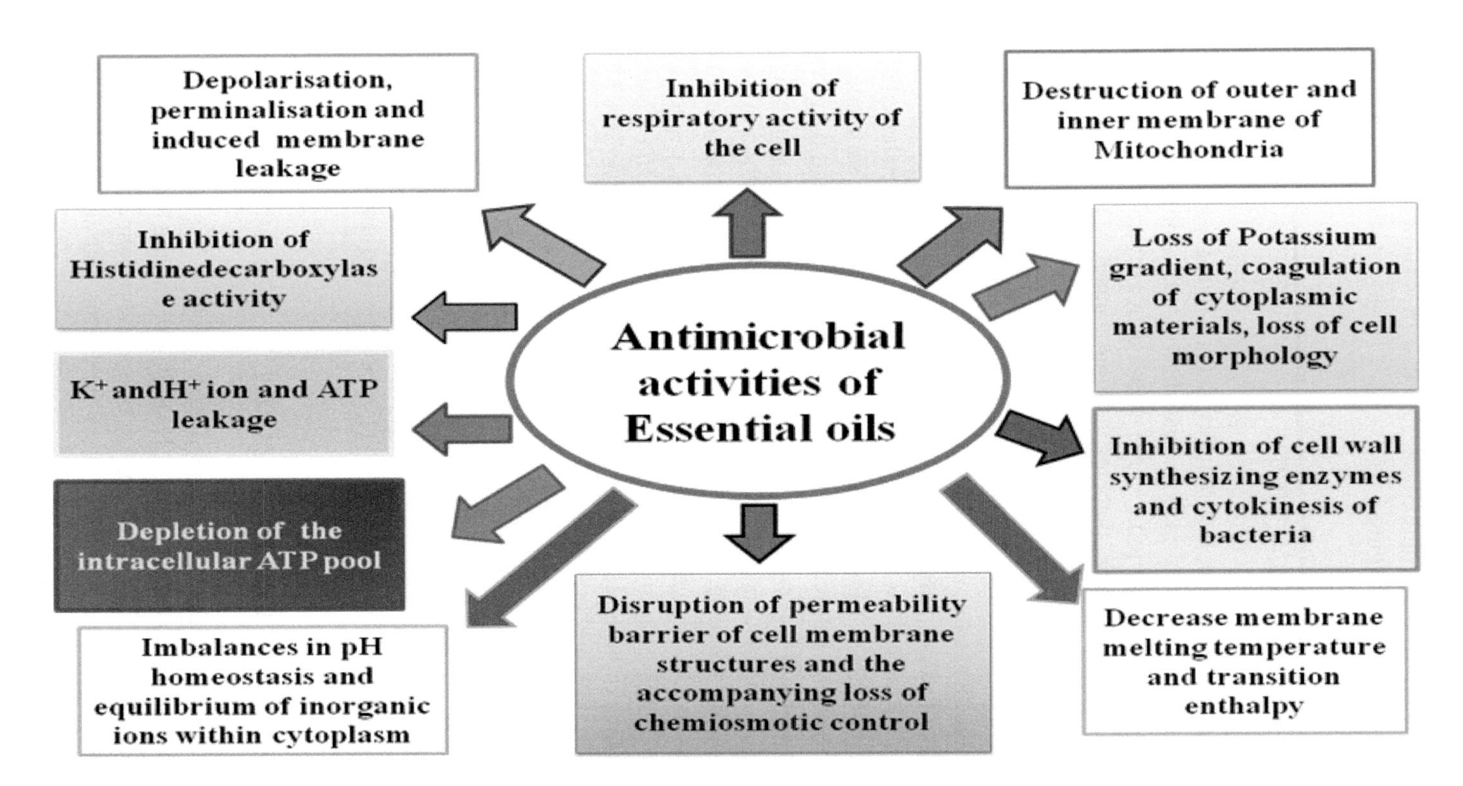


**FIGURE 3.2** **(See color insert.)** Different modes of action of essential oils as antimicrobial agent.

tumors [22]. Perusal of literature indicated that EOs show various modes of anticancerous activities (Figure 3.3).

### 3.3.2.3 ANTIOXIDANT ACTIVITY

During cellular metabolism, free radicals and reactive oxygen species (ROS) are generated as by-products due to oxidative stress, physiological imbalances and/or environmental pollutions, which in turn cause several health problems like, ageing, atherosclerosis, diabetes, asthma, cancer, AD and Parkinson's disease (PD) [45, 117]. In several scientific studies, plant-based EOs reveal significant antioxidant potential [24, 49, 117, 118, 148].

For example, the EOs extracted from the leaves of *Psidium cattleianum* exhibited significant antioxidant activity [150]. The aromatic components of the Basil like linalool, euginol, 1,8-Cineole, epi-α-cadinol, α-bergamotene, and γ-cadinene have shown very good antioxidant activity when tested in the DPPH assay [69, 171]. In another study, α-pinene, β-pinene, sabinene, and bicyclegermacrene extracted from *Cinnamodendron dinisii* and β-myrcene, germacrene-D, and bicyclogermacrene of *Siparuna guianensis* have shown moderate antioxidant activities in β-carotene/linoleic acid method [4].

Terpin-4-ol, sabinene, and γ-terpinene were found to be the major constituents in *Myristica fragrans*, whereas (E)-caryophyllene, α-eudesmol, β-eudesmol and γ-eudesmol were encountered in *S. microphylla.* Essential oil constituents of both plants showed significant antioxidant activity by the DPPH and β-carotene/linoleic acid tests [103].

Many of biocomponents in EOs are obtained from *Origanum majorana, Tagetes filifolia, Bacopa monnierii, Salvia cryptantha* and *S. multicaulis, Achillea millefolium, Melaleuca officinalis, Melaleuca alternifolia, Curcuma zedoaria, Ocimum* sp., *Mentha* sp., *Curcuma longa*, etc.; and considerable antioxidant and radical scavenging potentiality was observed for their antioxidant activity [60, 69, 85, 110, 111, 116, 137, 169, 172].

### 3.3.2.4 ANTI-INFLAMMATORY ACTIVITY

There are several inflammatory mediators such as the tumor necrosis factor-α (TNF-α); interleukin (IL)-1b, IL-8, IL-10; and the PGE2. Numerous biocomponents in the EOs have been investigated for their inhibitory effects on the expression of these inflammatory mediators and on several inflammatory conditions also.

Shinde et al., [156] have established the analgesic potential and anti-inflammatory action of *Cedrus deodara* wood oil through carrageenan-induced rat paw edema model [156]. Following the same method, the anti-inflammatory efficacy of EOs obtained from *Croton cajucara* leaf was observed by Bighetti et al., [14]. *Ocimum sanctum* is known to possess activity against inflammatory reactions. Linolenic acid in *O. sanctum* has been found to be responsible for its anti-inflammatory activity [161]. In some other studies, various biocomponents in EOs from *Baphia nitida, Lavendula angustifolia*, *Mentha piperita,* and *Eucalyptus* sp. have also shown promising anti-inflammatory activities [60, 62, 121, 130, 159]. The tea tree oil (100%) has shown its potentiality in reducing the inflammation caused by histamine diphosphate in humans [87].

### 3.3.2.5 ANTI-VIRAL ACTIVITY

The EOs from *Santolina insularis* had shown very low $IC_{50}$ value in plaque-reduction assay ($IC_{50}$ values of 0.88 mg/mL for HSV-1 and 0.7 mg/mL for HSV-2). This study indicated that the EOs reduced cell-to-cell transmission of Herpes simplex virus type1 and type 2 (HSV-1 and HSV-2) [35]. Aromatic fraction of the root extract of *Cynanchum stauntonii* showed potent antiviral effect against influenza virus in both *in vitro* and *in vivo* experiments [181]. In another experiment, promising anti-viral effect of the EOs from *Artemisia arborescens* has been evaluated against HSV-1 [162]. In an *in vitro* study, the scientists identified the active compounds in EOs from *Cedrus libani* with high antiviral activity against HSV-1 ($IC_{50}$ values of cones and leaves extracts were 0.50 and 0.66 mg/ mL, respectively) without provoking a cytotoxic effect against normal cell line (Vero cells) [106]. In another study, the same group of scientists found that the EOs from *Laurus nobilis* and *Thuja orientalis* were very effective against SARS-corona virus ($IC_{50}$: 120 ± 1.2 mg/mL, and 130 ± 0.4 mg/mL, respectively) and against Herpes simplex virus type 1 ($IC_{50}$: 60 + 0.5 mg/mL and 1000 mg/mL, respectively) [107].

### 3.3.2.6 ANTI-LICE ACTIVITY

Use of essential oil as lice eradicator has also been mentioned in traditional medicines. Different attempts have been made to find out the most potent essential oil components for minimizing the problem of lice. Bioactive compounds of tea tree oil showed their insecticidal activity through

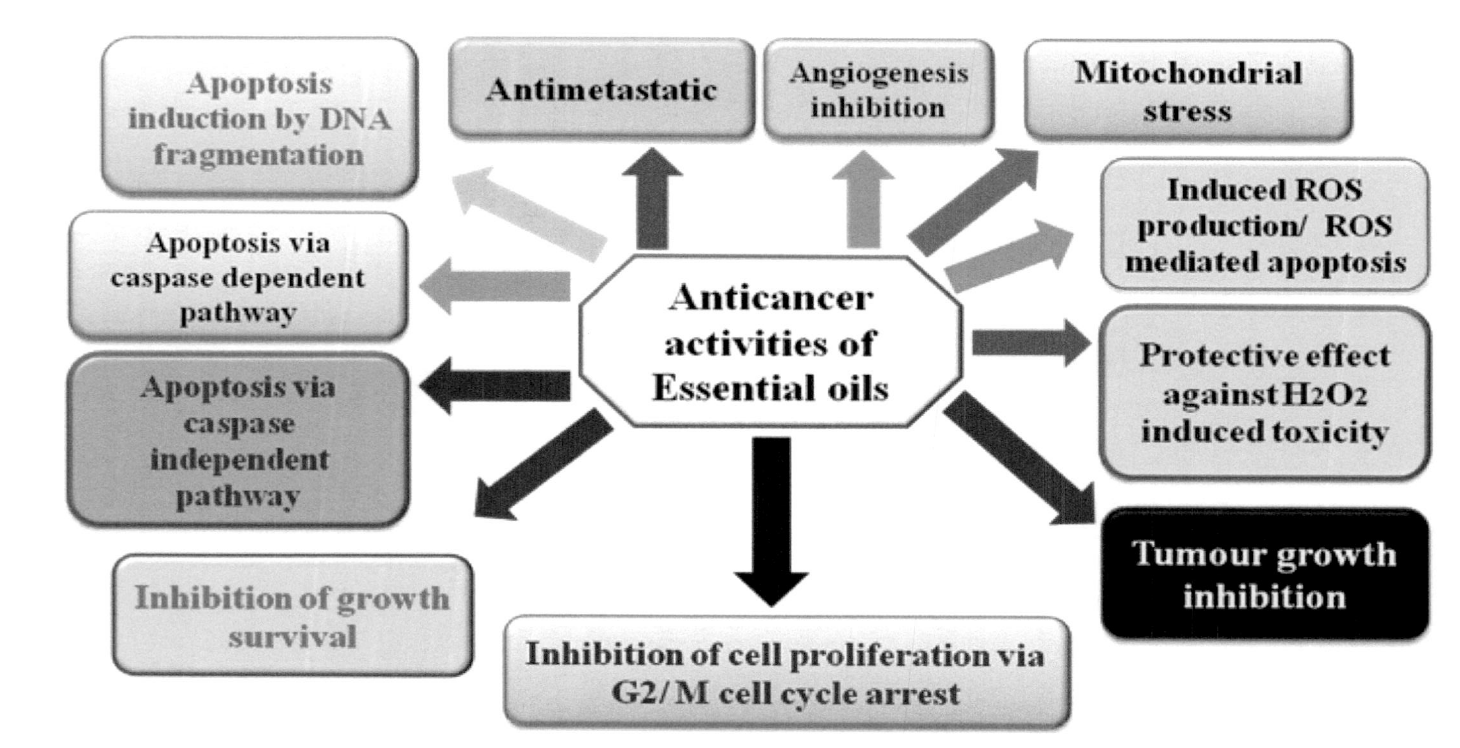


**FIGURE 3.3 (See color insert.)** Essential oils exhibiting different modes of anticancer action.

anticholinesterase action [42, 120]. Priestley et al., [139] studied the effect of monoterpenoid components in essential oil on ovicidal activity and range of lethality to the adult lice [139]. This study confirmed that (+)-Terpinen-4-ol is very effective for adult lice; and mono-oxygenated monocyclic compounds like nerolidol is most lethal to the lice eggs.

#### 3.3.2.7 ANTI-DANDRUFF ACTIVITY

Most common dandruff-causing microorganisms are *Malassezia furfur* and *Pityrosporum ovale*. Thymol in *Coleus amboinicus* and 1-8-Cineol in *Eucalyptus globulus* were found to have excellent anti-dandruff activity against *Malassezia furfur* [151]. Most of the antidandruff shampoos contain tea tree oil. In 2002, Satchell et al., [149] established its anti-dandruff efficacy against *Pityrosporum ovale* [149].

#### 3.3.2.8 INSECT REPELLANT ACTIVITY

Some plant-based insect repellants have better effectiveness than the synthetic ones. Growing research interests have been observed in relation to the repellant activity of EOs because of its volatile nature, which can create a vapor barrier preventing the insect from coming closer to the object.

Nerio et al., [124] reviewed in detail the EOs as insect-repellant based on research articles published till 2008 [124]. The EOs from *Cymbopogon citratus* and *Tagetes minuta* can be considered as potent natural insect repellant against adult sand flies according to recent findings of Kimutai et al., [86]. EOs extracted from freshly collected leaves of *Mentha piperita, Ocimum sanctum, Plectranthus amboinicus*, and *Eucalyptus globulus* have shown significant effectiveness against *Aedes aegypti* (the principal vector of diseases like dengue, yellow fever and chikungunya) [93]. Total of 55 essential oil compounds was from *Nepeta parnassica* by GC-MS, exhibiting positive results for ant (*Pogonomyrmex* sp.) toxicity and mosquito (*Culex pipiens f. molests*) repellency [57].

#### 3.3.2.9 ANTI-SPASMODIC ACTIVITY

Antispasmodics are used to get rid of spasms generally through their activity on smooth muscle. Peppermint oil has a traditional use to get relief from gastrointestinal spasmolysis [100]. Sousa et al., [165] considered rat trachea as an experimental model to find out the anti-spasmodic effect of Peppermint oil and they established its potentiality in relation to prostaglandins and nitric oxide (NO) synthase [165]. Strong anti-spasmodic activity was exhibited by essential oil from *Pycnocycla caespitosa*, which contains carvacrol, β-eudesmol, ρ-cymene, caryophyllene oxide, α-pinine and α-phelandrene as major biocomponents [146].

### 3.3.3 SAFETY ISSUES IN AROMATHERAPY

Some EOs can produce noxious effects at low concentrations because of their heavy load of toxic compounds, and certainly, these should be avoided in aromatherapy. The main points of concern regarding toxicity of EOs are its volatility, fat-solubility, and lipophilicity, which provide an easy access to pass across the cytoplasmic membrane.

Toxicity of non-toxic EOs may happen if one individual had gotten prior exposure to alike sensitive volatile compounds. Spectra of toxicity also depend on adulterants present in it and individuals' age; especially babies, young children, and elderly persons were found vulnerable. Number of aromatherapists has reported sensitive reaction after using massage oils, lotions, or creams that contain a certain percentage of essential oil constituents [32].

The most recent clinical review on hostile response to fragrances has pointed out examples of cutaneous reactions [59]. In the United States, about 6 million people got skin allergy after using some fragrance. Symptoms include:

- Headache;
- Dizziness;
- Nausea;
- Fatigue;
- Shortness of breath;
- Difficulty in concentrating.

High concentration of essential oil components (like phenols (e.g., eugenol, cinnamic aldehyde) and aldehydes (e.g., citral, citronellal)) may be responsible for skin reactions; and it is recommended that prior use of these constituents should be diluted enough to reduce its adverse effect.

It has been observed that fragrances are frequently associated with migraine headaches. According to the American Lung Association, asthma may be triggered by aroma compounds of perfumes and fragrances. Majority of the components of fragrances have been identified as respiratory irritants, and a few of them are known to be associated with respiratory sensitization. Such respiratory irritants are known to make the airways more susceptible to injury and allergens, which in turn exacerbate the conditions like asthma, hay fever, and other respiratory disorders [10, 119, 155].

### 3.3.4 PLACEBO EFFECT OF AROMATHERAPY

Potency of the placebo effect is related with either the patient's belief in the therapy or the patient or practitioner's belief in each other [177]. Studies so far made in this subject have confirmed the existence of placebo effect, because in the majority of cases, aromatherapy is administered to calibrate one's psychological balance. For example, breathing patterns of a patient under aromatherapy are changed intuitively due to the stimulation of emotion controlling center of the brain (the limbic system). Sweet odors help to raise the tidal volume of lung and to decrease the respiratory frequency, which causes harmonic slow and deep breathing. Induction of this meditationary breathing pattern helps to reduce pain [21]. It was also found that the level of opioids in the brain of rats was increased significantly when inhaled with certain EOs. It has also been observed that the pain-relieving factor of opioid was increased in the body by autosuggestion, belief, relaxation, etc. [19]. The increase of opioids basically reduces the pain and uneasiness associated with slow breathing and blood circulatory functions [19, 115]. There is a long debate among the scientists on the placebo effect in aromatherapy. Some scientists have a real faith for retaining complementary and alternative medicine, and others argued that scientific proof of placebo effect is necessary.

## 3.4 SUMMARY

This review confirms that aromatherapy is helpful in relieving the physical and psychological discomforts of the people. Aromatherapy does not heal the diseases only but also rejuvenates the whole body by the action of aromas. Many patients feel better, even if their disease is getting worse, due to their belief in such alternative therapy and this is a good example of "mind over matter," that is, a placebo effect. This therapy is not only preventive but also can be used in the acute and chronic stages of disease. Pharmaceutical industries are now engaged in the development of potent volatile oils, which can be used as an alternative medicine when safety and quality issues of it are studied properly. We can hope with certainty that Aromatherapy will bring new blessing and assurance in the sphere of treatment of diseases for the future generation.

Very few scientific clinical studies on the effectiveness of aromatherapy have been published till date, due to the non-acceptance of the doctrine of aromatherapy in its entirety by the scientists. It is well established that the EOs have a broad range of therapeutic activities, but the process of unveiling their exact mode of action and molecular mechanisms remains almost unattended. The systemic action of EOs exerts an effect on specific organs or tissues and stimulates or modulates any biochemical pathway to implement a specific therapeutic action. Without knowing the exact mode of action and molecular regulation, it is not possible to uplift or modernize the trends of aromatherapy.

The commercial importance of volatile oils is gradually diminishing day by day. Due to chemical evolution, industrialization, and ultramodern lifestyle, the people are gradually losing their interest and hope on aromatherapy. This can be explained with one suitable example. Very few plants are now cultivated for distillation, and many small factories involved in essential oil production have been closed down in the last few decades. The satisfactory growth of the essential oil industries is being hampered by excessive costs of labor and land and the pressure and demands of certain essential oil yielding plants. Overcoming all these problems is a very serious task for the world's scientific community. Also, proper identification tool is required to avoid the adulteration of EOs.

## KEYWORDS

- **aromatherapy**
- **essential oils**
- **pharmacology**
- **phytochemistry**
- **placebo effect**
- **plant sources**

## REFERENCES


1. Aeschbach, R., Löliger, J., Scott, B. C., Murcia, A., Butler, J., Halliwell, B., & Aruoma, O. I., (1994). Antioxidant actions of thymol, carvacrol, 6-gingerol, zingerone and hydroxytyrosol. *Food and Chemical Toxicology*, *32*, 31–36.
2. Aijaz Ahmad, A., Khan, A., Kumar, P., Bhatt, R. P., & Manzoor, N., (2011). Antifungal activity of *Coriaria nepalensis* essential oil by disrupting ergosterol biosynthesis and membrane integrity against *Candida*. *Yeast*, *28*(8), 611–617.
3. Amonkar, A. J., Padma, P. R., & Bhide, S. V., (1989). Protective effect of hydroxychavicol, a phenolic component of betel leaf, against the tobacco-specific carcinogens. *Mutation Research*, *210*(2), 249–253.
4. Andrade, M. A., Das Graças, C. M., De Andrade, J., Silva, L. F., Teixeira, M. L., Valério Resende, J. M., Da Silva Figueiredo, A. C., & Barroso, J. G., (2013). Chemical composition and antioxidant activity of essential oils from *Cinnamodendron dinisii* Schwacke and *Siparuna guianensis* Aublet. *Antioxidant*, *2*(4), 384–397.
5. Andrade, T. C., De Lima, S. G., Freitas, R. M., Rocha, M. S., Islam, T., Da Silva, T. G., & Militão, G. C., (2015). Isolation, characterization and evaluation of antimicrobial and cytotoxic activity of estragole, obtained from the essential oil of *Croton zehntneri* (Euphorbiaceae). *Anais Da Academia Brasileira De Ciências*, *87*(1), 173–182.
6. Antelme, R. S., & Rossini, S., (2001). *Sacred Sexuality in Ancient Egypt: The Erotic Secrets of the Forbidden Papyrus* (p. 112). Inner Traditions/Bear, Rochester, United States.
7. Arunasree, K. M., (2010). Anti-proliferative effects of carvacrol on a human metastatic breast cancer cell line, MDA-MB 231. *Phytomedicine*, *17*, 581–588.
8. Ashakirin, S. N., Tripathy, M., Patil, U. K., & Majeed, A. B. A., (2017). Chemistry and bioactivity of cinnamaldehyde: A natural molecule of medicinal importance. *International Journal of Pharmaceutical Sciences and Research*, *8*(6), 2333–2340.
9. Astani, A., & Schnitzler, P., (2014). Antiviral activity of monoterpenes beta-pinene and limonene against herpes simplex virus *in vitro*. *Iranian Journal of Microbiology*, *6*(3), 149–155.
10. Baldwin, C. M., Bell, I. R., & O'Rourke, M. K., (1999). Odor sensitivity and respiratory complaint profiles in a community-based sample with asthma, hay fever, and chemical odor intolerance. *Toxicology and Industrial Health*, *15*(3/4), 403–409.
11. Baratta, M. T., Dorman, H. J. D., Deans, S. G., Figueiredo, A. C., Barroso, J. G., & Ruberto, G., (1998). Antimicrobial and anti-oxidant properties of some commercial essential oils. *Flavor and Fragrance Journal*, *13*, 235–244.
12. Berwick, A., (1994). *Holistic Aromatherapy: Balance the Body and Soul with Essential Oils* (p. 118). Llewellyn Publications, St. Paul, MN.
13. Bezerra, D. P., Marinho, F. J. D. B., Alves, A. P. N. N., Pessoa, C., De Moraes, M. O., Pessoa, O. D. L., et al., (2009). Antitumor activity of the essential oil from the leaves of *Croton regelianus* and its component ascaridole. *Chemistry and Biodiversity*, *6*, 1224–1231.
14. Bighetti, E. J. B., Hiruma-Lima, C. A., Gracioso, J. S., & Brito, A. R. M. S., (1999). Anti-inflammatory and antinociceptive effects in rodents of the essential oil of *Croton cajucara* Benth. *Journal of Pharmacy and Pharmacology*, *51*(12), 1447–1453.
15. Bonamin, F., Moraes, T. M., Dos Santos, R. C., Kushima, H., Faria, F. M., Silva, M. A., et al., (2014). The effect of a minor constituent of essential oil from *Citrus aurantium*: the

role of β-myrcene in preventing peptic ulcer disease. *Chemico-Biological Interactions*, *212*, 11–19. doi: 10.1016/j.cbi.2014.01.009.
16. Bou, D. D., Lago, J. H. G., Figueiredo, C. R., Matsuo, A. L., Guadagnin, R. C., Soares, M. G., & Sartorelli, P., (2013). Chemical composition and cytotoxicity evaluation of essential oil from leaves of *Casearia sylvestris*, its main compound α-zingiberene and derivatives. *Molecules*, *18*, 9477–9487.
17. Braga, P. C., Dal Sasso, M., Culici, M., Galastri, L., Marceca, M. T., & Guffanti, E. E., (2005). Antioxidant potential of thymol determined by chemiluminescence inhibition in human neutrophils and cell-free systems. *Pharmacology*, *76*, 61–68.
18. Britto, A. C. S., De Oliveria, A. C., Henriques, R. M., Cardoso, G. M., Bomfim, D. S., Carvalho, A. A., et al., (2012). *In vitro* and *in vivo* antitumor effects of the essential oil from the leaves of *Guatteria friesiana*. *Planta Medica*, *78*(5), 409–414.
19. Buchbauer, G., (2010). Biological activities of essential oils. In: Hüsnü, K., (ed.), *Handbook of Essential Oils: Science, Technology, and Applications. 2nd Edition.* K. Husnu Can Baser and Gerhard Buchbauer. Boca Raton - FL: CRC Press; October 27, 2015; 1112 Pages.
20. Buck, L., & Axel, R., (1991). A novel multigene family may encode odorant receptors: A molecular basis for odor recognition. *Cells*, *65*(1), 175–187.
21. Busch, W., Magerl, U., Kern, J., Haas, G. H., & Eichhammer, P., (2012). The effect of deep and slow breathing on pain perception, autonomic activity, and mood processing-an experimentalstudy. *Pain Medicine*, *13*(2), 215–218.
22. Carnesecchi, S., Bras-Goncalves, R., Bradaia, A., Zeisel, M., Gosse, F., Poupon, M. F., & Raul, F., (2004). Geraniol, a component of plant essential oils, modulates DNA synthesis and potentiates 5-fluorouracil efficacy on human colon tumor xenografts. *Cancer Letter*, *215*, 53–59.
23. Carson, C. F., & Riley, T. V., (1995). Antimicrobial activity of the major components of the essential oil of *Melaleuca alternifolia*. *Journal of Applied Bacteriology*, *78*, 264–269.
24. Cavar, S., Maksimovic, M., Vidic, D., & Paric, A., (2012). Chemical composition and antioxidant and antimicrobial activity of essential oil of *Artemisia annua* L. from Bosnia. *Industrial Crops and Products, 37*, 479–485.
25. Chang, M. C., Uang, B. J., Tsai, C. Y., Wu, H. L., Lin, B. R., Lee, C. S., Chen, Y. J., Chang, C. H., Tsai, Y. L., Kao, C. J., & Jeng, J. H., (2007). Hydroxychavicol, a novel betel leaf component, inhibits platelet aggregation by suppression of cyclooxygenase, thromboxane production and calcium mobilization. *British Journal of Pharmacology*, *152*(1), 73–82.
26. Chaouki, W., Leger, D. Y., & Liagre, B., (2009). Citral inhibits cell proliferation and induces apoptosis and cell cycle arrest in MCF-7cells. *Fundamental & Clinical Pharmacology, 23*, 549–556.
27. Chen, W., & Viljoen, A. M., (2010). Geraniol: A review of a commercially important fragrance material. *South African Journal of Botany*, *76*(4), 643–651.
28. Chen, Y., Zeng, H., Tian, J., Ban, X., Ma, B., & Wang, Y., (2013). Antifungal mechanism of essential oil from *Anethum graveolens* seeds against *Candida albicans*. *Journal of Medical Microbiology*, *62*, 1175–1183.
29. Chouhan, S., Sharma, K., & Guleria, S., (2017). Antimicrobial activity of some Essential oils—present status and future perspectives. *Medicines*, *4*(58), 2–21.
30. Cimanga, K., Kambu, K., Tona, L., Apers, S., De Bruyne, T., Hermans, N., Totté, J., Pieters, L., & Vlietinck, A. J., (2002). Correlation between chemical composition and

antibacterial activity of essential oils of some aromatic medicinal plants growing in the Democratic Republic of Congo. *Journal of Ethnopharmacology, 79,* 213–220.

31. Cox, S. D., Mann, C. M., Markham, J. L., Bell, H. C., Gustafson, J. E., Warmington, J. R., & Wyllie, S. G., (2000). The mode of antimicrobial action of the essential oil of *Melaleuca alternifolia* (tea tree oil). *Journal of Applied Microbiology*, *88*(1), 170–175.
32. Crawford, G. H., Katz, K. A., Ellis, E., & James, W. D., (2004). Use of aromatherapy products and increased risk of hand dermatitis in massage therapists. *Archives of Dermatology*, *140*(8), 991–996.
33. Cristani, M., D'Arrigo, M., Mandalari, G., Castelli, F., Sarpietro, M. G., Micieli, D., et al., (2007). Interaction of four monoterpenes contained in essential oils with model membranes: Implications for their antibacterial activity. *Journal of Agricultural and Food Chemistry, 55,* 6300–6308.
34. Dahham, S. S., Tabana, Y. M., Ahamed, M. B. K., & Abdul, M. A. M. S., (2015). *In vivo* anti-inflammatory activity of β-caryophyllene, evaluated by molecular imaging. *Molecules & Medicinal Chemistry*, *1*, E-article 1001. doi: 10.14800/mmc.1001.
35. De Logu, A. G., Loy, M. L., Pellerano, L. B., & Schivo, M. L., (2000). Inactivation of HSV-1 and HSV-2 and prevention of cell-to-cell virus spread by *Santolina insularis* essential oil. *Antiviral Research*, *48*(3), 177–185.
36. De Oliveira, T. M., De Carvalho, R. B., Da Costa, I. H., De Oliveira, G. A., De Souza, A. A., De Lima, S. G., & De Freitas, R. M., (2015). Evaluation of p-cymene, a natural antioxidant. *Pharmaceutical Biology, 53*(3), 423–428.
37. Demirci, F., Berber, H., & Baser, K. H. C., (2007). Biotransformation of *p*-cymene to thymoquinone. *Book of Abstracts of the 38*[th] *ISEO, SL-1*, p. 6.
38. Derengowski, L. S., De-Souza-Silva, C., Braz, S. V., Mello-De-Sousa, T. M., Báo, S. N., Kyaw, C. M., & Silva-Pereira, I., (2009). Antimicrobial effect of farnesol, a *Candida albicans* quorum sensing molecule, on *Paracoccidioides brasiliensis* growth and morphogenesis. *Annals of Clinical Microbiology and Antimicrobials*, *8*(13), p. 8, E-article, doi: 10.1186/1476–0711–8–13.
39. Derwich, E., Chabir, R., Taouil, R., & Senhaji, O., (2011). *In-vitro* antioxidant activity and GC/MS studies on the leaves of *Mentha piperita* (Lamiaceae) from Morocco. *International Journal of Pharmaceutical Sciences and Drug Research*, *3*(2), 130–136.
40. Devereux, P., (1997). *The Long Trip: A Prehistory of Psychedelia* (p. 82). London, Penguin Arkana.
41. Dhar, S. K., (1995). Anti-fertility activity and hormonal profile of trans-anethole in rats. *Indian Journal of Physiology and Pharmacology*, *39*(1), 63–67.
42. Di-Campli, E., Di Bartolomeo, S., Delli, P. P., Di Giulio, M., Grande, R., & Nostro, A., (2012). Activity of tea tree oil and nerolidol alone or in combination against *Pediculus capitis* (head lice) and its eggs. *Parasitology Research*, *111*(5), 1985–1992.
43. Djilani, A., & Dicko, A., (2012). The therapeutic benefits of essential oils. In: Bouayed, J., (ed.), *Nutrition* (pp. 155–178). *Well-Being and Health*. In-Tech, Croatia.
44. Dunning, T., (2013). Aromatherapy: Overview, safety and quality issues. *OA Alternative Medicine, 1*(1), 1–6.
45. Edris, A. E., (2007). Pharmaceutical and therapeutic potentials of essential oils and their individual volatile constituents: A review. *Phytotherapy Research*, *21*, 308–323.
46. Enomoto, S., Asano, R., Iwahori, Y., Narui, T., Okada, Y., Singab, A. N., & Okuyama, T., (2001). Hematological studies on black cumin oil from the seeds of *Nigella sativa* L. *Biological and Pharmaceutical Bulletin, 24*, 307–310.

47. Esposito, E. R., Bystrek, M. V., & Klein, J. S., (2014). An elective course in Aromatherapy science. *American Journal of Pharmaceutical Education*, *78*(4), 1–9.
48. Fang, H. J., Su, X. L., Liu, H. Y., Chen, Y. H., & Ni, J. H., (1989). Studies on the chemical components and anti-tumor action of the volatile oils from *Pelargonium graveolens*. *Yao Xue Xue Bao*, *24*, 366–371.
49. Ferguson, L. R., & Philpott, M., (2008). Nutrition and mutagenesis. *Annual Review of Nutrition*, *28*, 313–329.
50. Fisher, K., & Phillips, C., (2008). Potential antimicrobial uses of essential oils in food: Is citrus the answer? *Trends in Food Science and Technology*, *19*, 156–164.
51. Fitzgerald, D. J., Stratford, M., Gasson, M. J., Ueckert, J., Bos, A., & Narbad, A., (2004). Mode of antimicrobial action of vanillin against *Escherichia coli*, *Lactobacillus plantarum* and *Listeria innocua*. *Journal of Applied Microbiology*, *97*, 104–113.
52. Fogang, H. P. D., Maggi, F., Tapondjou, L. A., Womeni, H. M., Papa, F., Quassinti, L., et al., (2014). *In vitro* biological activities of seed essential oils from the Cameroonian spices *Afrostyrax lepidophyllus* Mildbr. and *Scorodophloeus zenkeri* harms rich in sulfur-containing compounds. *Chemistry and Biodiversity*, *11*, 161–169.
53. Fujita, K. I., & Kubo, I., (2004). Potentiation of fungicidal activities of trans-anethole against *Saccharomyces cerevisiae* under hypoxic conditions. *Journal of Bioscience and Bioengineering*, *98*(6), 490–492.
54. Galeotti, N., Ghelardini, C., Mannelli, L., Mazzanti, G., Baghiroli, L., & Bartolini, A., (2001). Local anaesthetic activity of (+)- and (−)-menthol. *Planta Medica*, *67*, 174–176.
55. Ghelardini, C., Galeotti, N., Di Cesare, M. L., Mazzanti, G., & Bartolini, A., (2001). Local anaesthetic activity of beta-caryophyllene. *Farmaco, 56*(5–7), 387–389.
56. Gill, A. O., & Holley, R. A., (2006). Disruption of *Escherichia coli*, *Listeria monocytogenes* and *Lactobacillus sakei* cellular membranes by plant oil aromatics. *International Journal of Food Microbiology*, *108*, 1–9.
57. Gkinis, G., Tzakou, O., Iliopoulou, D., & Roussis, V., (2003). Chemical composition and biological activity of *Nepeta parnassica* oils and isolated nepetalactones. *Zeitschrift für Naturforschung C*, *58*, 681–686.
58. Gogoi, P., Baruah, P., & Nath, S. C., (2008). Microbiological research. effects of *Citrus sinensis* (L.) Osbeck epicarp essential oil on growth and morphogenesis of *Aspergillus niger* (L.) Van Tieghem. *Microbiological Research*, *163*, 337–344.
59. Guin, J. D., (1982). History, manufacture, and cutaneous reactions to perfumes. In: Frost P, Horwitz SW (eds) *Principles of cosmetics for the dermatologist.* Mosby, St. Louis, Calif.; pages 111–129.
60. Gulluce, M., Sahin, F., Sokmen, M., Ozer, H., Daferera, D., Sokmen, A., Polissiou, M., Adiguzel, A., & Ozkan, H., (2007). Antimicrobial and antioxidant properties of the essential oils and methanol extract from *Mentha longifolia* L. sp. *longifolia*. *Food Chemistry*, *103*, 1449–1456.
61. Gupta, S. C., Prasad, S., Sethumadhavan, D. R., Nair, M. S., Mo, Y. Y., & Aggarwal, B. B., (2013). Nimbolide, a limonoid triterpene, inhibits growth of human colorectal cancer xenografts by suppressing the proinflammatory microenvironment. *Clinical Cancer Research*, *19*(16), 4465–4476.
62. Hajhashemi, V., Ghannadi, A., & Sharif, B., (2003). Anti-inflammatory and analgesic properties of the leaf extracts and essential oil of *Lavandula angustifolia* Mill. *Journal of Ethnopharmacology*, *89*, 67–71.

63. Hamid, A. A., Aiyelaagbe, O. O., & Usman, L. A., (2011). Essential oils: Its medicinal and pharmacological uses. *International Journal of Current Research*, *33*, 86–98.
64. Haque, E., Irfan, S., Kamil, M., Sheikh, S., Hasan, A., Ahmad, A., Lakshmi, V., Nazir, A., & Mir, S. S., (2016). Terpenoids with antifungal activity trigger mitochondrial dysfunction in *Saccharomyces cerevisiae*. *Microbiology*, *85*, 436–443.
65. Helander, I. M., Alakomi, H. L., Latva-Kala, K. S., Mattila-Sandholm, T., Pol, I., Smid, E. J., Gorris, L. G. M., & Wright, A. V., (1998). Characterization of the action of selected essential oil components on gram-negative bacteria. *Journal of Agricultural Food Chemistry*, *46*, 3590–3595.
66. https://ayurvedicoils.com/tag/pharmacological-effects-of-b-pinene, https://ayurvedicoils.com/tag/menthone (Accessed on 29 July 2019).
67. Hua, Y., Zhang, J., Kong, W., Zhao, G., & Yang, M., (2017). Mechanisms of antifungal and anti-aflatoxigenic properties of essential oil derived from turmeric (*Curcuma longa* L.) on *Aspergillus flavus*. *Food Chemistry*, *220*, 1–8.
68. Huang, Y., Zhao, J., Zhou, L., Wang, J., Gong, Y., Chen, X., Guo, Z., Wang, Q., & Jiang, W., (2010). Antifungal activity of the essential oil of *Illicium verum* fruit and its main component trans-anethole. *Molecules*, *15*(11), 7558–7569.
69. Hussain, A. I., Anwar, F., Hussain, S. S. T., & Przybylski, R., (2008). Chemical composition, antioxidant and antimicrobial activities of basil (*Ocimum basilicum*) essential oils depends on seasonal variations. *Food Chemistry*, *108*, 986–995.
70. Hwang, E., & Shin, S., (2015). The effects of aromatherapy on sleep improvement: A systematic literature review and meta-analysis. *Journal of Alternative and Complementary Medicine*, *21*(2), 61–68.
71. Hyldgaard, M., Mygind, T., & Meyer, R. L., (2012). Essential oils in food preservation: Mode of action, synergies and interactions with food matrix components. *Frontiers in Microbiology*, *3*, 12–18.
72. Işcan, G., Kirimer, N., Kürkcüoğlu, M., Başer, K. H., & Demirci, F., (2002). Antimicrobial screening of *Mentha piperita* essential oils. *Journal of Agricultural and Food Chemistry*, *50*(14), 3943–3946.
73. Jabra-Rizk, M. A., Meiller, T. F., James, C. E., & Shirtliff, M. E., (2006). *Effect of farnesol on Staphylococcus aureus biofilm formation and antimicrobial susceptibility. Antimicrobial Agents and Chemotherapy*, *50*(4), 1463–1469.
74. Jäger, W., Nasel, B., Nasel, C., Binder, R., Stimpfl, T., Vycudilik, W., & Buchbauer, G., (1996). Pharmacokinetic studies of the fragrance compound 1, 8-cineol in humans during inhalation. *Chemical Senses*, *21*(4), 477–480.
75. Jahangir, T., Khan, T. H., Prasad, L., & Sultana, S., (2005). Alleviation of free radical mediated oxidative and genotoxic effects of cadmium by farnesol in Swiss albino mice. *Redox Report*, *10*(6), 303–310.
76. Jayaprakasha, G. K., Murthy, K. N. C., Uckoo, R. M., & Patil, B. S., (2012). Chemical composition of volatile oil from *Citrus limettioides* and their inhibition of colon cancer cell proliferation. *Industrial Crops and Products*, *45*, 200–207.
77. Jimbo, D., Kimura, Y., Taniguchi, M., Inoue, M., & Urakami, K., (2009). Effect of aromatherapy on patients with Alzheimer's disease. *Psychogeriatrics*, *9*, 173–179.
78. Juergens, U. R., Stober, M., Schmidt-Schilling, L., Kleuver, T., & Vetter, H., (1998). The anti-inflammatory activity of L-menthol compared to mint oil in human monocytes *in vitro*: A novel perspective for its therapeutic use in inflammatory diseases. *European Journal of Medical Research*, *3*, 539–545.

79. Kaefer, C. M., & Milner, J. A., (2008). The role of herbs and spices in cancer prevention. *Journal of Nutrition and Biochemistry*, *19*, 347–361.
80. Karkabounas, S., Kostoula, O. K., Daskalou, T., Veltsistas, P., Karamouzis, M., Zelovitis, I., et al., (2006). Anticarcinogenic and antiplatelet effects of carvacrol. *Experimental Oncology*, *28*, 121–125.
81. Kathirvel, P., & Ravi, S., (2012). Chemical composition of the essential oil from basil (*Ocimum basilicum* Linn.) and its *in vitro* cytotoxicity against HeLa and HEp-2 human cancer cell lines and NIH 3T3 mouse embryonic fibroblasts. *Natural Product Research*, *26*, 1112–1118.
82. Khalil, A. A., Rahman, U., Khan, M. R., Sahar, A., Mehmood, T., & Khan, M., (2017). Essential oil eugenol: Sources, extraction techniques and nutraceutical perspectives. *RSC Advances*, *7*(52), 32669–32681.
83. Khayyat, S. A., & Al-Zahrani, S. H., (2014). Thermal, photosynthesis and antibacterial studies of bioactive safrole derivative as precursor for natural flavor and fragrance. *Arabian Journal of Chemistry*, *7*, 800–804.
84. Kianinodeh, F., Tabatabaei, S. M., Alibakhshi, A., Mahshid, G., & Kaveh, T., (2017). Anti-tumor effects of essential oils of red clover and ragweed on MCF-7 breast cancer cell line. *Multidisciplinary Cancer Investigation, 1*(4), 17–23.
85. Kim, E. Y., Baik, I. H., Kim, J. H., Kim, S. R., & Rhyu, M. R., (2004). Screening of the antioxidant activity of some medicinal plants. *Korean Journal of Food Science and Technology*, *36*, 333–338.
86. Kimutai, A., Ngeiywa, M., Mulaa, M., Njagi, P. G., Ingonga, J., Nyamwamu, L. B., Ombati, C., & Ngumbi, P., (2017). Repellent effects of the essential oils of *Cymbopogon citratus* and *Tagetes minuta* on the sandfly, *Phlebotomus duboscqi. BMC Res Notes*, *10*(1), 98. doi: 10.1186/s13104–017–2396–0.
87. Koh, K. J., Pearce, A. L., Marshman, G., Finlay-Jones, J. J., & Hart, P. H., (2002). Tea tree oil reduces histamine-induced skin inflammation. *Brazilian Journal of Dermatology*, *147*, 1212–1217.
88. Koparal, A. T., & Zeytinoglu, M., (2003). Effects of carvacrol on a human nonsmall cell lung cancer (NSCLC) cell line, A549. *Cytotechnology*, *43*, 149–154.
89. Krishna, A., Tiwari, R., & Kumar, S., (2000). Aromatherapy-an alternative health care through essential oils. *Journal of Medicinal and Aromatic Plant Science*, *22*, 798–804.
90. Ku, C. M., & Lin, J. Y., (2015). Farnesol, a sesquiterpene alcohol in herbal plants, exerts Anti-inflammatory and Antiallergic effects on ovalbumin-sensitized and -challenged Asthmatic mice. *Evidence-Based Complementary and Alternative Medicine.* 387357. doi:10.1155/2015/387357.
91. Kumar, A., Malik, F., Bhushan, S., Sethi, V. K., Shahi, A. K., Kaur, J., Taneja, S. C., Qazi, G. N., & Singh, J., (2008). An essential oil and its major constituent isointermedeol induce apoptosis by increased expression of mitochondrial cytochrome and apical death receptors in human leukemia HL-60 cells. *Chemico Biological Interaction*, *171*, 332–347.
92. Kumar, A. M., & Devaki, T., (2015). Geraniol, a component of plant essential oils– a review of its pharmacological activities. *International Journal of Pharmacy and Pharmaceutical Sciences*, *7*(4), 67–70.
93. Lalthazuali, & Mathew, N., (2017). Mosquito repellent activity of volatile oils from selected aromatic plants. *Parasitology Research*, *116*(2), 821–825.

94. Lambert, R. J. W., Skandamis, P. N., Coote, P. J., & Nychas, G. J. E., (2001). A study of the minimum inhibitory concentration and mode of action of oregano essential oil, thymol and carvacrol. *Journal of Applied Microbiology, 91*, 453–462.
95. Laude, E. A., Morice, A. H., & Grattan, T. J., (1994). The antitussive effects of menthol, camphor and cineole in conscious guinea-pigs. *Pulmonary Pharmacology*, *7*, 179–184.
96. Lee, J. H., Kim, C., Kim, S. H., Sethi, G., & Ahn, K. S., (2015). Farnesol inhibits tumor growth and enhances the anticancer effects of bortezomib in multiple myeloma xenograft mouse model through the modulation of STAT3 signaling pathway. *Cancer Letter*, *360*(2), 280–293.
97. Lee, S. H., Kim, J. Y., Yeo, S., Kim, S. H., & Lim, S., (2015). Meta-analysis of massage therapy on cancer pain. *Integrative Cancer Therapies*, *5*, 245–256.
98. Legault, J., Dahl, W., Debiton, A. P., & Madelmont, J. C., (2003). Antitumor activity of balsam fir oil: Production of reactive oxygen species induced by α-humulene as possible mechanism of action. *Planta Medica*, *69*(5), 402–407.
99. Legault, J., & Pichette, A., (2007). Potentiating effect of beta-caryophyllene on anticancer activity of alpha-humulene, isocaryophyllene and paclitaxel. *Journal of Pharmacy and Pharmacology*, *59*(12), 1643–1647.
100. Leicester, R., & Hunt, R., (1982). Peppermint oil to reduce colonic spasm during endoscopy. *Lancet*, *2*(8305), 989.
101. Li, W. R., Shi, Q. S., Liang, Q., Xie, X. B., Huang, X. M., & Chen, Y. B., (2014). Antibacterial activity and kinetics of *Litsea cubeba* oil on *Escherichia coli*. *PLoS One*, *9*(11), 1–6.
102. Li, Y. L., Yeung, C. M., Chiu, L. C. M., Cen, Y. Z., & Ooi, V. E. C., (2009). Chemical composition and antiproliferative activity of essential oil from the leaves of a medicinal herb, *Schefflera heptaphylla*. *Phytotherapy Research*, *23*, 140–142.
103. Lima, R. K., Cardoso, M. G., Andrade, M. A., Guimarães, P. L., Batista, L. R., & Nelson, D. L., (2012). Bactericidal and antioxidant activity of essential oils from *Myristica fragrans* Houtt and *Salvia microphylla* H. B. K. *Journal of the American Oil Chemists' Society, 89*, 523–528.
104. Lin, J. P., Li, Y. C., Lin, W. C., Hsieh, C. L., & Chung, J. G., (2001). Effects of (−)-menthol on arylamine N-acetyltransferase activity in human liver tumor cells. *The American Journal of Chinese Medicine*, *29*, 321–329.
105. Liu, S. H., Lin, T. H., & Chang, K. M., (2013). The physical effects of aromatherapy in alleviating work-related stress on elementary school teachers in Taiwan. *Evidence-Based Complementary and Alternative Medicine* (p. 13). E-article. doi: 10.1155/2013/853809.
106. Loizzo, M. R., Saab, A., & Tundis, R., (2008a). Phytochemical analysis and *in vitro* evaluation of the biological activity against *Herpes simplex* virus type 1 (HSV-1) of *Cedrus libani* A. Rich. *Phytomedicine*, *15*(1/2), 79–83.
107. Loizzo, M. R., Saab, A. M., & Tundis, R., (2008b). Phytochemical analysis and *in vitro* antiviral activities of the essential oils of seven Lebanon species. *Chemistry & Biodiversity*, *5*, 461–472.
108. Loutrari, H., Magkouta, S., & Pyriochou, A., (2006). Mastic oil from *Pistacia lentiscus* var. *chia* inhibits growth and survival of human K562 leukemia cells and attenuates angiogenesis. *Nutrition and Cancer*, *55*(1), 86–93.
109. Madrid, A., Espinoza, L., Pavéz, C., Carrasco, H., & Hidalgo, M. E., (2014). Antioxidant and toxicity activity *in vitro* of twelve safrole derivatives. *Journal of the Chilean Chemical Society*, *59*(3), 2598–2601.

110. Maestri, D. M., Nepote, V., Lamarque, A. L., & Zygadlo, J. A., (2006). Natural products as antioxidants. In: Filippo, I., (ed.), *Phytochemistry: Advances in Research* (pp. 105–135). Res. Signpost, Trivandrum, India.
111. Maheshwari, R. K., Singh, A. K., Gaddipati, J., & Srimal, R. C., (2006). Multiple biological activities of curcumin: A short review. *Life Sciences*, *78*, 2081–2087.
112. Manniche, L., (1999). *Sacred Luxuries: Fragrance, Aromatherapy and Cosmetics in Ancient Egypt*. Cornell University Press, New York.
113. Marchand, L., (2014). Integrative and complementary therapies for patients with advanced cancer. *Annals of Palliative Medicine, 3*(3), 160–171.
114. Masaoka, Y., Takayama, M., Yajima, H., Kawase, A., Takakura, N., & Homma, I., (2013). Analgesia is enhanced by providing information regarding good outcomes associated with an odor: Placebo effects in aromatherapy? *Evidence-Based Complementary and Alternative Medicine* (p. 10). E-article. doi: 10.1155/2013/921802.
115. Masaoka, Y., Yajima, H., Takayama, M., Kawase, A., Takakura, N., & Homma, I., (2010). Olfactory stimuli modifies pain and unpleasantness: Investigating respiration and brain areas estimated by a dipole tracing method. *Journal of Japanese Society of Aromatherapy*, *9*(1), 23–29.
116. Mau, J. L., Lai, E. Y., Wang, N. P., Chen, C. C., Chang, C. H., & Chyau, C. C., (2003). Composition and antioxidant activity of the essential oil from *Curcuma zedoaria*. *Food Chemistry*, *82*, 583–591.
117. McCord, J. M., (2000). The evolution of free radicals and oxidative stress. *American Journal of Medicine*, *108*, 652–659.
118. Miguel, M. G., (2010). Antioxidant and anti-inflammatory activities of essential oils: A short review. *Molecules*, *15*, 9252–9287.
119. Millqvist, E., & Lowhagen, O., (1996). Placebo-controlled challenges with perfume in patients with asthma like symptoms. *Allergy*, *51*(6), 434–439.
120. Mills, C., Cleary, B. J., Gilmer, J. F., & Walsh, J. J., (2004). Inhibition of acetylcholine esterase by tea tree oil. *Journal of Pharmacy and Pharmacology*, *56*, 375–379.
121. Moreno, L., Bello, R., Primo-Yufera, E., & Esplugues, J., (2002). Pharmacological properties of the methanol extract from *Mentha suaveolens* Ehrh. *Phytotherapy Research*, *16*, 10–13.
122. Murakami, C., Cordeiro, I., Scotti, M. T., Moreno, P. R. H., & Young, M. C. M., (2017). Chemical composition, antifungal and antioxidant activities of *Hedyosmum brasiliense* mart. ex Miq. (chloranthaceae) essential oils. *Medicines (Basel)*, *4*(3), E55–E66.
123. Nakamura, C. V., Ueda-Nakamura, T., Bando, E., Negrão, M. A. F., Garcia, C. D. A., & Dias, F. F. B. P., (1999). Antibacterial activity of *Ocimum gratissimum* L. essential oil. *Memories do Instituto Oswaldo Cruz*, *94*, 675–678.
124. Nerio, L. S., Olivero-Verbela, J., & Stashenkob, E., (2010). Repellent activity of essential oils: A review. *Biores Tech*, *101*(1), 372–378.
125. Nguefack, J., Leth, V., Amvam, Z. P. H., & Mathur, S. B., (2004). Evaluation of five essential oils from aromatic plants of Cameroon for controlling food spoilage and mycotoxin producing fungi. *International Journal of Food Microbiology, 94*(3), 329–334.
126. Nissen, L., Zatta, A., Stefanini, I., Grandi, S., Sgorbati, B., Biavati, B., & Monti, A., (2010). Characterization and antimicrobial activity of essential oils of industrial hemp varieties (*Cannabis sativa* L.). *Fitoterapia*, *81*(5), 413–419.

127. Okokon, J. E., & Dar, A., (2013). Immuno modulatory, cytotoxic and antileishmanial activity of phytoconstituents of *Croton zambesicus*. *Phytopharmacology Journal*, *4*(1), 31–40.
128. Okuyama, E., Umeyama, K., Saito, Y., Yamazaki, M., & Satake, M., (1993). Ascaridole as a pharmacologically active principle of "Paico," a medicinal Peruvian plant. *Chemical and Pharmaceutical Bulletin*, *41*(7), 1309–1311.
129. Oliveira, C. B. S., Meurer, Y. S. R., & Medeiros, T. L., (2016). Anti-toxoplasma activity of estragole and thymol in murine models of congenital and noncongenital toxoplasmosis. *Journal of Parasitology*, *102*(3), 369–376.
130. Onwukaeme, N. D., (1995). Anti-inflammatory activities of flavonoids of *Baphia nitida* Lodd. (Leguminosae) on mice and rats. *Journal of Ethnopharmacology*, *46*, 121–124.
131. Owolabi, M. S., Ogundajo, A. L., Dosoky, N. S., & Setzer, W. N., (2013). The cytotoxic activity of *Annona muricata* leaf oil from Badagary, Nigeria. *The American Journal of Essential Oil and Natural Product*, *1*(1), 1–3.
132. Ozkan, A., Erdogan, A., Sokmen, M., Tugrulay, S., & Unal, O., (2010). Antitumoral and antioxidant effect of essential oils and *in vitro* antioxidant properties of essential oils and aqueous extracts from *Salvia pisidica*. *Biologia*, *65*(6), 990–996.
133. Pavela, R., Maggi, F., Cianfaglione, K., Bruno, M., & Benelli, G., (2018). Larvicidal activity of Essential oils of five Apiaceae taxa and some of their main constituents against *Culex quinquefasciatus*. *Chemistry & Biodiversity*, *15*(1), 80–90.
134. Pereira, F. O., Mendes, J. M., Lima, I. O., Mota, K. S., Oliveira, W. A., & Lima, E. O., (2015). Antifungal activity of geraniol and citronellol, two monoterpenes alcohols, against *Trichophyton rubrum* involves inhibition of ergosterol biosynthesis. *Pharmaceutical Biology*, *53*(2), 228–234.
135. Perry, N., & Perry, E., (2006). Aromatherapy in the management of psychiatric disorders clinical and neuropharmacological perspectives. *CNS Drugs, 20*, 257–280.
136. Pinheiro, M. A., Magalhães, R. M., Torres, D. M., Cavalcante, R. C., Mota, F. S., & Oliveira, C. E. M., (2015). Gastroprotective effect of alpha-pinene and its correlation with antiulcerogenic activity of essential oils obtained from *Hyptis* species. *Pharmacognosy Magazine*, *11*(41), 123–130.
137. Politeo, O., Jukic, M., & Milos, M., (2007). Chemical composition and antioxidant capacity of free volatile aglycones from basil (*Ocimum basilicum* L.) compared with its essential oil. *Food Chemistry*, *101*, 379–385.
138. Potawale, S. E., Luniya, K. P., Mantri, R. A., Mehta, U. K., Waseem, M., Sadiq, M., Vetal, Y. D., & Deshmukh, R. S., (2008). *Chenopodium ambrosioides*: An ethnopharmacological review. *Pharmacologyonline*, *2*, 272–286.
139. Priestley, C. M., Burgess, I. F., & Williamson, E. M., (2006). Lethality of essential oil constituents towards the human louse, *Pediculus humanus* and its eggs. *Fitoterapia, 77*(4), 303–309.
140. Priyadarsini, S. C., Murugan, R. S., Sripriya, P., Karunagaran, D., & Nagini, S., (2010). The neem limonoids azadirachtin and nimbolide induce cell cycle arrest and mitochondria-mediated apoptosis in human cervical cancer (HeLa) cells. *Free Radical Research*, *44*(6), 624–634.
141. Promoting Healthy Life. http://www.who.int/whr/2002/en/ (Accessed on 29 July 2019).
142. Rajput, S. B., & Karuppayil, S. M., (2013). Small molecules inhibit growth, viability and ergosterol biosynthesis in *Candida albicans*. *Springer plus*, *2*(1), 2–6.
143. Rao, V. S. N., Menezes, A. M. S., & Viana, G. S. B., (1990). Effect of myrcene on nociception in mice. *Journal of Pharmacy and Pharmacology*, *42*(12), 877–878.

144. Rashid, S., Rather, M. A., Shah, W. A., & Bhat, B. A., (2013). Chemical composition, antimicrobial, cytotoxic and antioxidant activities of the essential oil of *Artemisia indica* Willd. *Food Chemistry*, *138*, 693–700.
145. Rosato, A., Vitali, C., De Laurentis, N., Armenise, D., & Antonietta, M. M., (2007). Antibacterial effect of some essential oils administered alone or in combination with norfloxacin. *Phytomedicine*, *14*, 727–732.
146. Sadraei, H., Asghari, G., & Alipour, M., (2016). Anti-spasmodic assessment of hydroalcoholic extract and essential oil of aerial part of *Pycnocycla caespitosa* Boiss. & Hausskn on rat ileum contractions. *Research in Pharmaceutical Science*, *11*(1), 33–42.
147. Sánchez, E., García, S., & Heredia, N., (2010). Extracts of edible and medicinal plants damage membranes of *Vibrio cholerae*. *Applied and Environmental Microbiology*, *76*, 6888–6894.
148. Sanchez-Vioque, R., Polissiou, M., Astraka, K., Mozos-Pascual, M., Tarantilis, P., Herraiz-Penalver, D., & Santana-Meridas, O., (2013). Polyphenol composition and antioxidant and metal chelating activities of the solid residues from the essential oil industry. *Industrial Crops and Products*, *49*, 150–159.
149. Satchell, A. C., Saurajen, A., Bell, C., & Barnetson, R. S., (2002). Treatment of dandruff with 5% tea tree oil shampoo. *Journal of the American Academy of Dermatology*, *47*, 852–855.
150. Scur, M. C., Pinto, F. G. S., Pandini, J. A., Costa, W. F., Leite, C. W., & Temponi, L. G., (2016). Antimicrobial and antioxidant activity of essential oil and different plant extracts of *Psidium cattleianum* Sabine. *Brazilian Journal of Biology*, *76*, 101–108.
151. Selvakumar, P., Edhayanaveena, B., & Prakash, S. D., (2012). Studies on the antidandruff activity of the essential oil of *Coleus amboinicus* and *Eucalyptus globulus*. *Asian Pacific Journal of Tropical Disease*, *2*(2), S715–S719.
152. Shahsavari, R., Ehsani-Zonouz, A., Houshmand, M., Salehnia, A., Ahangari, G., & Firoozrai, M., (2009). Plasma glucose lowering effect of the wild *Satureja khuzestanica* Jamzad essential oil in diabetic rats: Role of decreased gluconeogenesis. *Pakistan Journal of Biological Sciences*, *12*(2), 140–145.
153. Sharma, M., Agarwal, S. K., Sharma, P. R., Chadha, B. S., Khosla, M. K., & Saxena, A. K., (2010). Cytotoxic and apoptotic activity of essential oil from *Ocimum viride* towards COLO 205 cells. *Food and Chemical Toxicology*, *48*, 336–344.
154. Sheikh, B. A., Ayyasamy, L. P., & Chandramohan, R. R., (2015). Trans-anethole, a terpenoid ameliorates hyperglycemia by regulating key enzymes of carbohydrate metabolism in streptozotocin induced diabetic rats. *Biochimie*, *112*, 57–65.
155. Shim, C., & Williams, M. H., (1986). Effect of odors in asthma. *American Journal of Medicine*, *80*(1), 18–22.
156. Shinde, U. A., Phadke, A. S., Nair, A. M., Mungantiwar, A. A., Dikshit, V. J., & Saraf, M. N., (1999). Studies on the anti-inflammatory and analgesic activity of *Cedrus deodara* (Roxb.) Loud. wood oil. *Journal of Ethnopharmacology*, *65*(1), 21–27.
157. Sikkema, J., De Bont, J. A. M., & Poolman, B., (1995). Mechanisms of membrane toxicity of hydrocarbons. *Microbiology Reviews*, *59*, 201–222.
158. Sikkema, J., De Bont, J. A., & Poolman, B., (1994). Interactions of cyclic hydrocarbons with biological membranes. *Journal of Biological Chemistry*, *269*, 8022–8028.
159. Silva, J., Abebe, W., Sousa, S. M., Duarte, V. G., Machado, M. I. L., & Matos, F. J. A., (2003). Analgesic and anti-inflammatory effects of essential oils of *Eucalyptus*. *Journal of Ethnopharmacology*, *89*, 277–283.

160. Silva, M. T., Ribeiro, F. P., Medeiros, M. A., Sampaio, P. A., Silva, Y. M., Silva, M. T., Quintans, J. S., Quintans-Júnior, L. J., & Ribeiro, L. A., (2015). The vasorelaxant effect of p-cymene in rat aorta involves potassium channels. *Scientific World Journal* (p. 8). E-article ID: 458080. doi: 10.1155/2015/458080.
161. Singh, S., & Majumdar, D. K., (1997). Evaluation of anti-inflammatory activity of fatty acids of *Ocimum sanctum* fixed oil. *Indian Journal of Experimental Biology*, *35*, 380.
162. Sinico, C., De Logu, A., Lai, F., Valenti, D., Manconi, M., Loy, G., Bonsignore, L., & Fadda, A. M., (2005). Liposomal incorporation of *Artemisia arborescens* L. essential oils and *in vitro* antiviral activity. *European Journal of Pharmaceutics and Biopharmaceutics*, *59*(1), 161–168.
163. Smith, C. A., Collins, C. T., & Crowther, C. A., (2011). Aromatherapy for pain management in labor. *The Cochrane Database of Systematic Reviews, 8*, 79–86.
164. Smith-Palmer, A., Stewart, J., & Fyfe, L., (1998). Antimicrobial properties of plant essential oils and essences against five important food-borne pathogens. *Letters in Applied Microbiology, 26*, 118–122.
165. Sousa, A. A., Soares, P. M., Almeida, A. N., Maia, A. R., Souza, E. P., & Assreuy, A. M., (2010). Antispasmodic effect of *Mentha piperita* essential oil on tracheal smooth muscle of rats. *Journal of Ethnopharmacology*, *130*(2), 433–436.
166. Sousa, O. V., Silvério, M. S., Del-Vechio-Vieira, G., Matheus, F. C., Yamamoto, C. H., & Alves, M. S., (2008). Antinociceptive and anti-inflammatory effects of the essential oil from *Eremanthus erythropappus* leaves. *Journal of Pharmacy and Pharmacology*, *60*(6), 771–777.
167. Suppakul, P., Miltz, J., Sonneveld, K., & Bigger, S. W., (2003). Antimicrobial properties of basil and its possible application in food packaging. *Journal of Agricultural and Food Chemistry*, *51*, 3197–3207.
168. Svoboda, K. P., & Deans, S. G., (1995). Biological activities of essential oils from selected aromatic plants. *Acta Horticulturae*, *390*, 203–209.
169. Tepe, B., Donmez, E., Unlu, M., Candan, F., Daferera, D., Vardar-Unlu, G., Polissiou, M., & Sokmen, A., (2004). Antimicrobial and antioxidative activities of the essential oils and methanol extracts of *Salvia cryptantha* (Montbret et Aucherex Benth.) and *Salvia multicaulis* (Vahl). *Food Chemistry*, *84*, 519–525.
170. Tisserand, R., Gattefossé, R. M., & Davies, L., (2012). *Gattefossé's Aromatherapy*. London: Ebury Digital. E-book.
171. Tomaino, A., Cimino, F., Zimbalatti, V., Venuti, V., Sulfaro, V., De Pasquale, A., & Saija, A., (2005). Influence of heating on antioxidant activity and the chemical composition of some spice essential oils. *Food Chemistry*, *89*, 549–554.
172. Tripathi, R., Mohan, H., & Kamat, J. P., (2007). Modulation of oxidative damage by natural products. *Food Chemistry*, *100*(1), 81–90.
173. Ultee, A., Bennik, M. H. J., & Moezelaar, R., (2002). The phenolic hydroxyl group of carvacrol is essential for action against the food-borne pathogen *Bacillus cereus*. *Applied and Environmental Microbiology*, *68*, 1561–1568.
174. Ultee, A., Kets, E. P. W., & Smid, E. J., (1999). Mechanisms of action of carvacrol on the food-borne pathogen *Bacillus cereus*. *Applied and Environmental Microbiology*, *65*, 4606–4610.
175. Vijayalaxmi, A., Bakshi, V., Begum, N., Kowmudi, V., Naveen, K. Y., & Reddy, Y., (2016). Anti-arthritic and anti inflammatory activity of β Caryophyllene against

Freund's complete adjuvant induced arthritis in Wistar Rats. *Journal of Bone Reports and Recommendations*, *1*, 9. doi: 10.4172/2469–6684.10009.

176. Walia, M., Mann, T. S., Kumar, D., Agnihotri, V. K., & Singh, B., (2012). Chemical composition and *in vitro* cytotoxic activity of essential oil of leaves of *Malus domestica* growing in Western Himalaya (India). *Evidence Based Complement and Alternative Medicine*. doi: 10.1155/2012/649727.
177. Weil, A., (1983). *Health and Healing* (p. 215). Houghton Mifflin, Boston, MA.
178. Wendakoon, C. N., & Morihiko, S., (1995). Inhibition of amino acid decarboxylase activity of *Enterobacter aerogenosa* by active components in spices. *Journal of Food Protection, 58*, 280–283.
179. Wiirzler, L. A. M., Silva-Filho, S. E., Aguiar, R. P., Cavalcante, H. A. O., & Cuman, R. K. N., (2016). Evaluation of anti-inflammatory activity of Estragole by modulation of Eicosanoids production. *International Journal of Pharma and Chemical Research*, *2*(1), 7–13.
180. Xiao, Y., Yang, F., Li, S. P., Hu, G., Lee, S. M. Y., & Wang, Y. T., (2008). Essential oil of *Curcuma wenyujin* induces apoptosis in human hepatoma cells. *World Journal of Gastroenterology*, *14*, 4309–4318.
181. Yang, Z. C., Wang, B. C., Yang, X. S., & Wang, Q., (2005). Chemical composition of the volatile oil from *Cynanchum stauntonii* and its activities of anti-influenza virus. *Colloids and Surfaces B: Biointerfaces*, *43*, 198–202.
182. Yu, J. Q., Liao, Z. X., Cai, X. Q., Lei, J. C., & Zou, G. L., (2007). Composition, antimicrobial activity and cytotoxicity of essential oils from *Aristolochia mollissima*. *Environment Toxicology and Pharmacology*, *23*, 162–167.
183. Yutani, M., Hashimoto, Y., Ogita, A., Kubo, I., Tanaka, T., & Fujita, K., (2011). Morphological changes of the filamentous fungus *Mucor mucedo* and inhibition of chitin synthase activity induced by anethole. *Phytotherapy Research*, *25*, 1707–1713.
184. Zanetti, M., Ternus, Z. R., Dalcanton, F., De Mello, M. M. J., De Oliveira, D., Araujo, P. H. H., Riella, H. G., & Fiori, M. A., (2015). Microbiological characterization of pure geraniol and comparison with bactericidal activity of the cinnamic acid in gram-positive and gram-negative bacteria. *Journal of Microbial and Biochemical Technology*, *7*(4), 186–193.
185. Zhu, R., Liu, H., Liu, C., Wang, L., Ma, R., Chen, B., Li, L., Niu, J., Fu, M., Zhang, D., & Gao, S., (2017). Cinnamaldehyde in diabetes: A review of pharmacology, pharmacokinetics and safety. *Pharmacological Research*, *122*, 78–89.

## CHAPTER 4

# EXTRACTION AND THERAPEUTIC POTENTIAL OF ESSENTIAL OILS: A REVIEW

SUJATHA GOVINDARAJ


## ABSTRACT

Essential oils (EOs) are fragrant and unbalanced compounds and are deposited in exceptional fragile secretory structures, such as:

- Secreting organs;
- Trichomes;
- Secretory canals;
- Secretory craters or resin canals.

EOs remain aquaphobic, solvable in alcohol, non-ionic or faintly polar solvents, waxes, and oils, however, somewhat dissolvable in water; and for the most part, they are dull or light yellow in color. Oils held inside plant cells are released by force from different fragments of the plant material: the leaves, blossoms, organic products, swards, roots, lumber, bay, and gums. Mining of volatile constituents from plants can be attained through diverse approaches, of which hydro-distillation and steam distillation are widely preferred. Other approaches embody aqueous infusion, solvent extraction, hot or cold pressing, supercritical fluid extraction (SFE), enfleurage, and phytonic method. Mono- and sesqui-terpene hydrocarbons and oxygenated materials are chief compounds present in EOs. They are utilized as a part of flavorings, aromas, in fragrant healing, antimicrobials, repellents, pharmaceutical preparations, and in numerous different ways. The best method to use is by outer application or inward breath. However, some can be exceptionally gainful when taken inside. The present review focuses on the extraction and therapeutic properties of EOs.

## 4.1 INTRODUCTION

There are two types of plant-based oils:

1. Non-volatile oils; and
2. Essential oils (EOs) (volatile oils).

Non-volatile constituents include glycerol and fatty acid esters (triglycerides or triacylglycerols), whereas volatile oils are a composite mixture of unstable and semi-volatile carbon-based mixtures instigating from the solitary botanic basis that regulates the odor of plants and also the savor and cologne of the plants [37]. EOs are usual products created by numerous volatile complexes and are otherwise referred to as concentrates, unbalanced oils, etheric oils or aetheroleum. In accordance with the International Standard Organization on EOs (ISO 9235: 2013) and conjointly with the European Pharmacopoeia, an essential oil is the product attained from plant raw material by hydrodistillation (HD), steam distillation or dry distillation or by an adequate power-driven methodology.

EOs are concerted volatile aromatic compounds synthesized by plants and are simply gaseous essences that offer plants their marvelous scents. Each of these complicated treasurable liquids is mined from a specific species of flora. Distinct sections of the biosphere with specific environmental conditions and adjoining plants and animals have resulted in the origin of each plant species. EOs are often stated as the "life strength" of plants. From roots, stems, leaves, fruit rinds, seeds, barks, and resins, the highly concentrated volatile oils are obtained in contrast to fatty oils. The volume of EOs discovered in these plants ranges from 0.01 to 10% of the whole plant. That is why; heaps of plant matter is required for simply a few pounds of oil. These oils are characteristically used for the flavor and healing or odoriferous properties, in an exceedingly good collection of product like foods, medicines, and cosmetics. The ingredients of plant EOs fall primarily into two different biochemical categories:

1. Terpenes; and
2. Phenylpropanoids.

Although terpenes and their oxygenated derivatives (terpenoids) are common and plentiful, yet few species comprise great amounts of shikimates; notably, phenylpropanoids; and once the compounds are existing, they deliver a precise aroma and savor to the plant [5, 50, 58]. Terpenes and terpenoids are resulted from the condensation of isoprene (2-methyl-1,3-butadiene), a 5-carbon unit with two unsaturated bonds, and are persistently named as isoprenoids (Figure 4.1), which possess numerous isomeric cyclic or linear structures, and varied grades of unsaturation, replacements, and ventilated derivatives, typically referred to as terpenoids [7]. Isoprene units are coupled in one path, where the divided end of the chain is mentioned as the head of the molecule and the supplementary end as a tail. Hence, the preparation of the structure is termed head-to-tail linking (Figure 4.2). This design of coupling is elucidated by the biogenesis of terpenoids [27]. Further, terpenes are the utmost vital and greatest assorted category of volatile organic compounds (VOCs).

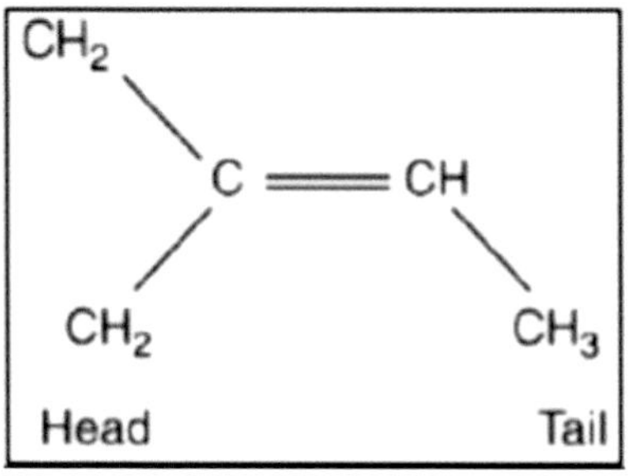

**FIGURE 4.1** Isoprene.

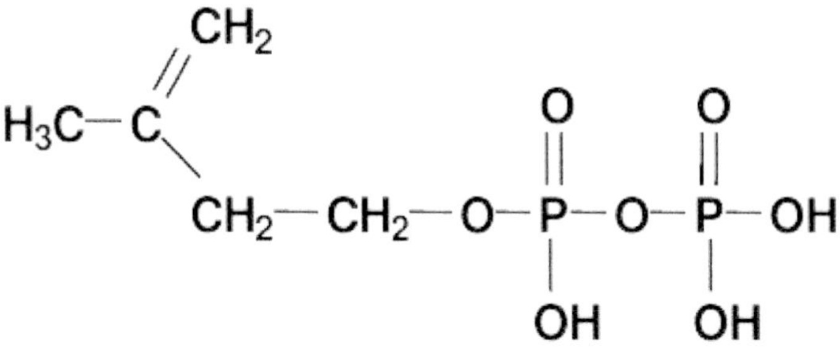

**FIGURE 4.2** Isopentenyl pyrophosphate.

Terpenes are categorized based on the number of isoprene units in their construction, e.g., hemiterpenes (1 unit), monoterpenes (2 units), sesquiterpenes (3 units), diterpenes (4 units) and more. Most volatile oils are highly advanced blends of monoterpenes ($C_{10}H_{16}$) and sesquiterpenes ($C_{15}H_{24}$), and embody biogenetically connected phenols (phenylpropanes and cinnamates), in conjunction with carbohydrates, alcohols, ethers, aldehydes, and ketones that are responsible for their characteristics. Besides, often trace quantities of heavier terpenes, like diterpenes, will likewise be present in EOs with four isoprene units; however, these generally do not boost the aroma of EOs, just like the diterpenes present in ginger oil [24]. The mevalonate pathway leading to sesquiterpenes, the methylerythritol pathway leading to mono- and diterpenes, and also the shikimic acid pathway leading to phenylpropenes are three important biosynthetic pathways for originating the necessary components of volatile oils [4].

Plant terpenoids are made from isopentenyl pyrophosphate (IPP) or isopentenyl diphosphate (IPD) (Figure 4.2) and its isomer, dimethylallyl diphosphate (DMAD). These thus referred to as "active isoprene units" that are resultant from the mevalonic acid and methylerythritol phosphate synthesis routes. IPD afterward reacts with DMAD to form geranyl diphosphate. This $C_{10}$ compound is the creator of monoterpenoids, as they comprise one pair of five-carbon units [50].

Aromatic plants have been used since early times for their preserving and therapeutic characteristics, and these enhance the fragrance and flavor to food. The pharmaceutical resources of aromatic plants are ascribed to volatile oils. The word 'essential oil' was introduced in the sixteenth century by Paracelsus Von Hohenheim, who termed the active part of a drug '*Quinta essential*' [19]. Several procedures have been established to extract EOs from various segments of the aromatic plant. The Persian physician Avicenna (980–1037 A.D.) executed the first modern distillation of volatile oils, and he extracted the *quintessence* from rose petals by means of 'enfleurage' method. The finding and successive usage of a beautiful fragrance substance gradually directed him to write a book on the *curative characteristics of the volatile oil of rose*.

In the early twentieth century, the French Chemist Rene-Maurice Gattefosse coined the "Aromatherapy." From 2000 plants, about 3000 EOs have been extracted, of which nearly 300 are saleable [13]. These volatile oils are deposited in oil canals, resin canals, glands, or trichomes (glandular hairs) of plants [7].

Volatile oils can be obtained from numerous parts of plants, consisting of grasses (lemongrass), foliage (Tulsi), florae (rose), dried flower buds (clove), fruits (lemon), seeds (fennel), roots (vetiver), rhizomes (ginger), bark (cinnamon), wood (Sandal), gum (frankincense), bulbs (garlic) and tree blooms (ylang–ylang) [8, 22, 50]. The varied extraction methods of EOs include HD, steam distillation, expression, enfleurage, organic solvent extraction, high-pressure solvent mining, microwave diffusion and gravity, microwave-assisted distillation, ultrasonic extraction, supercritical fluid extraction (SFE), solvent-free microwave extraction (SFME) and also the phytonic method [14, 40].

EOs are recommended for the medication in a wide range of health issues since ancient times. Antimicrobial, anticancer, antiviral, anti-inflammatory, antimutagenic, antidiabetic, and antiprotozoal activities are allocated to EOs. The in-depth phytochemical examination has paved the way to the categorization and documentation of chief components of volatile oils that are of extensive interest, notably to beautifying and medicinal industries. The various strategies of mining the EOs and their biologically active properties and medicinal ability are conferred in this chapter.

The review in this chapter focuses on the extraction and therapeutic properties of EOs.

## 4.2 MINING OF ESSENTIAL OILS (EOS)

### 4.2.1 STEAM DISTILLATION

Arabian physician Avicenna is attributed for the introduction of the distillation method for EOs. For natural aromatic compounds, steam distillation is an exceptional method of distillation. It was once a well-adorned protocol for isolation of organic compounds but is now outdated by vacuum distillation. Steam distillation remains vital in many manufacturing sectors. Water or steam distillation was introduced as separation by conventional distillation (at 1 atm), and will decompose organic compounds at elevated temperatures (Figure 4.3). The vapor carries tiny amounts of the volatilized complexes to the condensation flask, wherever the condensed section separates and allows easy collection. This method effectively permits for distillation at lower temperatures, lessening the corrosion of the required product. If the materials to be distilled are terribly subtle to warmth, steam distillation could be used at reduced pressure, thus dropping the operation temperature furthermore. The vapors are condensed after distillation [3]. Typically, the immediate product may be a two-stage system of water and therefore the organic distillation, giving separation of the parts by decantation, separating, or alternative appropriate ways. The principle of operation of this system indicates that the collective vapor pressure matches the surrounding pressure at around 100°C so that the volatile parts with the boiling points starting from 150–300°C are vaporized at a temperature near to that of water. Besides, this method can also be executed at low pressure depending on the volatile oils extraction issues.

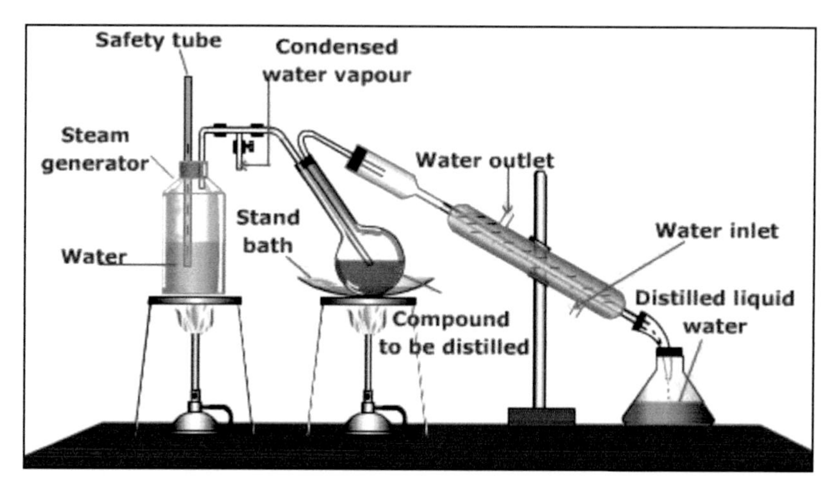


**FIGURE 4.3** **(See color insert.)** Schematic apparatus of steam distillation.

## *4.2.2 HYDRODISTILLATION (HD)*

Water or HD is the simplest strategy for separation of volatile oils [35]. HD is typically used to isolate EOs from the medicinal and aromatic plants. In HD, the EOs are volatilized by warming a blend of water or other solvent and herbal materials take part after by the liquefaction of the fumes in a condenser. The system includes a condenser and a decanter to gather the consolidate and to isolate EOs from water, appropriately (Figure 4.4).

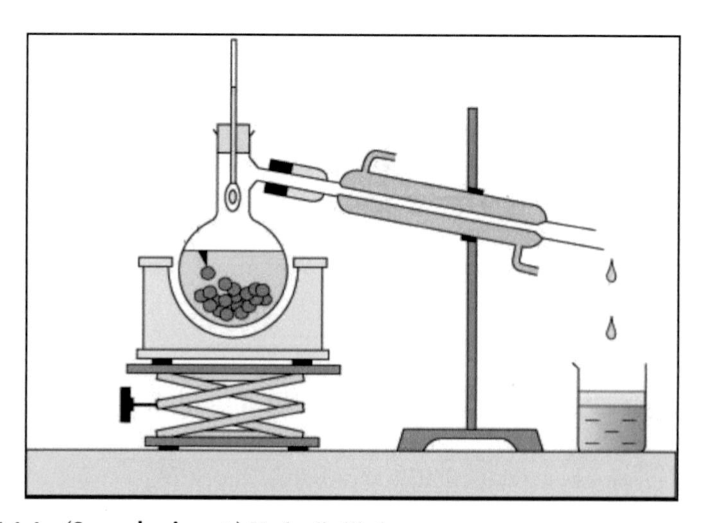

**FIGURE 4.4** **(See color insert.)** Hydrodistillation apparatus.

The principle of operation of separation relies on the isotropic distillation. Hydro-refining (HD) is a variation of steam distillation that was used in the French Pharmacopeia for the isolation of EOs from dehydrated plants and the superior control of volatile oils in the laboratory. Water immersion, through vapor injection and vapor inoculation, are three types of HD. The distillation time relies upon the plant material under test. Continued distillation yields only a trace volume of essential oil, but it also adds certain unwanted compounds and oxidation products.

### *4.2.3 ORGANIC SOLVENT EXTRACTION*

Solvent extraction (also called liquid-liquid removal or separation) is used to isolate a compound in line with the solubility of its components. This may be done by using two liquids that do not mix, for instance, water and a carbon-based solvent. Liquid extraction is employed to process perfumes, vegetable oils, or biodiesel. During this method, one of the elements of a mix dissolves in a specific liquid, and the other different component is separated as a residue by filtration (Figure 4.5). Solvent extraction is employed on subtle herbs to supply greater quantities of volatile oils at a lower price [9]. The superiority and amount of the mined blend are regulated by the sort of additional warmth as the technique is restricted by the complex solvability within the precise solvent used. Though the procedure is comparatively easy and fairly-effective, yet it experiences certain drawbacks as extended separation time, comparatively extraordinary solvent intake and frequent unsatisfying replicability [11] (Figure 4.6).

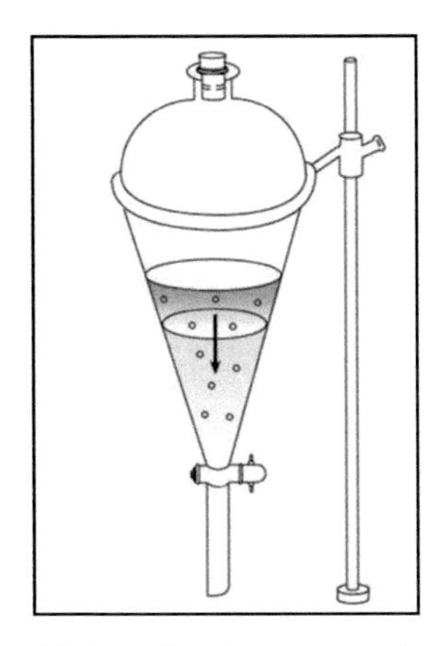

**FIGURE 4.5 (See color insert.)** Organic solvent extraction.

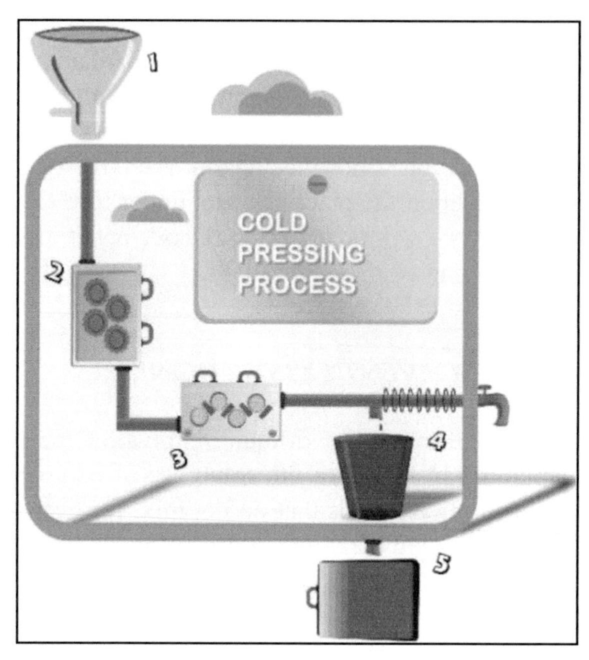


**FIGURE 4.6 (See color insert.)** Expression methods (*Legends:* 1–Filtering; 2–Milling; 3–Pressing; 4–filtering; 5–decantation).

### 4.2.4 EXPRESSION

Expression is an extraction method for citrus EOs such as lemon, sweet orange, tangerine, and lime. It is also known as cold pressing. The seeds are crumpled and pressurized so as to pull out the oil. However, the rubbing produced by the pressure might upsurge the heat of the product that is often not elevated. For example, to get jojoba oil without destructing the characteristics, it should not exceed 45°C. For extra virgin olive oil, it should not exceed 25°C. The method consists of the sieving phase, during which the seeds are passed over a sequence of sections by air propulsion systems. This method also removes any impurities. Granite millstones or trendy stainless-steel presses located in commercial operations are used for grinding fruits, seeds, and nuts into a paste. A rotating screw is used to agitate the semi-solid paste, which separates the oil from solid components and clumps. Pressure is used to pull out the oil. The pressure applied increases the temperature of the dough, which is one of the foremost vital moments of the entire method (Figure 4.6).

The tiny pieces or skin/rind or flesh of the fruit are separated from the pressed oil through a series of filters. This methodology of oil extraction needs the use of plenty of fruits and seeds (e.g., up to 30 kg of seeds for one liter of oil). The discarded components can be used as animal feed or fertilizers. Once the sieving method is complete, a decantation method is administrated. Easy vigor of gravity is used for separating the remaining sediments. Hence, 100% of clean native oil can be attained (Figure 4.7).

**FIGURE 4.7 (See color insert.)** Enfleurage method.

### *4.2.5 ENFLEURAGE*

The cold-fat extraction method (effleurage) is a process that uses odorless fats that are solid at room temperature to capture the fragrant compounds exuded by plants. The process can be "cold" enfleurage or "hot" enfleurage.

The fat used should be comparatively stable against rancidity. It is a technique for flowers, which endure developing and providing off their fragrance even during the collection (e.g., Jasmine, and tuberose). The method is of two types: "cold" enfleurage and "hot" enfleurage. In cold enfleurage, an oversized framed plate of glass (chassis) is spread with a layer of animal fat, typically lard or tallow (from pork or beef, correspondingly), and permitted to settle. Petals or whole flowers are placed flat for 1–3 days so that its scent diffuses into the fat. The method is then continued by exchanging the spent botanicals with new ones until the fat has reached the desired degree of fragrance saturation. In hot enfleurage, solid fats are heated, and plant matter is stirred into the fat (Figure 4.7).

Expended plant materials are recurrently strained from the fat and substituted with new material till the fat is saturated with cologne. This technique is taken into consideration as the oldest procedure for conserving plant fragrance substances. Enfleurage pomade is the one in which the fat is saturated with fragrance. It may be sold out as such, or it may be washed in ethanol to draw the scented particles into alcohol. This is further separated from the fat and endorsed to vaporize abandoning the utter of plant material. The spent fat is typically used in the preparation of soaps, because it is even scented [6].

### 4.2.6 MICROWAVE-ASSISTED DISTILLATION

With the assistance of microwave, extraction will stay accomplished in minutes rather than hours with varied benefits that are in line with the natural chemistry and separation values. During this methodology, plant resources are mined in microwave reactor either using organic solvents or water at diverse situations suitable for the investigative practices (Figure 4.8).

The very first Microwave-Assisted Extraction (MAE) of volatile oils was planned as a compacted air microwave distillation (CAMD) [10]. Based on the rule of steam distillation, the dense air is inoculated into the separator, and the plant matrices are submerged in water and warmed by microwave. Outside the microwave reactor, the water and EOs are condensed and unglued. The CAMD may be accomplished precisely in 5 min, and there is no variation in quantifiable and qualitative results amongst extracts of CAMD and standard extraction such as steam distillation for 90 min. To attain a good class of volatile oils, vacuum microwave hydro-distillation (VMHD) has also been used to evade hydrolysis [31]. Fresh plant materials are subjected to microwave irradiation to discharge the extracts; lowering the pressure to

100–200 milli-bars and allowing vaporization of the azeotropic water-oil mix at a temperature below 100°C. This operation may be recurrent stepwise with an optimum microwave power that is contingent on the specified yield. A chilling system present outside the microwave oven permits the continual condensation of the vaporized water-oil mix at atmospheric pressure. To maintain the suitable moisture of plant materials, the undue water is refluxed to the reactor.

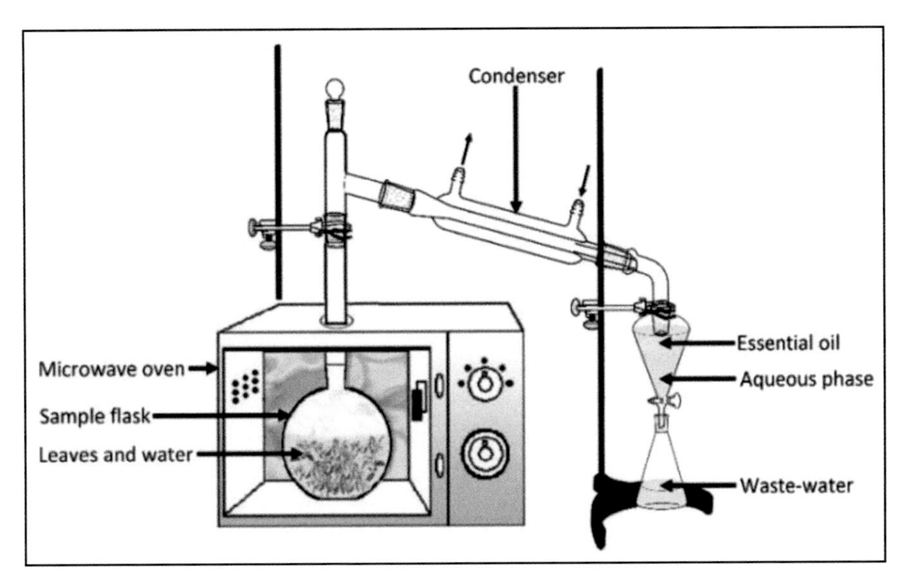


**FIGURE 4.8 (See color insert.)** Microwave-assisted distillation.

### *4.2.7 MICROWAVE HYDRODIFFUSION AND GRAVITY (MHG)*

This system is a microwave-induced hydro diffusion of herbal resources at ambient pressure, where all mines comprising of volatile oils and water plunge out of the microwave reactor under gravity into a nonstop condensation system over a perforated Pyrex support (Figure 4.9).

MHG is neither an altered MAE that practices organic solvents, nor an enhanced HD that needs high energy and water consumption, nor an SFME (solvent-free microwave extraction) that evaporates the EOs with the water in place exclusively. Additionally, MHG derivants like vacuum MHG and microwave dry-diffusion and gravity (MDG) have established later considering energy redeeming, clarity of end-products, and post-treatment of wastewater [14, 57].

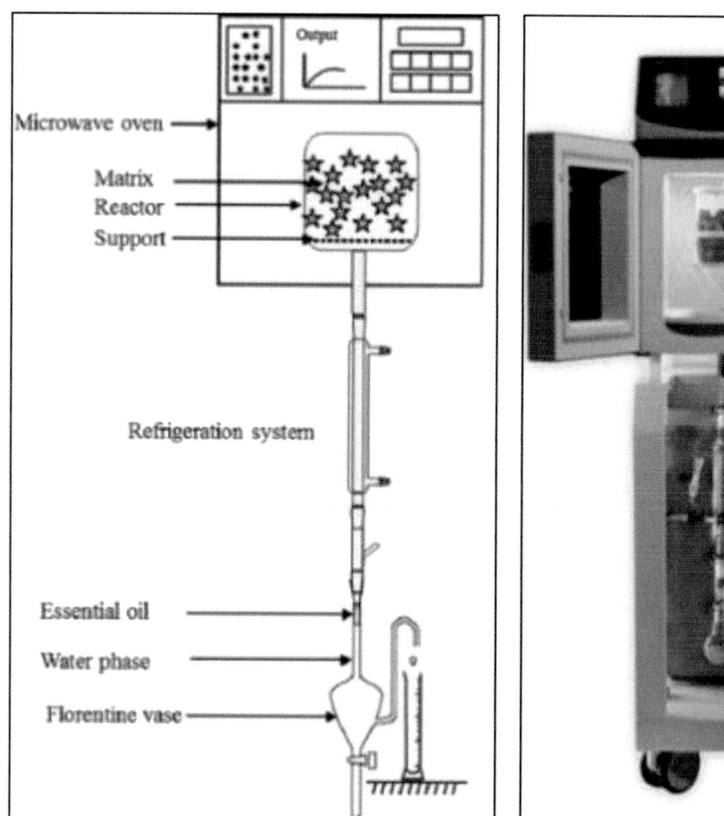

**FIGURE 4.9 (See color insert.)** Microwave hydrodiffusion and gravity (MHG).

### *4.2.8 HIGH-PRESSURE SOLVENT EXTRACTION*

Pressurized liquid extraction (PLE) is analogous to Soxhlet extraction, excepting in the separation method that the solvent state within the PLE cell reaches the supercritical area that ends up with additional effectual extractions. The raised temperature permits the sample to develop highly solvable and accomplishes a greater diffusion rate, whereas the raised pressure retains the solvent underneath its sweltering point. At increased pressures and temperatures, solvents will enter solid samples at high efficiency that reduces solvent usage (Figure 4.10).

PLE requires less extraction time and organic solvents compared to Soxhlet extraction. To perform a PLE, the sample is mixed with sodium sulfate, is loaded within the separation cell and covered with two filtration finish fixtures [15]. The PLE system afterward leads to pressurizing and heating of the samples. The pressure is kept at 1500–3000 psi, at a temperature of 70–200°C. The separated solvent comprising the goal analytes is then mechanically moved to a concentration/evaporation container wherever it is carried to final volume directly in a gas chromatography (GC) or liquid chromatography (LC) ampoule. The ampoule will then be moved to the investigative instrument for the final examination.

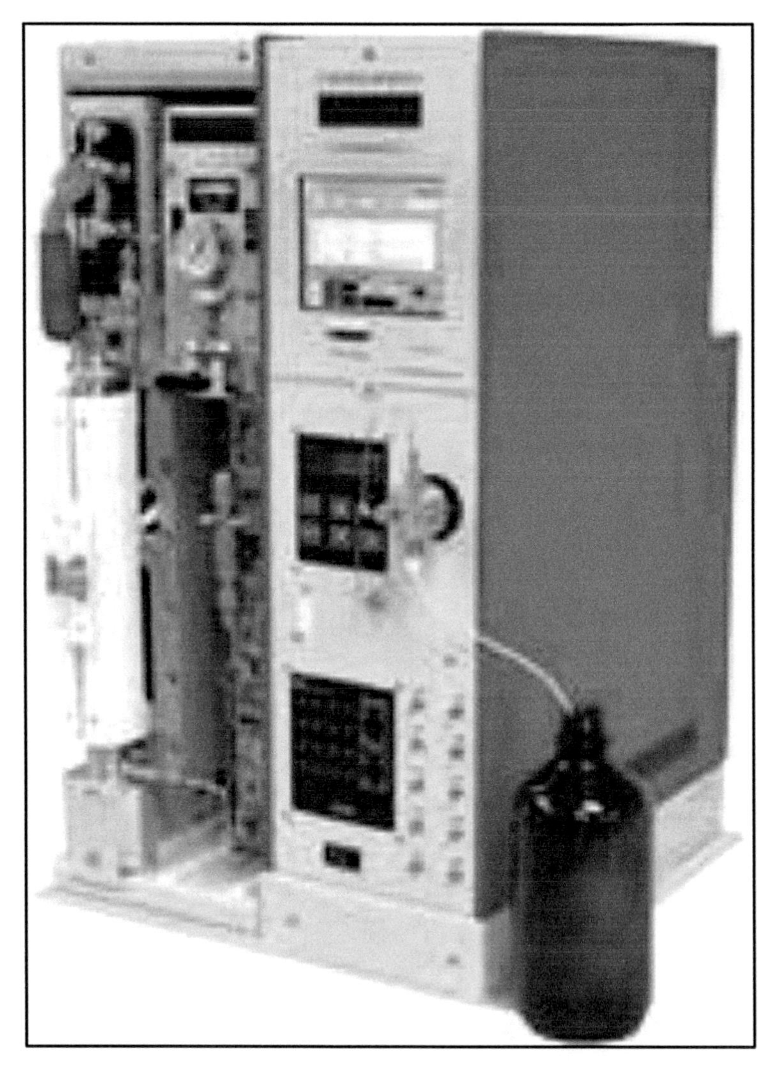

**FIGURE 4.10 (See color insert.)** High-pressure solvent extraction.

### 4.2.9 SUPERCRITICAL CARBON DIOXIDE EXTRACTION

SFE is a method for separation of one part (the extractant) from the other (the matrix) with supercritical fluids as the mining solvent. Extraction is sometimes from a solid medium; however, it can even be from liquids. The most important attention in the past decade was the use of supercritical $CO_2$, as it bears a close-ambient critical temperature (31°C) so that the biological matter is often treated at temperatures around 35°C. The mass of the supercritical $CO_2$ at about 200 bar pressure is nearer to hexane, and also the association characters are like hexane; therefore, it performs as a non-polar solvent. Supercritical fluid is any substance that bears temperature and pressure high enough beyond its analytical point. It will spread across solids like a mixture of gas and melted materials (Figure 4.11).

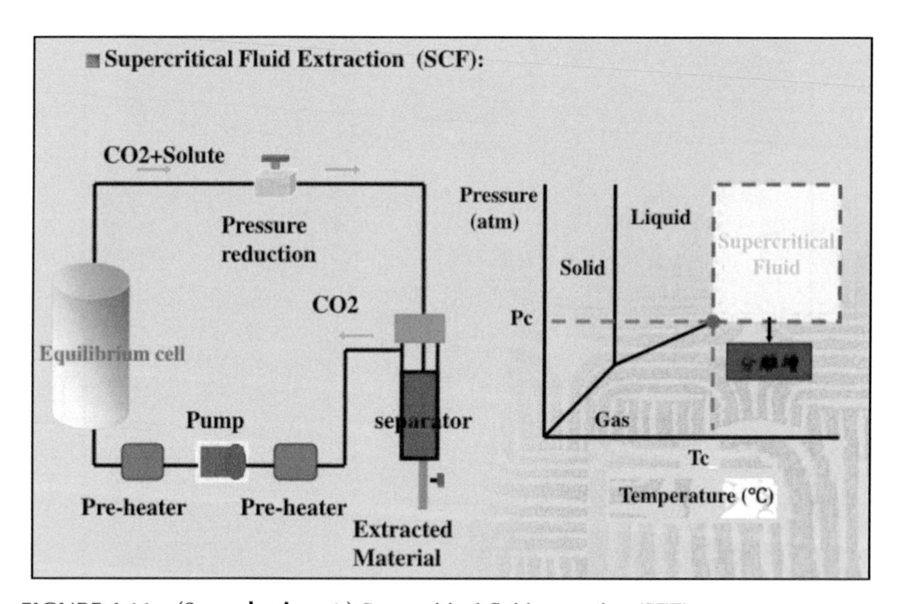


**FIGURE 4.11 (See color insert.)** Supercritical fluid extraction (SFE).

Minute changes in pressure or temperature result in huge changes in density, permitting many properties of a supercritical critical fluid to be "fine-tuned." Supercritical fluids are appropriate as an alternative for organic solvents in an extensive series of commercial and laboratory processes (Figure 4.11).

The higher diffusion coefficient, lower viscous, and higher yield are generated in this extraction technique. Several EOs are extracted with $CO_2$

extraction, which is not possible to be extracted with steam distillation. Still, this method is costly owing to the price of this instrument and is not also easy to be handled. Supercritical extracts are of excellent quality, with higher functional and biological activities.

### 4.2.10 ULTRASONIC EXTRACTION

For better extraction harvests and minor energy utilization, ultrasound aided extraction method has been established to boost the potency and to lessen the separation time. The downfall of cavitation bubbles created through ultrasonication provides an upsurge to micro-jets to extinguish volatile oil glands so that mass transfer and production of plant volatiles is enabled. The crater result is powerfully reliant on the operational limits (e.g., ultrasonic rate and power, temperature, treatment time, etc.) that are vital in effective designing and procedure of sono-reactors (Figures 4.12 and 4.13).

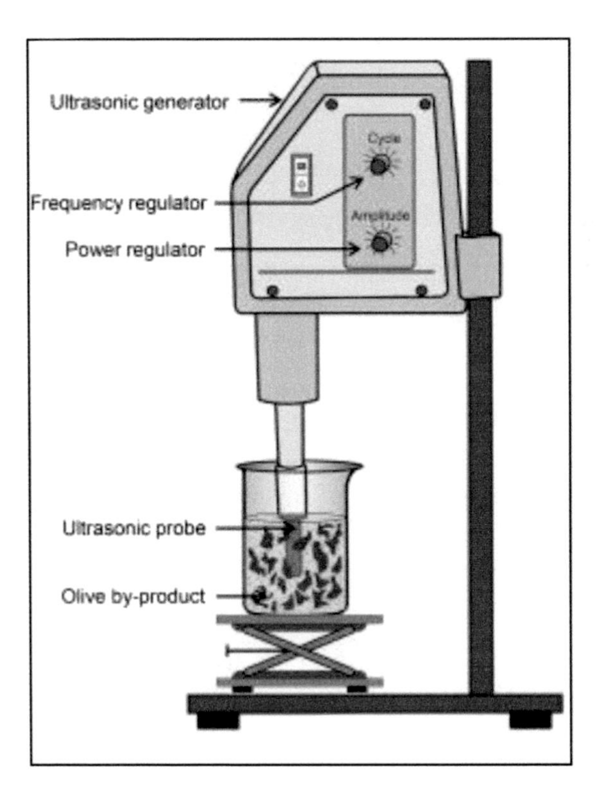


**FIGURE 4.12 (See color insert.)** Ultrasonic extraction.

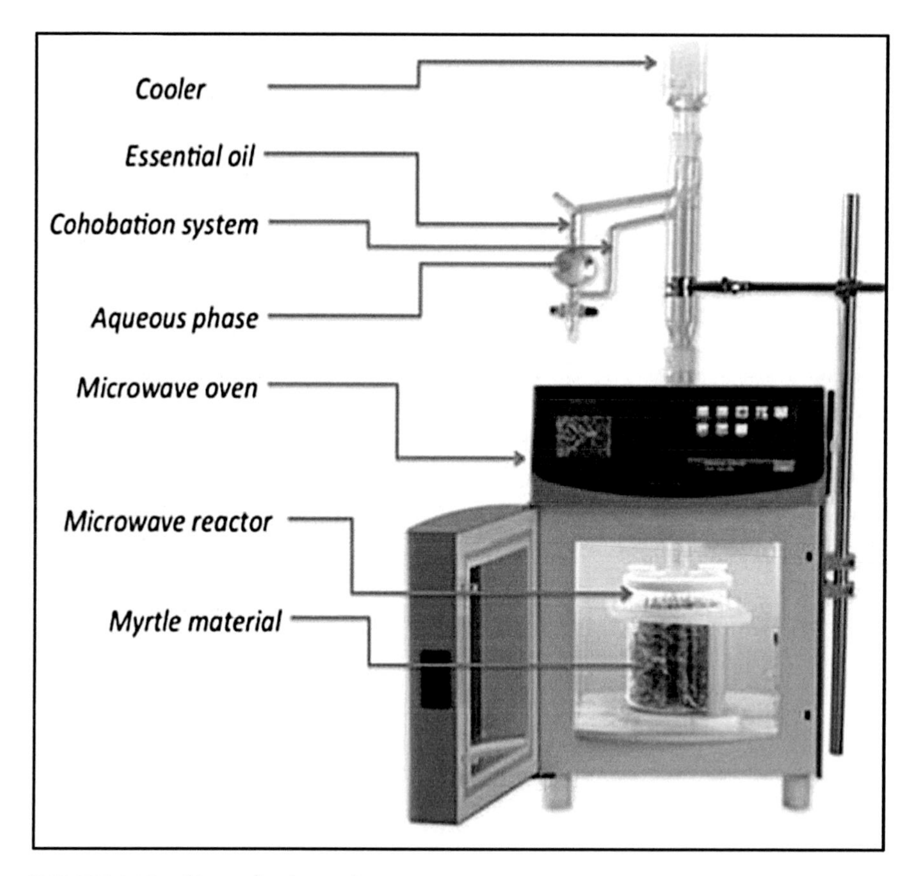


**FIGURE 4.13 (See color insert.)** Solvent-free microwave extraction.

In accumulation to the upgrading of yield, the volatile oils attained by Ultrasound-Assisted Extraction (UAE) exhibited low thermal destruction with a good quality and flavor [44]. However, the quality of sonotrode ought to be carefully decided because of metallic pollution, which can quicken oxidization and then decrease EOs' constancy [43]. Its efficiency has been up to 44% for separation yield of EOs from Japanese citrus that equaled to the regular approaches [29].

### *4.2.11 SOLVENT-FREE MICROWAVE EXTRACTION (SFME)*

SFME is a procedure to extract volatile oil in which the sample is extracted devoid of any solvent. Microwave distillation of the plant matter is carried out at atmospheric pressure devoid of solvents (Figure 4.13). During SFME procedure, the sample material is moistened with water for 1–2 hours before extraction, and the excess water is decanted. After that, the dampened resources are introduced into the microwave oven cavity, and the mined EOs are collected by a condenser as per the preset protocol. The radiation power, temperature, and separation time are managed by the panel within the instrument. The extracted volatile oils are dried using anhydrous sodium sulfate and maintained in 4°C darkness (Figure 4.14).

### *4.2.12 THE PHYTONIC PROCESS*

The most recent procedure for extracting EOs with non-CFCs (non-chlorofluorocarbons, R134a) is the Phytonic process. It is also called as florasol extraction and the resulting oils are known as phytols. This extraction methodology uses benign gaseous solvents. The extraction of aromatic oils and biologically active compounds from plant materials is founded on the distinctive properties of solvents used. The method is based on the solvent and florasol. Selectivity of solvents is needed for the extraction process, which produces free-flowing and wax-free clear oils (Figure 4.14) [2].

The solvents should be non-flammable, non-toxic, and non-explosive. The solvents used in the process are ozone-friendly, and they produce no harm to the environment. Vacuum distillation is not required to release solvents, and thus the highly volatile notes are protected. As a result, the end products are like the natural ones. The energy requirements are low for the process, and it also requires very low electricity, and no fossil fuels nor wood or coal is needed. There are not any "waters" to be eliminated. The used-up plant matter can be composted or could be used as fodder for cattle.

## 4.3 THERAPEUTIC POTENTIAL OF ESSENTIAL OILS (EOS)

Essential oil action starts after it enters the human body through three possible means viz., straight incorporation, inhalation, or ingestion:

- **Incorporation Through the Skin:** Essential oil complexes are solvable in fat, and therefore these possess the capability to pervade the skins of the peel previously being caught by the micro-circulation and exhausted into the complete movement that influences entire target tissues [4].

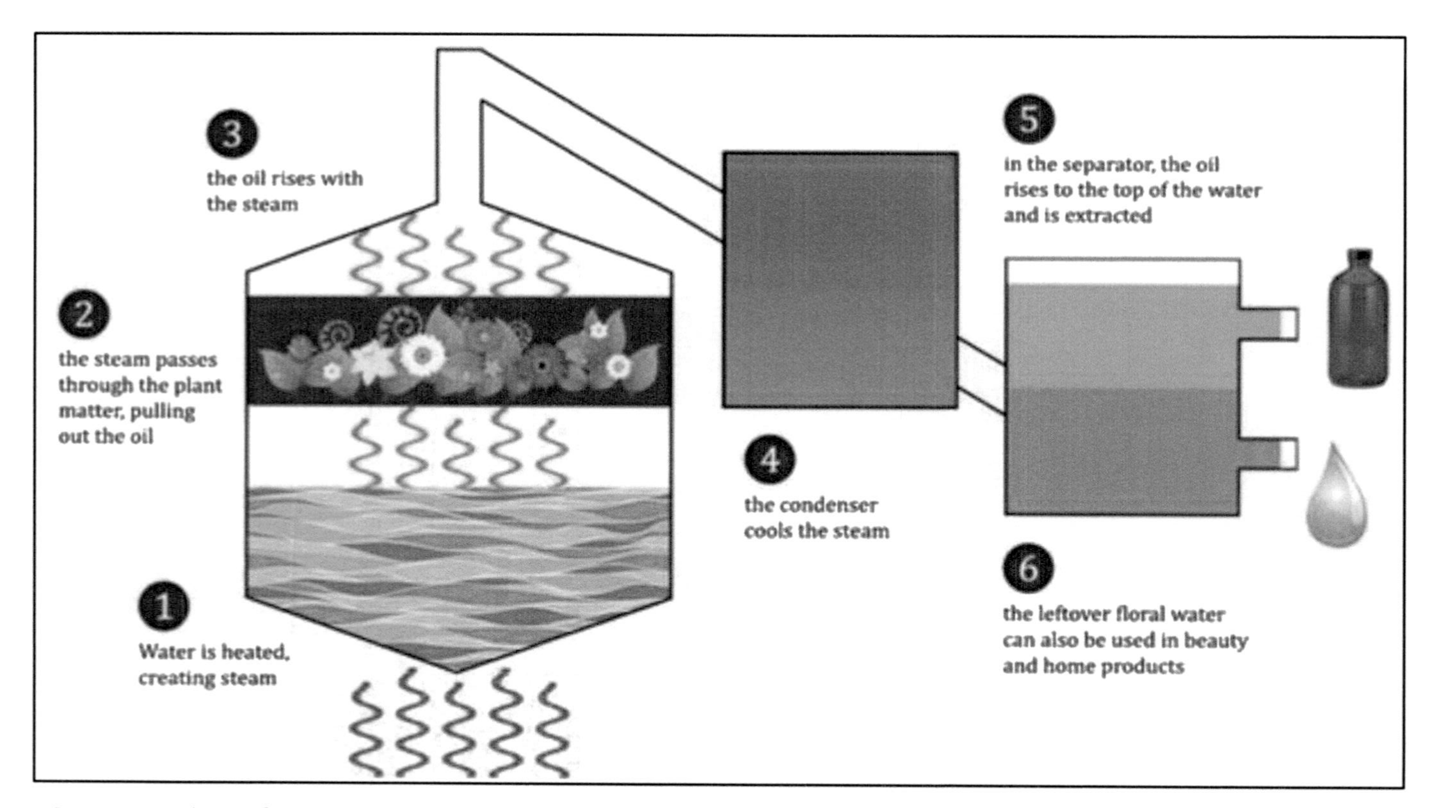


**FIGURE 4.14 (See color insert.)** The Phytonic process.

- **Inhalation:** The other way for EOs to enter the human body is by means of inhalation. Because of their volatile nature, EOs are easily inhaled across the respiratory tract and lungs [38]. The respiratory path offers the fastest route of entry, which is tracked by the dermal pathway.
- **Ingestion:** Care should be given during oral ingestion owing to possible poisonous action of EOs. Eaten essential oil mixtures and/or their metabolites will be engrossed and carried to the body by the blood and afterward to other segments of the body.

  As the volatile oils enter the human body, they interconnect with numerous functions through three modes of action.
- **Biochemical (Pharmacological):** Interaction with the bloodstream and chemicals like hormones and enzymes like farnesene.
- **Physiological:** By exerting an action on specific physiological function.
- **Psychological:** Through inhalation, the brain's olfactory area (limbic system) experiences a mechanism activated by the volatile oil molecules, and therefore the biochemical and neurotransmitter messengers offer alterations within the psychological and emotive actions of the individual [46]. Lavender and lemon EOs are examples for its tranquilizing and comforting characteristics.

## 4.4 ESSENTIAL OILS (EOS) AND THEIR MEDICINAL SIGNIFICANCE

### *4.4.1 CLOVE OIL*

The term 'Clove' originates from the French word 'Clou' and the English word 'Clout,' that means 'nail' – from the likelihood of the blossom of the clove to a broad-headed nail. Clove bud oil is the greatest supply of eugenol (about 80%) (Table 4.1) [12]. Clove (*S. aromaticum*) holds antibacterial drug properties. It is employed in tooth creams, dental pastes, mouth rinses, and throat sprigs to cleanse microorganism. It is used to release discomfort after sore gums and promotes complete tooth health. In odontology, eugenol is employed for provisional filling of cavities. Clove is an anodyne for dental emergencies. Cloves are aphrodisiac. Aromatherapists use clean essential oil to treat the indications of joint pains and inflammatory disease. Clove is employed for relieving flatulence, to extend acid within the abdomen and to boost bodily functions. The glop of the powder is applied with honey for treating inflammatory disease.

**TABLE 4.1** Essential Oils and Their Active Principles

| Plant Source | Active Compound | Molecular Formula | Molecular Mass |
|---|---|---|---|
| Clove – *Syzygiumaromaticum* (L.) | Eugenol | $C_{10}H_{12}O_2$ | 164.2 g/mol |
| Eucalyptus – *Eucalyptus globulus* | Eucalyptol | $C_{10}H_{18}O$ | 154.249 g/mol |
| Basil – *Ocimum basilicum* L. | Linalool | $C_{10}H_{18}O$ | 154.25 g/mol |
| Peppermint – *Mentha piperita* L. | Menthol | $C_{10}H_{20}O$ | 156.27 g/mol |
| Lemongrass – *Cymbopogon citratus* L. | Citral | $C_{10}H_{16}O$ | 152.237 g/mol |
| Oregano – *Origanumvulare* L. | Carvacrol | $C_{10}H_{14}O$ | 150.217 g/mol |
| Mugwort – *Artemisia vulgaris* L. | Camphor | $C_{10}H_{16}O$ | 152.23 g/mol |
| Pignut – *Hyptis suaveolens* (L.) Poit. | 1,8-Cineole | $C_{10}H_{18}O$ | 154.249 g/mol |
| Coriander – *Coriandrum sativum* L. | Linalool | $C_{10}H_{18}O$ | 154.25 g/mol |
| Jasmine – *Jasminum officinale* L. | Jasmineoil | $C_{10}H_{16}$ | 136.238 g/mol |

*Source:* PUBCHEM.

Cream of clove powder in water stimulates quicker curing of cuts and bites. *S. aromaticum* will effectively cure several gastric problems. It possesses healthful potentials to treat gas problems, loose stools, stomach upset, and biliousness. *S. aromaticum* are helpful in releasing the signs of diarrhea, irritation, and puking. *S. aromaticum* essential oil improves the physique by cleansing the blood and facilitates to contest against varied diseases. Essential oil is efficient in treating Athlete's foot and fungal infection in nails. The aromatic essential oil, once inhaled, will facilitate or soothe metabolic conditions like cold, cough, asthma, bronchitis, and inflammation. It conjointly helps in clearing the nasal tract. *S. aromaticum* can effectively fight against lung cancer and dermal cancer.

Eugenol aids in diminishing the destructive properties of atmospheric trashes, which will cause cancer of system *alimentarium*. Essential oil kindles blood movement and circulation creating it helpful for the individuals having cold extremities. It profits sugar patients by scheming the blood sugar levels. Eugenol is influential in averting blood clots. Intake of a clove bud decreases the want for alcohol. Muscular spasms are usually eased, once the oil of clove is used as a bandage close to the distressed space. It also helps to avoid the collapse in retina of the eye that reduces macular deterioration and assists vision in old age. The fundamental means is through the interference of the failure of omega-3 fatty acid that protects vision in old individuals. Scientists discovered that inhaling the spicy fragrance of cloves decreases somnolence, irritation, and headache.

One droplet of essential oil applied to the roof of the mouth will promptly relieve headaches, and it promotes memory retaining. It is counseled for releasing brain fog, weariness, and cheerless state of mind. Analysis has revealed that essential oil is a good dipteran repulsive too. A drop of essential oil is four hundred times influential as an anti-oxidant compared to wolfberries or blueberries [36].

### 4.4.2 EUCALYPTUS OIL

Bluegum is aimmortal tall tree that is indigenous to Australia and Tasmania [52]. Eucalyptus (*Eucalyptus globulus* L.) oils comprise of more than 100 diverse compounds. Many of these natural complexes have their own exclusive qualities, and their efficacy can be significantly improved as they unite synergistically to create superior therapeutic or other consequences. The 1,8-cineole (eucalyptol), ß-pinene, α- gurjunene, globulol, α-terpinen, ß-terpinen and γ-terpinen-4-ol, alloaromadendrene, and pipertone are present in both leaves and in shoots. In Eucalyptus, eucalyptol is a vital component of EOs [54]. Though there are 18 different compounds found in EOs of Eucalyptus, yet around 79.85% is represented by eucalyptol (Table 4.1).

High amounts of oxygenated monoterpenes are found to be present in Eucalyptus, but it varied among different species of the genus. The composition of EOs varied depending on the geographical location and seasons, which had a great influence on the biological activities [52]. The plant was expended traditionally as an antiseptic, for treating respiratory tract infections, colds, sore throats, chest infections like bronchitis, pneumonia, colds [56]. The fragrant components of EOs are used as antipyretic, anti-inflammatory, analgesic, etc. [47]. It also possesses antibacterial properties. In the 19$^{th}$ century, it was used to sterilize urinary catheters. It was also discovered that leaf of *E. globulus* was found to very effective against *E. coli* and *S. aureus*. It also promotes oral health. Chewing gum mixed with eucalyptus oil lowered the plaque formation and promoted healthier gums. It is found as an ingredient in cough syrups, rubs, and vapor baths [32].

### 4.4.3 BASIL OIL

Basil (*Ocimum basilicum* L.) has been used since ancient times for embalming and as preserving herb. It denotes a symbol of bereavement in Greece, and it is referred to as basilisk on photon, which means glorious, regal or majestic herb. In Italy, it is a symbol of love, and in France, it is named as an herb of royalty. It was used to gain strength during the times of fasting by the Jewish people, whereas Africans entitle it for its protection against scorpion bites. But a European group considered it as a sign of satan [28]. The 1,8-cineole, linalool, eugenol, limonene, camphor, bergamotene, methyl chavicol, geraniol, etc. are the principal constituents of *O. basilicum* oils (Table 4.1).

Scientists have found that basil possesses diverse essential oil components, which are rich in phenolic and polyphenolic components (like flavonoids and anthocyanins). Antimicrobial, antiviral, and antioxidant attributes of basil are owed to the existence of vicenin and orientin. In ayurvedic medication, it is used for treating cold, cough, headaches, and stomach related disorders. It is also a powerful insect repellant. The presence of citronellol, eugenol, and limonene in basil leaves contributes to the anti-inflammatory activity. Leaves, when crushed and applied on poisonous bites, can expel the venom from site. Seeds of the plant help in treating diabetes, obesity, and urinary infections. Basil also stimulates breast milk secretion and blood circulation after childbirth in women. For killing the harmful oral bacteria, it is used as a mouthwash or gargle. β-caryophyllene in basil treats arthritis and inflammatory bowel diseases. Leaf also holds anti-aging properties [17].

### 4.4.4 PEPPERMINT OIL

Peppermint (*Mentha piperita* L.) is a significant medicinal herb that is well known in the traditions of the Eastern and Western world. It is commonly found in Europe and North America. The name is derived from the Greek word *Mintha* that is the name of a mythological nymph that has metamorphosed into a plant and *piper* is a Latin word meaning pepper. Peppermint oil bears menthol smell, and it is pale yellow in color, and its viscosity is watery. Since ancient times, it was used for various ailments like cold, cough, neuralgia, headache, spasms, etc. Menthol is the active gradient of peppermint oil (Table 4.1); but other constituents like cineole, menthone, isomenthone, isopulegol, pulegone, carvone, limonene are also present [1].

India is the major manufacturer of mint oil in the biosphere. It is used in traditional medicine as flavoring agents. It possesses antiseptic, analgesic, anti-catarrhal, rubefacient, antiemetic, carminative, diaphoretic, stimulant, antimicrobial, antipruritic properties. The vapor of the oil is inhaled for respiratory congestion. Infusion of peppermint oil in tea treats bronchitis and oral mucosal inflammation. It is also used to treat various digestive problems such as flatulence, diarrhea, nausea, vomiting, morning sickness, anorexia, and colic in babies. It is used in curing irritating bowel symptoms, gall bladder, Crohn's illness, and biliary tract complaints. The oil is applied externally for myalgia, neuralgia, migraine, and headaches [23, 42].

### 4.4.5 LEMONGRASS OIL

Lemongrass (*Cymbopogon citratus* L.) is resultant from the Greek words '*kymbhe*,' which means boat; and 'pogon,' which means beard. The oil is a yellow or amber color liquid, which is extracted from the air-dried leaves by distillation. It contains about 75–85% of aldehydes, mainly citral (65–85%) (Table 4.1) and linalool, citronellol, geraniol, myrcene [53]. It is used in the manufacturing of cosmetics, dermal care products, and tonics. Lemongrass oil mixed with geranium or bergamot oil is an effective air freshener and deodorizer.

It is also a powerful insect repellant due to the presence of high citral and geraniol contents. It can be sprayed around the home or rubbed as a diluted mixture on the skin. It owns a clean and calming aroma, which helps in relieving stress, anxiety, irritability, insomnia, and drowsiness. It helps in relaxing muscles, relieves muscle pain, rheumatism, migraines, stomach, tooth, and headaches. It is a powerful analgesic and reduces pain and inflammation. It also strengthens hair follicles and promotes growth. It can kill both internal and external fungal infections, ringworms, and athlete's foot. It is a strong antipyretic agent that reduces fever levels. It promotes nutrient assimilation and lifts the functioning of the digestive system, thus lessening bowel problems and digestive complaints. It is a very good diuretic and eliminates toxins from the body [33].

### 4.4.6 OREGANO OIL

*Origanum vulgare* (Oregano) is also known as origanum or wild marjoram that is a perpetual herb. It is well-known for its flavorful leaves and flowering tops. The name is resultant from the Greek word 'oros' means mountain, and 'ganos' means joy. Essential oil of oregano comprises of monoterpenoids and monoterpenes. About 60 various components are found in the oil, and the chief ones are carvacrol and thymol reaching up to 80%, whereas the remaining compounds are p-cymene, caryophyllene, spathulenol, germacrene-D, β-fenchyl alcohol and terpineol (Table 4.1) [55].

It is used as an antiseptic and in treating respiratory, gastrointestinal, and urinary disorders, skin diseases (acne and dandruff) and menstrual cramps. Carvacrol has antimicrobial products, which is effective against 41 food pathogen of *Listeria monocytogens*. It also treats osteoporosis, arteriosclerosis, and breast cancer, cold aches, muscle cramps, pimples, dandruff, bronchitis, tooth problems, swellings, headache, cardiac problems, allergies, intestinal worms, earache, painful throat, and tiredness. The oil is also a powerful insect repellant. A few drops of oil added to water is used to treat nail or foot fungus. Inhalation of the oil is used as a cure for sinus and cold infections [30].

### 4.4.7 MUGWORT OIL

*Artemisia vulgaris L.* (Mugwort) grows in Europe, Asia, North, and Central America and North Africa. The name is given after the queen Artemisia of Asia Minor, and the epithet *vulgaris* means usual, common or vulgare [16]. The chief compounds of *A. vulgaris* are flavonoids, coumarins, volatile oil, sesquiterpene lactones, inulin, etc. The main components of EOs are camphor, camphene, α-thujone, germacrene-D, 1,8-cineole, and β-caryophyllene [18, 25]. The composition of EOs varies depending on different geographical locations and based on the method of extraction. Indian grown mugwort oil is rich in camphor, isoborneol, artemisia alcohol (Table 4.1) [18, 21].

It was reported that mugwort oils are used for their insecticidal, antimicrobial, and anti-parasitical properties [25]. It is effective against mosquitoes and other insects. It also possesses fumigant and repellent activity against *Musca domestica*. The essential oil exhibited 90% mosquito repellency against *Aedes aegypti*, the yellow fever mosquito [39]. The plant showed repellant fumigant properties against the insect pest *Tribolium castaneum*. The main compounds of EOs include camphor, 1,8-cineole, α-thujone, camphene, and β-caryophyllene. Its various properties include pain-relieving, painkilling, antimicrobial, antispasmodic, CNS-stimulant, cancer preventive, anti-acne, anti-neuralgic, decongestant, expectorant, larvicidal, insect repellant, insectifuge, nematicide, pesticide, antiasthmatic, anti-inflammatory, antispasmodic, tranquilizing, and flavor and perfumery [48, 49].

### 4.4.8 PIGNUT OIL

Genus epithet *Hyptis* derived Greek term 'huptios' (turned back), a reference to the lower brink location of the flower. Species epithet *suaveolens* means 'fragrant' or 'sweetly scented,' describing the plant's minty scent. EOs are major compounds of *Hyptis suaveolens* L. (Pignut). Sabinene, 1,8-cineole, β-caryophyllene, cymene, pinenes, eugenol, humulene, etc. are present in pignut oil. However, the chief compounds are 1, 8-cineole (32%) (Table 4.1) and ß-caryophyllene (29%) [41].

Leaves and shoots are used in anti-rheumatic and anti-fever baths and consumed to treat spasms. Infusion of dried leaves drunk to treat fever, or is drunk as an ordinary beverage. Small amounts of roots chewed with betel nut are systemic treatment for wounds and skin diseases. Poulticed leaves are applied on the head as headache relief, or onto skin boils. Leaf juice is drunk with lime juice as a remedy for colic and stomachache. Leaves are also applied as snuff to treat nose bleeds. EOs from leaves have shown to possess anti-microbial and anti-fungal properties. In the Philippines, the strong-smelling leaves are often placed under beds and chairs to drive bedbugs out.

Shoot tips are edible and sometimes used as a flavoring, and the decoction of roots is used as an appetizer. Plant is grown as green manure in parts of India and is used as cattle fodder. The extracted leaf compounds have insecticidal properties that potentially can be developed into bio-pesticides against lepidopteran pests. Root decoction is used as an emmenagogue and stimulant, whereas flowers possess antifungal, carminative, febrifuge, and stomachic properties. It is used as an external application for skin disorders such as dermatitis, eczema, headaches, boils, etc. Pounded fresh material is used as a poultice for snake bites. Juice of leaves is used for treating athlete's foot. Leaves are used for treating bed-sores, fungal skin infections, and bedbug repellent [51].

### 4.4.9 CORIANDER OIL

The name Coriandrum is derived from the Greek word 'koriannon' meaning bed bug as it produces a bed-bug like smell. Coriander (*Coriandrum sativum* L.) has been used for culinary purposes. The EOs of coriander consist of linalool, p-cymene, camphene, limonene, myrcene, ocimene, α- and β- phellandrene, α- and ß-pinenes, sabinene, α-terpinene, γ-terpinene, terpinolene, α-thujene. Linalool is the chief compound consisting of 60–70% (Table 4.1).

Coriander is one of the best spices, which is a common flavoring substance. The aerial part of the plant bears a nice aromatic odor. The tender plant is used in the preparation of chutneys, and specifically, foliage is used for flavoring curries, sauces, and soups; coriander oil and oleoresin are chiefly used in seasonings for sausages and other meat products. It also finds applications in bakeries, condiment, munching gums, and in curry mixtures [45]. In parts of Europe, it is referred as an anti-diabetic plant. In India, the plant is used traditionally for its anti-inflammatory possessions and cholesterol-lowering effects. It lowers bad cholesterol and prevents from cardiovascular disease, strokes, and heart attacks, also support smooth bowel and liver movements. Insulin secretion by the pancreas is promoted by coriander thus regulating sugar level in blood.

The leaves are anti-inflammatory, which is helpful in relieving arthritis pain and rheumatism. Juice of the plant mixed with turmeric powder is used for curing pimples, dry skin, and blackheads. It is excellent in treating conjunctivitis. It is a good source of calcium, and it strengthens bones by preventing osteoporosis. It helps in regulating enzymes and hormones of the body. Coriander seeds powder mixed with coconut oil and applied on the skull prevents hair fall and stimulates hair follicles. It is also an antiseptic, analgesic, aphrodisiac, antispasmodic, depurative, deodorant, lipolytic, and stimulant [20, 26].

### 4.4.10 JASMINE OIL

Jasmine (*Jasminum officinale* L.) is found in the wild in north India, south China, and Britain. The aromatic extract is obtained from the plant flowers. The essential oil obtained from jasmine includes linalool, nerol, farnesol, eugenol, terpineol, jasmone, trans-methyl jasmonate, jasmonic acid, jasmolactone, cis-and trans-ethyl jasmonate (Table 4.1).

It has been used traditionally as an antiseptic, aphrodisiac, stimulant, antidepressive, emollient, rejuvenating, and calming agent. Roots and leaves of jasmine are used as anthelminthic and which fights against ringworm and tapeworm. Extract of the flower is used to treat cough, and infusion of the greeneries is employed in curing sore eyes and wounds. It reduces stretch marks, prevents wrinkles, and has soothing property. However, extracting oil from jasmine is a tough task. It is widely used in perfume industries for producing quality perfumes. It possesses antidepressant, cicatrizant, analgesic, expectorant, anti-inflammatory, sedative, aphrodisiac, uterine tonic properties. For depression, nervous exhaustion, and stress-related conditions, jasmine promotes a feeling of optimism, confidence, euphoria, and it is good for apathy and listlessness. It is used in the treatment of cough, laryngitis, labor pains, skin problems, uterine problems, spasms, and sprains, dysmenorrhoea, and catarrh. When a few drops of oil are applied on the forehead, it will calm the mind and gives a feel of hope and happiness. It is also applied topically for treating skin disorders and muscle spasms. The inhalation of absolute oil helps in relieving stress and nervousness [34].

## 4.5 SUMMARY

Volatile or EOs find usage in various sectors such as in medical, beautifying products, agricultural, and bioactivity. Separation of EOs can be administrated by varied methodologies. Novel strategies avoid inadequacies of content facultative methods to reduced chemical hazard, separation time, and high energy input and procure yield quality of EOs. Exertions are being done to additionally discover the big vary of biological actions of volatile oils and its possible industrial applications. New approaches of therapy and chemo- interference are required within the arrival of several drug resistance connected with contagious and non-transmissible diseases. It is a must to increase the attention on the risks and advantages related to the healthful uses of volatile oils among the therapeutic and health care peoples further as amongst the patients is must. Use of plant molecules towards healing infectious and non-transmissible diseases are often an honest strategy.

## KEYWORDS

- **camphor**
- **enfleurage**
- **essential oils**
- **Listeria monocytogens**
- **Origanum vulgare**
- **isoprenes**
- **menthol**
- **repellant**
- **volatile oils**

## REFERENCES

1. Aishwarya, B., (2015). Therapeutic uses of peppermint: Review. *Journal of Pharmaceutical Science & Research*, *7*(7), 474–476.
2. Aromaworld 4-U. http://www.aromaworld4u.co/ExtractionMethods/Florasols.html (Accessed on 29 July 2019).
3. Badami, S., Rai, S. R., & Suresh, B., (2004). *In-vitro* antioxidant properties of Indian traditional paan and its ingredients. *Indian Journal of Traditional Knowledge*, *3*(2), 187–191.

4. Baser, K. H. C., & Buchbauer, G., (2010). *Handbook of Essential Oils: Science, Technology, and Applications* (1st edn.). CRC Press, Boca Raton, FL.
5. Baser, K. H. C., & Buchbauer, G., (2015). *Handbook of Essential Oils: Science, Technology, and Applications* (2nd edn., p. 1128). CRC Press, Boca Raton, FL.
6. *Bauer, K., Garbe, D., & Surburg, H., (1997). Common Fragrance and Flavor Materials: Preparation, Properties and Uses* (p. *290*). *WILEY-VCH.*
7. Berger, R. G., (2007). *Flavors and Fragrances: Chemistry, Bioprocessing and Sustainability* (p. 648). Springer Science and Business Media, Heidelberg.
8. Chami, F., Chami, N., Bennis, S., Trouillas, J., & Remmal, A., (2004). Evaluation of carvacrol and eugenol as prophylaxis and treatment of vaginal candidiasis in an immunosuppressed rat model. *Journal of Antimicrobial Chemotherapy*, *54*, 909–914.
9. Chrissie, W., (1996). *The Encyclopedia of Aromatherapy* (pp. 16–21). Healing Arts Press, Vermont.
10. Craveiro, A. A., Matos, F. J. A., Alencar, J. W., & Plumel, M. M., (1989). Microwave oven extraction of an essential oil. *Flavor Fragrance Journal*, *4*, 43–44.
11. Dawidowicz, A. L., Rado, E., Wianowska, D., Mardarowicz, M., & Gawdzik, J., (2008). Application of PLE for the determination of essential oil components from *Thymus Vulgaris* L. *Talanta*, *76*, 878–884.
12. Deyama, T., & Horiguchi, T., (1971). Studies on the components of essential oil of clove (*Eugenia caryophylatta* Thumberg). *Yakugaku Zasshi*, *91*, 1383–1386.
13. Djilani, A., & Dicko, A., (2012). The therapeutic benefits of essential oils. In: Bouayed, J., & Bohn, T., (eds.), *Nutrition, Well-Being and Health* (pp. 155–178). Intech, Croatia.
14. Farhat, A., Fabiano-Tixier, A. S., Visinoni, F., Romdhane, M., & Chemat, F., (2010). A surprising method for green extraction of essential oil from dry spices: Microwave dry-diffusion and gravity. *Journal of Chromatography*, *1217*(47), 7345–7350.
15. Fluid Management Systems. http://www.fms-inc.com/Products/PressurizedLiquid Extraction.aspx (Accessed on 29 July 2019).
16. Gledhill, D., (1990). *The Names of Plants* (2nd edn., p. 150). Cambridge University Press, Cambridge, MA.
17. *Global Food Book*. https://globalfoodbook.com/health-benefits-of-basil-*Ocimum basilicum* (Accessed on 29 July 2019).
18. Govindaraj, S., & Kumari, B. D. R., (2013). Composition and larvicidal activity of *Artemisia vulgaris* L. stem essential oil against *Aedes aegypti*. *Jordan Journal of Biological Sciences*, *6*(1), 11–16.
19. Guenther, E., (1953). *The Essential Oils: Individual Essential Oils of the Plant Families Gramineae, Lauraceae, Burseraceae, Myrtaceae, Umbelliferae and Geraniaceae* (p. 752). D. Van Nostrand Co., New York, volume IV.
20. *Guyanun Limited*. Available at: http://www.gyanunlimited.com/health/wonder-health-benefits-and-medicinal-uses-of-coriander-cilantro/9565/ (Accessed on 29 July 2019).
21. Haider, F., Dwivedi, P. D., Naqvi, A. A., & Bagchi, G. D., (2003). Essential oil composition of *Artemisia vulgaris* harvested at different growth periods under indo-Gangetic plain conditions. *Journal of Essential Oil Research*, *15*(6), 376–378.
22. Hammer, K. A., & Carson, C. F., (2011). Antibacterial and antifungal activities of essential oils. In: Thormar, H., (ed.), *Lipids and Essential Oils as Antimicrobial Agents* (pp. 255–306). John Wiley & Sons, Ltd, UK.
23. Hoffman, D., (1996). *The Complete Illustrated Holistic Herbal* (p. 256). Element Books Inc., Rockport, MA.

24. Hunter, M., (2012). *Essential Oils: Art, Agriculture, Science* (p. 780). Nova Science Publishers, Inc., Hauppauge, NY.
25. Judzentiene, A., & Buzelyte, J., (2006). Chemical compositions of essential oils of *Artemisia vulgaris* L. (mugwort) from North Lithuania. *Chemija, 17*(1), 12–15.
26. Laribi, B., Kouki, K., M'Hamdi, M., & Bettaieb, T., (2015). Coriander (*Coriandrum sativum* L.) and its bioactive constituents. *Fitoterapia, 103*, 9–26.
27. Mann, J., Davidson, R. S., Hobbs, J. B., Banthorpe, D., & Harborne, J. B., (1994). *Natural Products: Their Chemistry and Biological Significance* (p. 455). Longman Scientific and Technical, Harlow.
28. Marwat, S. K., & Khan, M. A., (2011). Interpretation and medicinal potential of Ar-Rehan (*OcimumbasilicumL.*)–A review. *American-Eurasian Journal of Agricultural and Environmental Sciences, 10*, 478–484.
29. Mason, T., Chemat, F., & Vinatoru, M., (2011). The extraction of natural products using ultrasound and microwaves. *Current Organic Chemistry, 15*, 237–247.
30. *Medical News Today*. https://www.medicalnewstoday.com/articles/266259.php (Accessed on 29 July 2019).
31. Mengal, P., Behn, D. B. M., & Monpon, B., (1993). VMHD: Extraction of essential oil by microwave. *Parfums, Cosmétiques, Arômes, 114*, 66–67 (in French).
32. Mercola, J., (2018). *Herbs and Spices*. Available at: https://articles.mercola.com/herbs-spices/eucalyptus.aspx (Accessed on 29 July 2019).
33. Mercola, J., (2018). *Herbal Oils*. https://articles.mercola.com/herbal-oils/lemongrass-oil.aspx (Accessed on 29 July 2019).
34. Mercola, J., (2018). *Herbal Oils*. https://articles.mercola.com/herbal-oils/jasmine-absolute-oil.aspx (Accessed on 29 July 2019).
35. Meyer-Warnod, B., (1984). Natural essential oils: Extraction processes and application to some major oils. *Perfumer and Flavorist, 9*, 93–104.
36. Milind, P., & Deepa, K., (2011). Clove: A champion spice. *International Journal of Research in Ayurvedha and Pharmacy, 2*(1), 47–54.
37. Moghaddam, M., & Mehdizadeh, L., (2017). Chemistry of essential oils and factors influencing their constituents. In: *Soft Chemistry and Food Fermentation* (pp. 379–419). Elsevier.
38. Moss, M., Cook, J., Wesnes, K., & Duckett, P., (2003). Aromas of rosemary and lavender essential oils differentially affect cognition and mood in healthy adults. *International Journal of Neuroscience, 113*, 15–38.
39. Nentwig, G., (2003). Use of repellents as prophylactic agents. *Parasitology Research, 90*, S40–S48.
40. Okoh, O., Sadimenko, A., & Afolayan, A., (2010). Comparative evaluation of the antibacterial activities of the essential oils of *Rosmarinus officinalis* L. obtained by hydrodistillation and solvent free microwave extraction methods. *Food Chemistry, 120*(1), 308–312.
41. Peerzada, N., (1997). Chemical composition of the essential oil of *Hyptis Suaveolens*. *Molecules, 2*, 165–168.
42. Peirce, A., (1999). *The American Pharmaceutical Association Practical Guide to Natural Medicines* (p. 752). William Morrow and Company, Inc., New York.
43. Pingret, D., Fabiano-Tixier, A. S., & Chemat, F., (2013). Degradation during application of ultrasound in food processing: A review. *Food Control, 31*, 593–606.

44. Porto, C., Decorti, D., & Kikic, I., (2009). Flavor compounds of *Lavandula angustifolia L.* to use in food manufacturing: Comparison of three different extraction methods. *Food Chemistry*, *112*, 1072–1078.
45. Sharma, M. M., & Sharma, R. K., (1999). Coriander. *Handbook of Herbs and Spices* (pp. 1–6). Woodhead Publishing Ltd., Cambridge, UK.
46. Shibamoto, K., Mochizuki, M., & Kusuhara, M., (2010). Aromatherapy in anti-ageing medicine. *Anti-Ageing Medicine*, *7*, 55–59.
47. Silva, J., Abebe, W., Sousa, S. M., Duarte, V. G., Machado, M. I., & Matos, F. J., (2003). Analgesic and anti-inflammatory effects of essential oils of eucalyptus. *Journal of Ethnopharmacology*, *89*(2/3), 277–283.
48. Sujatha, G., Ranjitha-Kumari, B. D., Cioni, P. L., & Flamini, G., (2008). Mass propagation and essential oil analysis of *Artemisia vulgaris* L. *Journal of Bioscience and Bioengineering*, *105*(3), 176–183.
49. Teixiera Da Silva, J. A., (2004). Mining the essential oils of Anthemideae. *African Journal of Biotechnology*, *3*(12), 706–720.
50. Tisserand, R., & Young, R., (2013). *Essential Oil Safety: A Guide for Health Care Professionals* (p. 784). Elsevier health sciences (Churchill Livingstone), United Kingdom.
51. *Useful Tropical Plant Database*. http://tropical.theferns.info/viewtropical.php?id=Hyptis+suaveolens (Accessed on 29 July 2019).
52. Vecchio, M. G., Loganes, C., & Minto, C., (2016). Beneficial and healthy properties of Eucalyptus plants: A great potential use. *The Open Agricultural Journal*, *10*, 52–57.
53. Virmani, O. P., Srivastava, R., & Datta, S. G., (1979). Oil of lemongrass, Part 2: West Indian. *World Crops*, *31*(3), 120–121.
54. Vishin, A. P., & Sachin, A. N., (2014).A review on *Eucalyptus globulus*: A divine medicinal herb. *World Journal of Pharmacy and Pharmaceutical Sciences*, *3*(6), 559–567.
55. *Wikipedia*. https://en.wikipedia.org/wiki/Oregano (Accessed on 29 July 2019).
56. Williams, L. R., Stockley, J. K., Yan, W., & Home, V. N., (1998). Essential oils with high antimicrobial activity for therapeutic use. *International Journal of Aromatherapy*, *8*(4), 30–39.
57. Zill-e-Huma, H., Abert-Vian, M., Elmaataoui, M., & Chemat, F., (2011).A novel idea in food extraction field: Study of vacuum microwave hydrodiffusion technique for by-products extraction *Journal of Food Engineering*, *105*, 351–360.
58. Zuzarte, M., & Salgueiro, L., (2015). Essential oils chemistry. In: De Sousa, D. P., (ed.), *Bioactive Essential Oils and Cancer* (pp. 19–28). Springer, Switzerland.

# PART II

# Disease Management with Medicinal Plants

CHAPTER 5

# THERAPEUTIC POTENTIAL OF *OCIMUM TENUIFLORUM* L.: METABOLIC AND MENTAL DISORDERS

NISHIKANT A. RAUT, DADASAHEB M. KOKARE, and
GAIL B. MAHADY

## ABSTRACT

*Ocimum tenuiflorum* (*tulsi*) is used traditionally in India for management of asthma, common cough, mild upper respiratory infection, and various metabolic and mental disorders. The pharmacological activities of *tulsi* are attributed to a wide range of phytochemical constituents present in the plant extracts. It is suggested that *tulsi* is beneficial in the treatment of both metabolic and neurological disorders. The review in this chapter summarizes the experimental research that supports the utilization of *tulsi* for the management of metabolic and mental stress. Based on the available animal experiments and clinical data, it is attempted to correlate these biological activities and their potential clinical implications. The discussion of the clinical implications may be beneficial in deciding alternative therapies for the control of metabolic and mental disorders.

## 5.1 INTRODUCTION

Psychological stress is known to be associated with poor quality of life, and chronic mental stress reduces brain performance, eventually triggering an impairment of high-level brain functions, particularly attention and decision processes and working memory [55]. Mental stress is associated with strong feelings of anxiety, burden, and apprehension that can usually be correlated with problems related to family, lifestyle, and work situations.

Under specific conditions, all living beings will experience stress, and the body's response to these stressful situations could be considered as life-saving. Both physical and mental stress induces many physical changes, including alterations in immune function, as well as inflammatory responses in both the brain and periphery. These changes are induced by several inflammatory neuropeptides, cytokines, and stress hormones. These molecules act together to facilitate the fight or flight response. When an individual is placed in a dangerous situation, increasing heart rate, respiration, muscular activity, oxygen consumption, and brain activity, then there is an increased immune response. All these changes occur as a part of the body's defense system against acute stress situations. During chronic stress situations, however, the neurotransmitters are life-saving and are able to suppress functions that are not required for immediate survival, including the immune system.

Under continuous stress situations, the immune system becomes suppressed, and there are multiple alterations in digestion, excretion, and reproduction. However, unlike acute stress under chronic stress situations, body functions do not return to normal due to alterations in signaling and inhibition of the immune system. Thus, chronic stress can ultimately induce severe physio-psychological health complications, including anxiety, depression, and metabolic disorders, including diabetes and cardiovascular diseases [32]. Thus, chronic stress situations are associated with increased chronic disease and overall mortality.

Chronic metabolic (physical) stress is associated with catabolic and hypermetabolic responses to severe injury or diseases, due to constant override of metabolic homeostasis. Therefore, chronic stress adversely alters metabolic regulatory systems making them vulnerable to serious physical alterations such as acute protein synthesis, increased hormone release, hyperglycemia, reliance on gluconeogenesis, and subsequent release of glucose. This is commonly known to shift the fluid balance and decrease urinary output [44]. Furthermore, chronic stress enhances hunger and promotes obesity, both of which are important risk factors for several metabolic diseases, such as cardiovascular disease, type-2 *diabetes mellitus*, and polycystic ovarian syndrome (PCOS) [55]. Furthermore, under chronic stress, the body may not heal appropriately following injury, due to improper utilization of nutrients, increased demand for calories and protein; and in these situations, the patient may be considered at high nutritional risk that may lead to protein-energy malnutrition.

Natural metabolic processes involve the digestion of food materials normally made up of protein, fats, and carbohydrates into simple useable molecules such as glucose and amino acids by enzymes in the gastrointestinal tract (GIT). The energy produced during the process is stored in the liver, muscles, and fat. Any disturbance in the proper digestion of food because of impairment of vital organs (liver and pancreas) may result in metabolic disorders. Based on the abnormalities, metabolic disorders can be classified into different groups such as those affecting metabolism of protein, carbohydrates, fat or inter-conversion of small molecules like amino acids, glucose, or fatty acids. In some of the metabolic disorders, energy-producing organelles of the cells, the mitochondria are also affected, causing mitochondrial dysfunction.

## 5.2 TREATMENT OF METABOLIC AND MENTAL STRESS

The initial management of stress should begin with a change in lifestyles, such as modifications in exercise and diet. Primary lifestyle changes include a balanced diet, physical activities, regular exercise, maintenance of a healthy weight, and minimizing poor health habits (such as smoking and excessive drinking of alcoholic beverages). Regular exercise trims down the release of stress hormones and related neurotransmitters. Several reports have suggested that exercise has strong antidepressant and anxiolytic effects, as well as aiding in sleep. However, sometimes changes in lifestyle are not sufficient to reduce stress, and thus pharmacotherapy becomes essential to deal with it.

There are abundant pharmacological agents available for the prevention and treatment of anxiety and depression caused by stress that are prescribed by physicians, including sedative-hypnotics, antidepressants, anxiolytics, tranquilizers, and beta-blockers [57, 59]. However, careful evaluation of the patient should be carried out prior to using medications for relieving stress, as many of these medicines are addictive in nature. Along with addiction, such medications may cause life-threatening adverse effects, including severe behavioral effects. Sedatives depress the central nervous system (CNS) causing sedation and relaxation; however, these drugs have a low margin between therapeutic effects and lethal doses.

Benzodiazepines are regularly prescribed CNS depressants that are considered relatively safe when compared with the barbiturates. Some commonly prescribed benzodiazepines are:

- Alprazolam;
- Clonazepam;
- Diazepam;

- Lorazepam;
- Triazolam;
- Temazepam;
- Chlordiazepoxide.

Alcohol, barbiturates, and benzodiazepines stimulate the function of gamma aminobutyric acid (GABA), an inhibitory neurotransmitter that is widely distributed in the brain and is responsible for reducing brain activity. In addition to the benzodiazepines, antihistamines, and some sleeping medications (such as glutethimide and methyprylon) are also prescribed. Almost all sedative drugs are addictive and develop dependence; and they may experience severe withdrawal symptoms on termination of the treatment.

## 5.3 NATUROPATHY FOR METABOLIC AND MENTAL STRESS

In many instances, patients may stop using their prescribed anxiolytics, antidepressants, and sedative drugs due to adverse effects of these drugs; and opt to use complementary and alternative medicines (CAM). CAM therapies are not considered conventional medicine practice but are rather considered a separate group of alternative remedial health care system, products, and/or practices. CAM therapies are classified into five subtypes according to the National Center for Complementary and Integrated Health (NCCIH, formerly NCCAM, USA) including [31]:

- Alternative or complete medical systems of therapy and its practice;
- Biologically-based systems including herbal medicines;
- Energy therapy;
- Manipulative, and body-based methods, including chiropractic and massage therapies;
- Mind-body interventions or techniques.

These therapies are labeled as alternative and complementary when they either replace or are used concurrently with other conventional treatments, respectively. Integrative medicine combines mainstream medical- with CAM-therapies in cases where there is high-quality scientific evidence of safety and effectiveness [31]. It should be stated that patients with serious mental health issues being treated with prescription medications usually cautioned against self-medicating with CAM treatments without the supervision of their physician or other healthcare professional as the combination may cause adverse events and serious drug interactions. Nevertheless, >70% of Americans are using CAM-related therapies and prescription drugs and may be exposed to a high risk of severe herb-drug interactions [54, 60].

## 5.4 AYURVEDA

Ayurveda is an ancient system of natural and holistic medicine in India and one of the world's oldest traditional medical systems. It remains one of the traditional healthcare systems in India [37]. As cited in *Sushruta Samhita*, the two major aims of Ayurveda are maintaining the health of people and to treat the diseases. The basic concepts of Ayurveda focus on preventative medicine so that patients remain in good health. The World Health Organization (WHO) defines health as, *"Health is a condition of complete physical, mental, and social wellbeing and not just the absence of disease or infirmity"* [6]. On the other hand, Ayurveda defines health as–

> *"Sama doshasamaagnishchasamadhatu mala kriyaaha |*
> *Prasannaatma mana indriyahaswasthaitiabhidheeyate||"*

Which means, *"Health is a condition wherein the Tridosha (Digestive fire), all the body tissues and components, all the physiological processes are in great unison and the soul, the sense organs and mind are in a state of total satisfaction (Prasanna) and content"* [37].

According to the definition of health by WHO, an individual is considered to be healthy if, he is in a mental, physical, and social state of wellbeing. The first two, physical and mental well-being, are quite easy to understand. However, social well-being refers to the positive interaction of an individual with the surrounding people. Whereas, the definition of health in Ayurveda is elaborative that considers balance in *Tridosha* (*Vata*, *Pitta*, and *Kapha*: three fundamental principles to regulate forces of nature for movement, metabolism, and growth, respectively) that refers physical wellbeing.

Any disturbance in the digestive fire (Agni) is considered as the basic cause for the majority of the diseases. As the *Agni* is directly related with the metabolism of the body. Imbalance in the working of this element may be considered as a diseased condition, and proper working is normal healthy condition and/or also refers to physical wellbeing. The remaining part of the definition of health based on Ayurveda refers to mental and social wellbeing along with spiritual wellbeing. Therefore, *Ayurveda* explains the meaning of health completely by considering not only the mental, physical, and social wellbeing but also the spiritual wellbeing [53]; and it four-dimensional.

According to Ayurveda, the mind (or *manas*) has three dimensions in terms of three *Gunas* (qualities of mind), such as *sattva*, *rajas*, and *tamas*. *Sattva* is the state of complete balance or termed as pure mind. The *rajas* signify activity and dynamism, whereas the *tamas* represents inertia and darkness. Cessation or merging of both, *rajas*, and *tamas,* the dangerous qualities of mind in each other and their disorders are considered as the basis of all mental illness. Therefore, *rajas*, and *tamas* are termed as *manasdoshas* while *sattva* is not at all the cause of illness [53].

Ayurveda assumes that mind is a sensory and motor organ and associated with touch sense and consciousness. The touch sense with related faculty in mind yields pleasurable and/or painful sensations. This aspect links body metabolism with the mind [53]. According to Ayurvedic practice, mental health is a state of intellectual, sensorial, and spiritual well-being. The mental has brought ill-health essentially because of unwholesome interaction between the individuals and their environment.

The concepts of Ayurveda about health and disease encourage the use of herbal compounds, special diets, and other unique health practices. The Indian religion is based on the spiritual scriptures *Vedas*, which are composed by ancient scientists called '*Rishis*.' Hence, various Indian traditional rituals use *tulsi* in worship based on its scientific and practical reasoning. In Indian culture, this plant is also considered extremely sacred and pious. *Oscimum tenuiflorum* Linn. (synonym: *Ocimum sanctum*) is known as *tulsi* (Figure 5.1). It has a deep spiritual and religious significance in India. In Sanskrit, it is believed that, *"tulana anaasti athaiva Tulsi,"* which means that *Tulsi* is incomparable for its benefits to mankind.

**FIGURE 5.1 (See color insert.)** A typical *tulsi* plant with flowers and seeds.

It is a principal herb of Ayurveda, therefore known as ‘The Incomparable One,’ ‘The Queen of Herbs,’ and ‘The Mother Medicine of Nature.’ In Ayurveda, tulsi is considered a *divyaaushadhi* (divine herb). The tulsi leaf has immense medicinal value to cure diverse ailments, including the common cold. Tulsi is grown in *Tulsi Vrindavan* (grove of *Tulsi*) or near almost every Hindu house. The plant is cultivated for obtaining essential oil and is also for medicinal and religious purposes.

Traditionally, *tulsi* is used as a nerve tonic to improve the complications related to the nervous system. Ayurveda describes *tulsi* as *Rasayana*, which has properties of immune stimulation or adaptogens. *Tulsi* is thought to impart immunity, avert aging, augment longevity, improve mental functions, and responsible for the addition of vigor and vitality, which are considered as *Ayurvedic Rasayana* for the body [4]. Recent studies report widespread use of *tulsi* as an anti-stress agent [47], an anticonvulsant [20], an anti-inflammatory [13], an antioxidant [22], as well as an antidepressant [35]. The primary phytochemical constituents of *tulsi* leaf are:

- Eugenol (volatile oil);
- Palmitic acids;
- Stearic acids;
- Linoleic acids;
- Linolenic acids
- Oleic acids (fixed oil);
- Luteolin;
- Orientin;
- Vicenin (flavonoids)
- Rosmarinic acid
- Ursolic acid (triterpene) [22].

‘Charak Samhita’ and ‘Sushruta Samhita,’ two of the most primeval manuscripts available especially describe Indian traditional medicines report several magnificent remedial effects of tulsi. According to the Charak manuscript, tulsi possess therapeutic effects for cough, dyspnoea, and pleurodyspnoea, hiccups, as well as reducing the effects of poisons. It is curative for ‘Kapha’ and ‘Vata’ while promotive for ‘Pitta’ and eliminates ‘Fetor’ [30]. The various parts of tulsi are used to treat bronchitis, earache, fever, gastritis, leucoderma, pain, inflammation, and vomiting, diseases of the eye, heart, and blood [23].

## 5.5 POTENTIAL OF TULSI FOR THE MANAGEMENT OF STRESS

Modern lifestyle is always connected with a high level of mental stress caused by personal, societal, and occupational reasons. Daily consumption of *tulsi* leaves helps to detoxify the body and helps to reduce stress by relaxing the mind. It has also reported to have antidepressant and memory-enhancing effects, as well as improvement in cognitive functions [9]. Numerous experiments have revealed the anti-stress activities of *O. tenuiflorum* in various animal models (pre-clinical) and few clinical studies [9].

Maity et al., [26] studied the effects of methanol extract of *tulsi* roots on swimming performance in mice. An increase in the swimming time confirmed the anti-stress effects of *tulsi* that were comparable to the anti-depressant drug, desipramine [26]. In an interesting study, the effect of *tulsi* was studied in an experimental animal model that looked at the effects of stress-induced by noise [42]. Exposure to noise alters the levels of neurotransmitters such as dopamine, epinephrine, norepinephrine, and serotonin in distinct regions of the brain. After treatment with *tulsi*, these altered levels of neurotransmitters were normalized [42]. Similarly, another study reported prophylactic use of *tulsi* against noise-induced stress [51]. Exposure to noise significantly reduces the content and the activities of acetylcholine (Ach) and acetylcholine esterase in different brain loci, such as corpus striatum, cerebral cortex, hypothalamus, and hippocampus.

Sembulingam et al., [51] reported that pretreatment with *tulsi* prevented the noise-induced changes that occurred in the mice central cholinergic system [51]. Samson et al., [47, 48] indicated that the neuronal modifications induced by noise were reduced by *tulsi* by increasing the synthesis of serotonin and not by reducing its degradation. The alterations in the neurotransmitters in different regions of the brain after exposure to noise were also reversed. These changes in the neurotransmitters were not limited to specific brain regions [47]. They further suggested that some of the constituents of *tulsi* may cross the blood-brain barrier and produce their action directly on the brain. They hypothesized that active constituents of *tulsi* (such as α-caryophyllene, 1,8-cineole, eugenol, flavonoids (like orientin, vicenin, etc.), linalool, methyl chavicol, methyl eugenol, and ursolic acid) produce this action by upregulating the synthesis and/or reuptake of neurotransmitters [47].

Pare and Glavin [34] demonstrated that the action of *tulsi* on dopaminergic neurons was blocked by haloperidol and sulfide, and that the dopamine D2 receptors played a role in *tulsi*'s pharmacological effects [34]. In addition, changes in the membrane of the hippocampus and sensorimotor cortex, which were induced by stress, were restored by *tulsi* treatment [52], indicating anti-stressor non-specific activity of *tulsi*.

Interestingly, noise stress is coupled with an increase in the levels of catalase (CAT), glutathione peroxidase (GPx), lipid peroxidation (LPO), superoxide dismutase (SOD) and oxidized glutathione (GSSG), as well as reduced level of reduced glutathione (GSH), indicating that it induces an oxidative state. GSH/GSSG ratio induced by noise exposure was restored in animals after *tulsi* treatment, and these effects were ascribed to the presence of phenolics and flavonoids in ethanol extract of *tulsi* [46].

Gupta et al., [15] isolated and characterized three new and two known compounds from an ethanol extract of *tulsi* and screened them for anti-stress activities. Ocimumoside-A and -B normalized some stress parameters such as adrenal hypertrophy, hyperglycemia, and plasma corticosterone and plasma creatine kinase. However, ocimumoside C was not effective [15].

Ahmad and his colleagues [2] investigated the role of ocimarin, ocimumoside-A, and -B in reducing chronic stress-induced central changes in a chronic unpredictable stress (CUS) model in rats [2]. Ocimumoside-A and -B pretreatments reduced CUS-elevated plasma corticosterone levels to control, butocimarin was ineffective. The levels of serotonin, dopamine, noradrenaline, activities of CAT and SOD and glutathione (GSH) content, or the metabolites of dopamine, GPx activity, and LPO by CUS was not affected by any of these compounds. However, the combination of ocimumoside-A and -B normalized parameters altered by CUS with a similar efficacy as of other anti-stress agents such as *Panax quinquefolium* or melatonin. Pretreatment with ocimarin was not effective against CUS-induced alterations. The authors suggested that the ocimumosides-A and -B may penetrate the blood-brain barrier and produce their effects. This was supported by a reduction in the activation of hypothalamus-pituitary-adrenocortical (HPA) axis as seen by a reduction in corticosterone levels and a restoration of the monoamines and the redox state following the pretreatment with these phytoconstituents [2].

Richard et al., [43] investigated the effects of *tulsi* in a chronic variable stress (CVS) model in rats. In CVS model, rats exposed to multiple stressors for 2 to 5 weeks to induce stress, is a commonly employed and very efficient model to examine the effects of interventions on stress and anti-stress [16, 19, 28, 41]. The anti-stress effect of *tulsi* in CVS was assessed with the forced swim test (FST) by recording immobility time. Along with this, they also conducted cell-based and cell-free *in-vitro* assay to understand the mechanism of the anti-stress effects of *tulsi*. The results of this study showed that the immobility time was reduced in FST, thereby demonstrating the anti-stress activities of *tulsi*. In terms of the mechanism for anti-stress activities, it was reported that there was an inhibition of corticol secretion and catechol-O-methyl transferase (COMT) activity following the treatment with *tulsi*.

In a randomized placebo-controlled, double-blind clinical trial conducted by Saxena et al., patients treated with a *tulsi* extract did exhibited an improvement in cognition and a reduction in forgetfulness, as compared with the placebo group [49]. Furthermore, it was reported that other stress-related symptoms (including sexual disorders, irregular sleeping habits, day time drowsiness, and frequent feeling of exhaustion) were all improved by treatment with *tulsi*. This study also reported a reduction in the severity of abnormal hearing and sensations, avoidance of familiar people, blurred vision, GI symptoms, headache, heart palpitations, and quarrelsome behavior [49].

## 5.6 USE OF *TULSI* FOR THE MANAGEMENT OF ANXIETY

A controlled clinical trial of *tulsi* was performed in the patients suffering with generalized anxiety disorders (GAD) [5]. Treatment with *tulsi* demonstrated significant anxiolytic and anti-stress effects of *tulsi,* with an improvement in the willingness in adjustment and attention [5]. It is suggested that regular administration of *tulsi* may regulate the hypothalamus-adrenocortical axis (HHA axis) in patients with stress disorders and maybe a safe alternative replacement for treatment with benzodiazepines.

Tabassum et al., [56] demonstrated the anxiolytic activity of *tulsi* to restrain stress conditions that are consistent with anti-stress effects [56]. However, the precise mechanism has not been reported for this activity.

## 5.7 USE OF *TULSI* FOR THE MANAGEMENT OF DEPRESSION

Symptoms of depression are correlated with a reduced level of monoamines, including dopamine, serotonin, and noradrenaline, which is consistent with the "Monoamine hypothesis," one of the most popular theories of depression. Many phytoconstituents of *tulsi* such as ursolic acid and others have been reported to increase levels of dopamine, serotonin, and noradrenaline in the brain [50]. However, the antidepressant effect of *tulsi* reported in the literature has not confirmed the accurate mechanism for antidepressant activity [35]. Similar work demonstrating antidepressant action of *tulsi* leaves using different animal paradigm has also been reported by Manu et al., [27].

## 5.8 USE OF *TULSI* FOR THE MANAGEMENT OF OTHER MENTAL DISORDERS

### 5.8.1 NEURO-PROTECTION

The *tulsi* has been reported to possess neuroprotective activity according to the research study by Yanpallewar et al., They demonstrated a neuroprotective effect of *tulsi* in a chronic cerebral hypoperfusion model that induced behavioral and structural changes [61]. Cerebrovascular insufficiency and neurodegenerative conditions induced by permanent occlusion of common carotid artery bilaterally were attenuated after treatment with *tulsi*. This activity was attributed to some of the chemical constituents of *tulsi,* including eugenol and its derivatives, which act as vasodilators by blocking voltage-dependent $Ca^{++}$ channels [18, 24, 33].

It is reported that *tulsi* protects the brain from extrapyramidal adverse effects induced by the neuro-psychiatric drugs [36]. Rajagopal et al., also observed the neuroprotective effect of *tulsi* against monosodium glutamate-induced neurotoxicity [40].

*Tulsi* has also been reported to have anticonvulsant effects. Epilepsy is a neurological disease caused by abnormal and excessive brain cell activities that cause seizures in specific individuals. There are several anticonvulsant agents available to treat epilepsy from both natural and synthetic sources. Jaggi et al., demonstrated that stem extract of *tulsi* had anticonvulsant effects [20]. Sakina et al., also reported anticonvulsant activity of ethanol extract of *tulsi* leaves by impacting dopaminergic neurons [45].

In Parkinson's disease (PD), dopaminergic neurons die due to increased oxidative stress [11]. However, Zhou et al., suggested that oxidative stress is not the only causative factor for neurodegeneration [62]. Venuprasad et al., suggested that *tulsi* prevented neuronal damage due to oxidative stress due to the antioxidant potential of flavonoids and polyphenols contents in *tulsi* [58].

Alzheimer's disease (AD) is a progressive and neurodegenerative disease that is age-associated and is mainly characterized by cognitive deficits; however, it is accompanied by behavioral disorders and changes in mood. AD induced dementia is a common form of accounting approximately ~70%cases in Western nations [3, 17]. *Tulsi* is reported to possess antioxidant and cognition-enhancing effects in cerebral-degenerative disorders.

Raghavendra and his colleagues [38] investigated the efficacy of *tulsi* in AD. *Tulsi* treatment significantly reduced the cognitive deficits observed after treatment of animals with neurotoxins (colchicine and ibotenic acid). The experimental animals treated with *tulsi* showed enhanced learning ability, including spatial and working memory. To determine if *tulsi* could reduce psychological parameters that were associated with learning disabilities, the effect of *tulsi* treatments were assessed on anxiety and depression induced by ibotenic acid and colchicine. These effects were evaluated using open field, elevated plus maze, and Porsalt's swim tests. The results showed that *tulsi* treatments were able to reduce secondary psychological effects associated with AD [38].

Similar to AD, drug-induced, and natural disruption of memory and learning abilities are also significant issues that are seen in patients of all age groups. It is suggested that nootropic agents ameliorate some of these drug-induced cognitive dysfunctions. *Tulsi* extracts possess nootropic activity and found to be useful for memory improvement and also in conditions of amnesia [21].

Antipsychotic activity of *tulsi* root extract was reported by Mukherjee et al., who suggested remedial use of root extract in the mental disturbances and stressful conditions [29]. It was also suggested to use *tulsi* roots for the induction of sleep in such mental disturbances due to insomnia [29]. In a study conducted to assess behavioral effects, *tulsi* leaves were reported to induce anxiolytic and antidepressant effects. Along with these effects, *tulsi* did not impair neuromuscular tone [7]. Similarly, Gradinariu et al., [14] also demonstrated antidepressant and anxiolytic effects of *tulsi* essential oil.

## 5.9 USE OF *TULSI* FOR THE MANAGEMENT OF METABOLIC DISORDERS

Under the Medicinal Plants Project of Central Drug Research Institute (CDRI) at Lucknow - India, Dhar, and his team [10] screened a variety of biological activities of several Indian Medicinal Plants. *Tulsi* was screened and found to possess hypoglycemic activity [10]. After the publication of Dr. Dhar reporting hypoglycemic activity of tulsi, several other scientists worked to establish the associated mechanism for the same.

Chattopadhyay et al., demonstrated that *tulsi* reduced blood glucose level in glucose-induced hyperglycemia and diabetic rats treated with insulin [8]. While one of the studies reported inhibition of intestinal absorption of glucose by *tulsi* leaf, the active constituents and exact mechanisms, by which it inhibits the glucose absorption, have not been described [12]. One of the critical factors greatly affected in diabetes is beta-cell function, and the quantity of insulin secreted. Considering this fact, Agrawal and his colleagues [1] indicated that *tulsi* treatment may improve the function of beta-cell and enhance the secretion of insulin from beta cells [1].

In 1964, a very interesting clinical trial was carried out in "People's Hospital" in Havana-Cuba to assess the hypoglycemic effect of decoction of *tulsi* (*Albahacamorada* in Caribbean language) as a supplement. The freshly prepared decoction (about 24 oz) was given orally to each set of the volunteers (normal and diabetic), and blood glucose levels were monitored. The diabetic patients under study were considered difficult cases are having a constant increase in blood sugar levels in spite of the increase in dosage of their regular insulin or tolbutamide; and therefore, this regimen was maintained during the study. Luthy and his coworkers observed positive results with the supplement of *tulsi* decoction along with their regular insulin or tolbutamide regimen. The diabetic patients were observed to be stabilized with their blood glucose levels and negative glucosuria without any side effects when compared with previous blood and urine reports [25].

Alterations in blood lipid profile in experimental normal rabbits after administration of mixed diet with 1–2% of *tulsi* was investigated [48]. This study showed that daily consumption of 1–2% of *tulsi* leaves along with the regular diet could significantly lower down the serum LDL-cholesterol, total cholesterol, phospholipid, and triglyceride levels and it significantly increased the HDL-cholesterol and total fecal sterol contents [48].

One report studied the effects of *tulsi* supplements on blood glucose levels, glycated protein, total amino acids, and serum lipid profile in non-insulin dependent *diabetes mellitus* patients. The patients and healthy volunteers were administered @ 1 g of dry powder of *tulsi* leave in empty stomach in the morning for 30 days without disturbing their regular medication and diet. The healthy volunteers comprising of 5 males and females have reported to have a slight increase in blood glucose levels, whereas diabetic volunteers were observed to decrease the hyperglycemia and hyperlipidemia. However, the study does not report the mechanism behind these effects [39].

## 5.10 SUMMARY

Ayurvedic treatment with *tulsi* has been used as an herbal nerve tonic to improve the complications related to nervous stress. *Tulsi* is thought to impart immunity, increase longevity, prevent aging, improve mental functions, and responsible for the addition of vigor and vitality to the body. Animal studies showed that *tulsi* and its chemical constituents could reduce both mental and physical stress in animal models, as well as have anti-depressant and anxiolytic effects; and may be of interest in Parkinson's and ADs. In addition, in terms of physiological stress, *tulsi* reduces blood sugar, cholesterol, and thus may be of benefit in type-2 diabetes and metabolic syndrome.

## KEYWORDS

- **Alzheimer's disease**
- **metabolic syndrome**
- **neurological stress**
- **neuroprotection**
- ***Ocimum sanctum***
- ***Ocimum tenuiflorum***
- ***tulsi***

## REFERENCES

1. Aggarwal, P., Rai, V., & Singh, R. B., (1996). Randomized placebo - controlled single blind trial of holy basil leaves in patients with non-insulindependent *diabetes mellitus*. *International Journal of Clinical Pharmacology and Therapeutics, 34*, 406–409.
2. Ahmad, A., Rasheed, N., Gupta, P., Singh, S., Siripurapu, K. B., Ashraf, G. M., et al., (2012) Novel ocimumoside A and B as anti-stress agents: Modulation of brain monoamines and antioxidant systems in chronic unpredictable stress model in rats. *Phytomedicine, 19*, 639–647.
3. Alzheimer's disease facts and figures: Alzheimer's association report, (2013). *Alzheimer's and Dementia, 9*(2), 208–245.
4. Bhargava, K. P., & Singh, N., (1981). Anti-stress activity of *Ocimum sanctum* Linn. *Indian Journal of Medical Research, 73*, 443–451.
5. Bhattacharyya, D., Sur, T. K., Jana, U., & Debnath, P. K., (2008). Controlled programmed trial of Ocimum sanctum leaf on generalized anxiety disorders. *Nepal Medical College Journal, 10*, 176–179.
6. Callahan, D., (1973). The WHO definition of 'health.' In: *The Concept of Health* (Vol. 1, No. 3, pp. 77–87). The Hastings Center Studies, Garrison, New York.
7. Chatterjee, M., Verma, P., Maurya, R., & Palit, G., (2011). Evaluation of ethanol leaf extract of Ocimum sanctum in experimental models of anxiety and depression. *Pharmaceutical Biology, 5*, 477–483.
8. Chattopadhyay, R. R., (1993). Hypoglycemic effect of *Ocimum sanctum* leaf extract in normal and streptozotocin diabetic rats. *Indian Journal of Experimental Biology, 31*(11), 891–893.

9. Cohen, M. M., (2014). Tulsi-*Ocimum sanctum*: A herb for all reasons. *Journal of Ayurveda and Integrative Medicine*, *5*, 251–259.
10. Dhar, M. L., Dhar, M. M., Dhawan, B. N., Mehrotra, B. N., & Ray, A. C., (1968). Screening of Indian plants for biological activity, Part I. *Indian Journal of Experimental Biology*, *6*(4), 232–247.
11. Dias, V., Junn, E., & Mouradian, M. M., (2013). The role of oxidative stress in Parkinson's disease. *Parkinson's Disease*, *4*, 461–491.
12. Giri, J. P., Suganthi, B., & Meera, G., (1987). Effect of tulsi on diabetes mellitus. *The Indian Journal of Nutrition and Dietetics*, *24*, 337–341.
13. Godhwani, S., Godhwani, J. L., & Was, D. S., (1988). *Ocimum sanctum*- A preliminary study evaluating its immunoregulatory profile in albino rats. *Journal of Ethnopharmacology*, *24*, 193–198.
14. Gradinariu, V., Cioanca, O., Hritcu, L., Trifan, A., Gille, E., & Hancianu, M., (2015). Comparative efficacy of *Ocimum sanctum* L. and *Ocimumbasilicum* L. essential oils against amyloid beta(1–42)-induced anxiety and depression in laboratory rats. *Phytochemistry Reviews*, *14*, 567–575.
15. Gupta, P., Yadav, D. K., Siripurapu, K. B., Palit, G., & Maurya, R., (2007). Constituents of *Ocimum sanctum* with antistress activity. *Journal of Natural Products*, *70*, 1410–1416.
16. Hill, M. N., Hellemans, K. G. C., Verma, P., Gorzalka, B. B., & Weinberg, J., (2012). Neurobiology of chronic mild stress: Parallels to major depression. *Neuroscience & Biobehavioral Reviews*, *36*, 2085–2117.
17. Huang, Y., & Mucke, L., (2012). Alzheimer mechanisms and therapeutic strategies. *Cell*, *148*(6), 1204–1222.
18. Huang, Y. C., Wu, B. N., Lin, Y. T., Chen, S. J., Chiu, C. C., Cheng, C. J., & Chen, I. J., (1999). Eugenedilol: A third-generation h-adrenoceptor blocker derived from eugenol, with a-adrenoceptor blocking and h-2-adrenoceptor agonist associated vasorelaxant activities. *Journal of Cardiovascular Pharmacology*, *34*, 10–20.
19. Ittiyavirah, S., & Anurenj, D. A., (2014). Adaptogenic studies of acetone extract of *Musa paradisiaca* L. fruit peels in albino Wistar rats. *International Journal of Nutrition, Pharmacology, Neurological Disease*, *4*, 88–94.
20. Jaggi, R. K., Madaan, R., & Singh, B., (2003). Anticonvulsant potential of holy basil, *Ocimum sanctum* Linn. and its cultures. *Indian Journal of Experimental Biology*, *11*, 1329–1333.
21. Joshi, H., & Parle, M., (2006). Evaluation of nootropic potential of *Ocimum sanctum* Linn. in mice. *Indian Journal of Experimental Biology*, *44*, 133–136.
22. Kelm, M. A., Nair, M. G., Strasburg, G. M., & DeWitt, D. L., (2000). Antioxidant and cyclooxygenase inhibitory phenolic compounds from *Ocimum sanctum* Linn. *Phytomedicine*, *7*, 7–13.
23. Kirtikar, K. R., & Basu, B. D., (1975). *Ocimum sanctum* Linn. In: Blatter, E., Caius, J. F., & Mhaskar, K. S., (eds.), *Indian Medicinal Plants* (Vol. III, pp. 1965–1968). M/s. Bishen Singh Mahendra Pal Singh Publisher, Dehradun.
24. Lin, Y. T., Wu, B. N., Horng, C. F., & Huang, Y. C., (1999). Selective beta1-adrenergic antagonist with tracheal and vascular smooth muscle relaxant properties. *Japanese Journal of Pharmacology*, *80*, 127–136.
25. Luthy, N., & Martinez-Fortun, O., (1964). A study of a possible oral hypoglycemic factor in *Albahaca morada* (*Ocimum sanctum* L.). *The Ohio Journal of Science*, *64*(3), 223–224.

26. Maity, T. K., Mandal, S. C., Saha, B. P., & Pal, M., (2000). Effect of *Ocimum sanctum* roots extract on swimming performance in mice. *Phytotherapy Research*, *14*, 120–121.
27. Manu, G., Hema, N. G., Parashivamurthy, B. M., & Kishore, M. S., (2014). Evaluation of effect *Ocimum sanctum* in experimental models of depression. *International Medical Journal*, *1*(9), 599–602.
28. Mizoguchi, K., Yuzurihara, M., Ishige, A., Aburada, M., & Tabira, T., (2003). *Saiko-ka-ryukotsu-borei-to*: A herbal medicine, ameliorates chronic stress-induced depressive state in rotarod performance. *Pharmacology Biochemistry and Behavior*, *75*, 419–425.
29. Mukherjee, J., Bhaumik, U., Mukherjee, P. K., & Saha, B. P., (2009). CNS activity of the methanol extract obtained from the roots of *Ocimum sanctum* Linn. *Pharmacology Online*, *2*, 673–685.
30. Nadkarni, K. M., & Nadkarni, A. K., (1954). *Ocimum sanctum* Linn. In: Nadkarni's, K. M., (ed.), *Indian Materia Medica, with Ayurvedic, Unani-Tibbi, Siddha, Allopathic, Homeopathic, Naturopathic & Home Remedies, Appendices & Indexes* (Vol. 1, pp. 865–867). Popular Prakashan, Panvel, Bombay.
31. NIH (National Institutes of Health), (2018). *Collection Development Manual*, NLM *(U. S. National Library of Medicine)*. https://www.nlm.nih.gov/tsd/acquisitions/cdm/subjects24.html (Accessed on 29 July 2019).
32. *NIH*. stress_factsheet_14289, https://www.nimh.nih.gov/health/publications/stress/index.shtml (Accessed on 29 July 2019).
33. Nishijima, T., Uchida, R., Kamayama, K., Kawakami, N., Ohkubo, T., & Kitamura, K., (1999). Mechanisms mediating vasorelaxing action of eugenol, a pungent oil, on rabbit arterial tissue. *Japanese Journal of Pharmacology*, *79*, 327–334.
34. Pare, W. P., & Glavin, G. B., (1986). Restraint stress in biomedical research: A review. *Neuroscience and Biobehavioral Reviews*, *10*, 339–370.
35. Pemminati, S., Gopalakrishna, H. N., Alva, A., Pai, M. R. S. M., Seema, Y., Raj, V., & Pillai, D., (2010). Antidepressant activity of ethanolic extract of leaves of *Ocimum sanctum* in mice. *Journal of Pharmacy Research*, *3*, 624–626.
36. Pemminati, S., Nair, V., Dorababu, P., & Gopalakrishna, H. N., (2007). Effect of ethanolic extract of leaves of *Ocimum sanctum* on haloperidol induced catalepsy in albino mice. *Indian Journal of Pharmacology*, *39*(2), 87–89.
37. Prajapati, N. D., Purohit, S. S., Sharma, A. K., & Kumar, T., (2003). *A Handbook of Medicinal Plants: A Complete Source Book* (pp. 12–18). Agrobios (India), New Delhi.
38. Raghavendra, M., Maiti, R., Kumar, S., & Acharya, S. B., (2009). Role of *Ocimum sanctum* in the experimental model of Alzheimer's disease in rats. *International Journal of Green Pharmacy*, 6–15.
39. Rai, V., Mani, U. V., & Iyer, U. M., (1997). Effect of *Ocimum sanctum* leaf powder on blood lipoproteins, glycated proteins and total amino acids in patients with non-insulin-dependent diabetes mellitus. *Journal of Nutritional & Environmental Medicine*, *7*, 113–118.
40. Rajagopal, S. S., Lakshminarayanan, G., & Rajesh, R., (2013). Neuroprotective potential of *Ocimum sanctum* (Linn) leaf extract in monosodium glutamate induced excitotoxicity, *African Journal of Pharmacy and Pharmacology*, *7*(27), 1894–1906.
41. Ramanathan, M., Balaji, B., & Justin, A., (2011). Behavioral and neurochemical evaluation of Perment an herbal formulation in chronic unpredictable mild stress induced depressive model. *Indian Journal of Experimental Biology*, *49*, 269–275.
42. Ravindran, R., Sheela Devi, R., Samson, J., & Senthilvelan, M., (2005). Noise stress induced brain neurotransmitter changes and the effect of *Ocimum sanctum* (Linn) treatment in albino rats. *Journal of Pharmacological Sciences, 98*, 354–360.

43. Richard, E. J., Illuri, R., Bethapudi, B., & Anandhakumar, S., (2016). Anti-stress activity of *Ocimum sanctum*: Possible effects on hypothalamic-pituitary-adrenal axis. *Phytotherapy Research*, *30*, 805–814.
44. Ryan, K. K., (2014). Stress and metabolic disease. In: *Sociality, Hierarchy, and Health: Comparative Biodemography: A Collection of Papers* (pp. 247–268). National Academy Press, Washington DC.
45. Sakina, M. R., Dandiya, P. C., & Hamdard, M. E., (1990). Preliminary psychopharmacological evaluation of *Ocimum sanctum* leaf extract. *Journal of Ethnopharmacology*, *28*, 143–150.
46. Samson, J., Sheeladevi, R., & Ravindran, R., (2007). Oxidative stress in brain and antioxidant activity of *Ocimum sanctum* in noise exposure. *Neuro Toxicology*, *28*, 679–685.
47. Samson, J., Sheeladevi, R., Ravindran, R., & Senthilvelan, M., (2006). Biogenic amine changes in brain regions and attenuating action of *Ocimum sanctum* in noise exposure. *Pharmacology, Biochemistry and Behavior*, *83*, 67–75.
48. Sarkar, A., Lavania, S. C., Pandey, D. N., & Pant, M. C., (1994). Changes in the blood lipid profile after administration of *Ocimum sanctum* leaves in the normal albino rabbits. *Indian Journal of Physiology and Pharmacology*, *38*(4), 311–312.
49. Saxena, R. C., Singh, R., Kumar, P., & Negi, M. P. S., (2012). Efficacy of an extract of *Ocimum tenuiflorum* (OciBest) in the management of general stress: A double-blind, placebo-controlled study. *Evidence-Based Complementary and Alternative Medicine*, 1–7.
50. Schildkraut, J. J., (1965). The catecholamine hypothesis of affective disorders: A review of supporting evidence. *American Journal of Psychiatry*, *122*(5), 509–522.
51. Sembulingam, K., Sembulingam, P., & Namsivayam, A., (2005). Effect of *Ocimum sanctum* Linn. on the changes in central cholinergic system induced by acute noise stress. *Journal of Ethnopharmacology*, *96*, 477–482.
52. Sen, P., Maiti, P. C., Puri, S., & Ray, A., (1992). Mechanism of anti-stress activity of *Ocimum sanctum* Linn, eugenol and *Tinosporamalabarica* in experimental animals. *Indian Journal of Experimental Biology*, *30*, 592–596.
53. Srikanth, N., Sudhakar, D., Padhi, M. M., & Lavekar, G. S., (2011). *Role of Ayurveda in Mental Health- An Appraisal of CCRAS Research Contributions* (pp. 1–11). Central council for research in Ayurveda and Siddha, New Delhi.
54. Stradford, D., (2012). Promoting wellness in mental health: The CAM approach in psychiatry. In: Stradford, D., Berger, C., Vickar, G., & Cass, H., (eds.), *The Flying Publisher Guide to Complementary and Alternative Medicine Treatments in Psychiatry* (pp. 13–21). The Flying Publisher, Bernd KampsSteinhauser Verlag.
55. Subhani, A. R., Kamel, N., & Mohamad, S. M. N., (2018). Mitigation of stress: New treatment alternatives. *Cognitive Neurodynamics*, *1*, 1–20.
56. Tabassum, I., Siddiqui, Z. N., & Rizvi, S. J., (2010). Effects of *Ocimum sanctum* and *Camellia sinensis* on stress induced anxiety and depression in male albino *Rattus norvegicus*. *Indian Journal of Pharmacology*, *42*, 283–288.
57. Tupper, T., & Gopalakrishnan, G., (2007). Prevention of diabetes development in those with the metabolic syndrome. *Medical Clinics of North America*, *91*, 1091–1105.
58. Venuprasad, M. P., Kumar, K. H., & Khanum, F., (2013). Neuroprotective effects of hydroalcoholic extract of Ocimum sanctum against $H_2O_2$ induced neuronal cell damage in SH-SY5Y cells via its antioxidative defense mechanism. *Neurochemical Research*, *38*, 2190–2200.

59. Welty, F. K., Alfaddagh, A., & Elajami, T. K., (2016). Targeting inflammation in metabolic syndrome. *Translational Research, 167*, 257–280.
60. Wicks, S. M., Lawal, T. O., Raut, N. A., & Mahady, G. B., (2017). Interactions of commonly used prescription drugs with food and beverages. In: *Sustained Energy for Enhanced Human Functions and Activity* (pp. 465–477). Academic Press (An imprint of Elsevier), London Wall, London, UK.
61. Yanpallewar, S. U., Rai, S., Kumar, M., & Acharya, S. B., (2004). Evaluation of antioxidant and neuroprotective effect of *Ocimum sanctum* on transient cerebral ischemia and long term cerebral hypoperfusion. *Pharmacology, Biochemistry and Behavior*, *79*, 155–164.
62. Zhou, C., Huang, Y., & Przedborski, S., (2008). Oxidative stress in Parkinson's disease: A mechanism of pathogenic and therapeutic significance. *Annals of the New York Academy of Sciences*, *1147*, 93–104.

# CHAPTER 6

# ANTIMICROBIAL COMPOUNDS FROM MEDICINAL PLANTS: EFFECTS ON ANTIBIOTIC RESISTANCE TO HUMAN PATHOGENS

OLUTOYIN OMOLARA BAMIGBOYE and
IDOWU JESULAYOMI ADEOSUN

## ABSTRACT

This chapter explores the potential of antimicrobial phytochemicals from medicinal plants as an alternative medicine to combat antibiotic-resistant strains of human pathogens. Increase in antibiotic resistance to certain human pathogens at global levels has necessitated research into safer and more acceptable alternatives, such as the use of medicinal plant and plant products with antimicrobial properties. More research is needed to establish the *in-vivo* effect in humans and establish their use as a therapeutic agent.

## 6.1 INTRODUCTION

Antimicrobial agents can reduce the menace of infectious diseases. Therefore, they play critical roles in human healthcare [22]. The most commonly known antimicrobials are antibiotics that target bacteria, while others are antiviral, antifungal, and antiparasitic [47]. Bacteria such as *Pseudomonas aeruginosa, Shigella boydii, Staphylococcus aureus, Klebsiella pneumonia,* and *Escherichia coli* result in numerous human infections [56], such as nosocomial infection; and health care providers are facing the challenge of combating the resistance of these species to antimicrobial agent [44].

Antibiotic resistance is as old as antibiotics itself. Alexander Fleming, in his acceptance speech, when receiving the novel award for the discovery of Penicillin, warned that bacteria could develop resistance to itself [11].

Resistance to antimicrobial agent occurs when it becomes ineffective in killing or preventing the growth of the microbe at the therapeutic dose [38]. A vast increase in multidrug-resistant organisms as well as strain, that is less susceptible to antibiotics, has greatly increased in number and the emergence of strains becoming resistant to antibiotics [15]. The severity of drug resistance has created an unprecedented challenge both to the scientific community and human populace.

Plant-based antimicrobials have exhibited immense potentials with no side effects in combating bacterial, fungal, protozoal, and viral diseases [12]. Plants have a wide variety of metabolites, such as terpenoids, alkaloids, tannins, and flavonoids, and these have shown to have antimicrobial properties [13].

Bioactive agents of plant origin are now being investigated for treating drug-resistant infection [8]. Exploration of the potential of the medicinal plant will bring out plant-based products that will alleviate the burden of antimicrobial resistance.

This chapter explores the antimicrobial potential of medicinal plants and phytochemicals as an alternative medicine to combat antibiotic-resistant strains of human pathogens.

## 6.2 ANTIMICROBIAL RESISTANCE AMONG PATHOGENS

Resistance to antimicrobials is a phenomenon, where antimicrobial agents are no longer effective on microbes through gene acquired by them. This creates challenges for medical intervention at different levels of health care, specifically in greater dimensions for chemical therapeutic and public health [54].

Bacteria produce antimicrobial compounds as an adaptive feature to be able to survive among other microbes. This has led to the production of many antibiotics by microorganisms, such as. Streptomyces. Likewise, antibiotic resistance must have arisen as an adaptation of bacteria for survival in a competitive environment using their innate genetic ability [26, 36]. Antimicrobial-resistant bacteria are therefore selected in an antibiotic rich environment, and these lead to an increase in transmission of diseases [33].

The increase and spread of antimicrobial resistance have also been attributed to include indiscriminate, inappropriate, and irrational use of antibiotics, intake of incomplete dosage of drug, transplantation of organs, agents capable of suppressing the immune system, HIV infection and intravenous catheters [12, 17, 28, 39, 40]. Moreover, transfer across the food chain can lead to antibiotic resistance. Antimicrobial agents are presently used to treat infections in plants, animals, as well as humans and sub-therapeutically in foods or growth promoters in animals [6]. Antibiotic use in both plant and animals has resulted to increase the prevalence of resistance of bacteria affecting human health. Bacteria make use of this opportunity to develop new resistant genes that could be transferred across the food chain [36]. Starr et al., reportedly isolated streptomycin-resistant bacteria from the intestine of turkey that was previously treated with streptomycin [49]. Bauer-Garland et al., also demonstrated that the use of antibiotics influences the transmission of resistant bacteria strain in poultry [9].

## 6.3 MECHANISM OF ANTIMICROBIAL RESISTANCE

Various mechanisms are involved in microbial resistance to the drug. These include:

- Inactivation of the drug through enzyme activities;
- Mutations that change target sites or cellular functions;
- Reduced cellular uptake and extrusion by efflux [12, 32, 41, 43, 48].

Bacteria and other microbes usually develop resistance as a means of adaptation to the antimicrobial agent, and this is quickly passed to the progeny. This makes the treatment of bacterial infection increasingly complicated [51].

The classical production of β-lactamase is a typical example of such a mechanism, which precedes the first antibiotic penicillin discovery [16]. They utilize biochemical pathways, such as:

- Inactivation of drugs with beta-lactamases.
- Acetylases, adenylases, and phosphorylases.
- Reduction of drug access to sites of action.
- Alteration of drug target sites.
- Bypassing of drug metabolism and developing tolerance.

Most antibiotics, including penicillin, ceflazidime, and ampicillin contain the β-lactam ring. Upon the production of β-lactamase enzyme by bacteria, the amide bond of the β-lactamase ring is hydrolyzed, and this renders the antibiotics ineffective [55].

Mutation or acquisition of new DNA is a genetic mechanism, in which microorganisms resist antimicrobials. Point mutation is usually involved, which can alter bacteria targets or its functions. This will lead to a decrease in its affinity for an antimicrobial agent and influence its efflux pump and protein production [35]. Changes in the genetic makeup of resistant bacteria happen swiftly so that the potency of the antibiotic may varnish within a short period of time [12]. Antibiotic effectiveness is usually lost quickly as genetic constitution changes rapidly in the resistant bacteria.

Antibiotic resistance can also occur when the penetrable ability of the bacteria outer membrane is reduced due to suppressed porin expression; hence, access of antibiotic to bacteria target is reduced. Antimicrobial agents make use of porin and other protein channels to reach the bacterial targets. Loss of certain protein from the outer membrane has been associated with antibiotic resistance in some bacteria species [32].

Active drug efflux is another prevalent channel of drug resistance in bacteria. Such resistant bacteria extrude antimicrobial agents to non-inhibitory levels through active efflux pumps in their structure [53]. Active efflux pump may be primary active transport, which effluxes drugs from cells through the hydrolysis of ATP or secondary active transport drug-using ion gradient or glycoprotein transporters [57]. Efflux primarily acts in removing substances that can disrupt the cytoplasmic membrane, including those with antimicrobial properties, thus, removing antimicrobial agents from the cell [10].

To date, there is no known available method to reverse antibiotic resistance in bacteria [12]. Drug resistance has become so severe that professionals and scientists must urgently search for alternative treatments to combat drug-resistant bacteria.

## 6.4 ANTIMICROBIAL BIOACTIVITIES OF MEDICINAL PLANTS

Medicinal plant has been defined by the WHO as one that can be used wholly or in parts for disease management [2]. Flowering plant species that have been reported are about 422,000, and up to 12% of the species are medicinal plants [27]. Plants are a potential source of antimicrobial agents in diverse countries [4]. Many workers have studied the antimicrobial properties of medicinal plants against bacterial, fungal, protozoal, and viral pathogens [5, 7, 20].

Gislene et al., [26] evaluated 10 plant extracts and their phytochemicals for activity against antibiotics susceptible and resistant microorganisms. The highest antimicrobial activity was also active against resistant bacteria. Extract of *Caryophyllus aromaticus* (clove) had the highest antimicrobial activity. *Syzygyum joabalamum* (joabalam) was also active against resistant bacteria. The two plant extracts also inhibited *Pseudomonas aeruginosa.* In West Africa (Nigeria), three medicinal plants were investigated by Falodun et al., [20] for *in vitro* antiplasmodial and antimicrobial effects. The methanol concentrate of *Picralima nitida* exhibited action against *Plasmodium falciparum*, which is either sensitive or resistant to chloroquine while ether extract showed activity against Methylin resistant *Staphylococcus aureus.* Extract of *Persea Americana* revealed antifungal activity against *Cryptococcus neoformans.*

Research conducted on antimicrobial properties of medicinal plant in Argentina using an extract from 122 known plant species [5] showed that *Staphylococcus aureus*, *Escherichia coli*, and *Aspergillus niger* were inhabited by some of the plant species. *Tarbebua impetiginosa* extract exhibited the most potent extract with the highest bioactivity. *Dabur* et al., [15] screened 77 extracts for antimicrobial activity. The water extract of *Acacia nilotica as well as Lantana camara* and *Saraca asoca* showed antibacterial activity. It was also effective against all the fungi pathogens,

Martinez et al., [34, 35] reported that *Pantherum argentatum* produced antibacterial compounds effective against *P. aeruginosa* and *Klebsiella* species and antifungal compounds against *Torulopsis hansemula and Candida albicans.*

Khan et al., [30] investigated the activity of three plant species against human pathogens. *Bergenia ciliata* extract showed higher activity than the traditional antibiotics cefloriaxone and erythromycin against *Bacillus subtilis*. Extracts of *S. album* and *J. officinale* showed variable antimicrobial activity. It was also effective against *S. aureus*, *P. aeruginosa. Proteus vulgaris,* and *E. coli.*

*S. surathence* have been reported to show antifungal activity against *A. fumigatus* [15]. Inhibitory effect of *S. asoca* flower and bud water-soluble extract was reported against *Shigella boydii* [37].

Farjana et al., [21] determined the antimicrobial action of extracts of guava (*Pisidium guajava*), green tea (*Camellia sinensis*), Neem (*Azardirachita indica*) and Marigold (*Calendula officinales*) against different species of bacterial including *Pseudomonas spp, Vibrio cholera, V. parahemolyticus, Klebsiella spp, E. coli. Salmonella spp* and *S. aureus*. The leaf extracts showed antibacterial activity against different bacteria species and can be utilized as an alternative to routine antimicrobial agents as a remedy for bacterial infection.

Ozcelik et al., [43] investigated the antiviral, antifungal, and antibacterial effects of 15 lipophilic extracts obtained from various parts of *Pistacia vera* against the standard and isolated strains of common human pathogens. The extract showed little antibacterial activity, but obvious antifungal activity as seed extract and kernel showed notable antiviral activity.

Table 6.1 summarizes medicinal plant extracts that possess antimicrobial activity against some human pathogens.

**TABLE 6.1** Effective Plant Extract Against Antibiotic Resistant Strain of Common Human Pathogens [5, 11, 15, 20, 21, 26, 30, 34, 35, 37]

| Human Pathogen | Antibiotic Resistant Strain | Effective Plant Extract |
|---|---|---|
| *Escherichia coli* | Timotheprine-R *E. coli*<br>Ampicillin-R *E. coli* | *Acacia nilotica, Picralima nitida, Lantana camara, Saraca asoca, Teucrium polium, Perganiumh amala, Prangos ferulace, Santalum album* |
| *Klebsiella pneumonia* | MDR- *K. pneumoniae*<br>XDR- *K. pneumoniae* | *Panthenum argentatum, Calendula officinales, Azaridachta indica* |
| *Pseudomonas aeruginosa* | PDR-*Pseudomonas*<br>MDR-*Pseudomonas* | *Syzygium joabalanum, Caryopsis ameranthus, Thymus vulgaris, P. hamela, T. polium, T. pratensis* |
| *Salmonella typhi* | MDR-*Salmonella* | *Acacia nilotica, Picralima nitida, Lantana camara, Saraca asoca* |
| *Shigella boydii* | Tetracyline-R *Shigella*<br>Fluoroquinoline-R *Shigella* | *Acacia nilotica, Picralim anitida, Lantana camara, Saraca asoca* |
| *Staphylococcus aureus* | Methicilin-R *Staphylococcus*<br>Penicillim-R *Staphylococcus*<br>Linezoid-R *Staphylococcus*<br>Vancomycin-R *Staphylococcus*<br>Ceftarolin-R *Staphylococcus* | *Persea americana, Acacia nilotica, Justicia zelenica, Picralima nitida, Lantana camara, Saraca asoca, Bergenia ciliate* |

## 6.5 PLANT-BASED BIOCOMPONENTS WITH ANTIMICROBIAL PROPERTIES

Plants possess a variety of phytochemicals with antimicrobial properties, which include alkaloids, tannins, flavonoids, and terpenoids [8, 19, 50]. Other plant metabolites with antimicrobial activity include:

- Phenols;
- Quinines;
- Lectins;
- Polypeptide;
- Coumarin;
- Essential oils (EOs) [12].

However, the mode of action and potency of these plant extracts in the most assessment are still needed to be scientifically proven [14, 45].

Flavonoids derived from plants can be found in fruits, vegetables, and tea and occur as a large group of phenyl chromones in nature. They have antibacterial, antiviral, and anti-inflammatory properties; and antidiabetic, antioxidant, and antimutagenic properties. They are active against cancerous growth (antiproliferative and anticarcinogenic) and protect the liver (hepatoprotective) [52].

Orhan et al., [42] reported that the flavonoids from six plant extracts showed antibacterial activity against isolated strains of *Staphylococcus aureus*, and *Pseudomonas aeruginosa,* and antifungal activity against *Candida krusei*. Some of the flavanones also inhibited Parainfluenzas, PI-3 virus and Herpes simplex virus I HSV-I. Plants have been shown to consistently synthesize aromatic phenolic substances or their oxidized derivatives [24]. This includes quinone, which is aromatic in nature and has a wide range of antimicrobial effect.

Terpenoids, as well as EOs, are toxic to viruses, bacteria, protozoa, and fungi [3, 7, 23, 25]. Alkaloids are effective against trypanosome and plasmodia, while lectins and polypeptides are inhibitory to bacteria and fungi [18]. Tannins were found to be toxic to filamentous fungi, yeast, and bacteria [46].

## 6.6 PLANT-BASED ANTIMICROBIAL AGENTS: SCOPE AND CHALLENGES

The challenge facing countries with diverse plant species of medicinal value is that of proper identification and extraction of active agents from the plant. Moreover, most of the medicinal plants have not been well studied [31, 29] for their beneficial effects; and *in-vivo* studies of various useful phytochemicals with antimicrobial properties have not been conducted.

There is a rapid loss of the natural habitat of some of these plants due to anthropogenic activities [1]. There is also a rapid extinction of plant species, and many scientists have observed that the risk of irretrievably losing potentially useful plant and plant product is great.

There is a general awareness that most antibiotics are overprescribed and used wrongly, and many people would rather take charge of their medication [13]. Many substantial plant products are being sold locally, and self-medication is common. This calls for urgent research efforts by scientists and medical professionals to stem the menacing tide of antibiotic-resistant pathogens.

## 6.7 SUMMARY

Indiscriminate use of antibiotics, self-medication, often accompanied with improper dosage and transfer across the food chain through antibiotic use in food animals both as therapeutic and growth promoters are some of the major causes of antimicrobial resistance. Several antibiotics are no longer potent in the treatment of certain diseases due to bacteria developing resistance to them.

The ability of bacteria to genetically develop drug resistance is a survival strategy to adapt to their environment. The effectiveness of phytochemicals as antimicrobial agents has been demonstrated by many investigators across various continents. In developing countries, little research has been done, and the plants are rapidly going into extinction. The urgent research focus should be beamed at this rich phytodiversity and their chemicals to maximize their potentials as antimicrobial agents.

## KEYWORDS


- **alkaloids**
- *Aspergillus niger*
- *Azaridachta indica*
- *Calendula officinalis*
- *Camelia sinensis*
- *Caryophyllus aromaticus*
- *Cephalosporium*
- *Escherichia coli*
- **flavonoids**
- *Jasminum officinalis*
- *Klebsiella pneumonia*
- *Lantana camara*
- **lectins**
- **metabolite**
- **multidrug resistant**
- *Penicillium*
- *Persea americana*
- *Picralima nitida*
- *Pisidium guajava*
- *plasmodium falciparum*
- **polypeptide**
- *Pseudomonas aeruginosa*
- **quinones**
- *Salmonella spp.*
- *Santalum album*
- *Saraca asoca*
- *Shigella boydis*
- *Staphylococcus aureus*
- *Streptomyces*
- *Syzygyum joabalanum*
- **tannins**
- **terpenoids**

## REFERENCES

1. Abd El-Ghani, M. M., (2016). Traditional medicinal plants of Nigeria: An overview. *Agricultural and Biology J. of North America*, *7*(5), 220–247.
2. Archaya, D., & Silvastava, A., (2008). *Indigenous Herbal Medicines: Tribal Formulations and Traditional Herbal Practices* (1st edn., p. 241). Havishker Publishers, New Delhi–India.
3. Ahmed, A. A., Mahmoud, A. A., William, H. J., & Scott, A. I., (1993). New sesquiterpene α-methylene lactones from the Egyptian plant *Jalconiacandicans*. *J. Nat. Prod, 56*, 1276–1280.
4. Alviano, D. S., & Alviano, C. S., (2009). Plant extracts: Search for new alternatives to treat microbial diseases. *Curr. Pharm. Biotechnol, 10*(1), 106–121.
5. Anesini, E., & Perez, C., (1993). Screening of plants used in Argentina folk medicine for antimicrobial activity. *J. Eth. Pharmacol, 39*, 119–128.
6. Angulo, F., Johnson, K., Tauxe, R., & Cohen, M., (2000). Origins and consequences of antimicrobial resistant non-typhoidal Salmonella: Implications for the use of fluoroquinolones in food animals. *Microbial Drug Resist, 6*, 77–83.
7. Ayafor, J. F., Tchuendim, M. H. K., & Nyasse, B., (1994). Novel bioactive diterprenoids from *Aframomumaulacocarpos*. *J. Nat. Prob. Cross-Ref Medline, 57*, 917–923.
8. Bankole, A. E., Adekunle, A. A., & Sowemimo, A. A., (2016). Phytochemical screening and *in vivo* antimalarial activity of extracts from three medicinal plants used in malaria treatment in Nigeria. *Braz. J. Microbiol, 115*, 299–305.
9. Bauer-Garland, J., Frye, J. G., Gray, J. T., Berrang, M. E., Harrison, M. A., & Fedorka-Cray, P. J., (2006). Transmission of salmonella serotype typhimurium, in poultry with or without antimicrobial selective pressure. *J. Appl. Micro, 101*, 1301–1308.
10. Bonomo, R. A., & Szabo, D., (2006). Mechanism of multidrug resistance in Acinetobacter species and Pseudomonas aureginosa. *J. Infect. Dis. Clin., 43*, 49–56.
11. Centre for Disease Control and Prevention (CDC), (2017). *Antibiotic and Antimicrobial Resistance* (p. 23). CDC, Atlanta, GA, USA.
12. Chandra, H., Bishnoi, P., Yadav, A., Patni, B., Mishra, A. P., & Nautiyal, A. R., (2017). Antimicrobial resistance and the alternative resources with special emphasis on plant-based antimicrobials - A review. *Plants, 6*(2), 16, E-article. doi: 10.3390/plants6020016.
13. Cowan, M. M., (1999). Plant products as antimicrobial agents. *Clinical Microbiology Reviews, 12*(4), 564–582.
14. Cruz, M. C., Santos, P. O., Barbosa, A. M., Melo, D. L., & Alviano, C. S., (2007). Antifungal activity of Brazilian medicinal plants involved in popular treatment of mycoses. *J. Ethnopharmacology, 11*(2), 409–412.
15. Dabur, R., Gupta, A., Mandal, T. K., & Singh, D. D., (2007). Antimicrobial activity of some Indian medicinal plants. *Afr. J. Trad. Compl. Med., 4*(3), 313–318.
16. Davies, J., (1994). Inactivation of antibiotics and the dissemination of resistant genes. *Science* (New York, NY), *264*(3), 75–82.
17. Dean, D. A., & Burchard, K. W., (1996). Fungal infection in surgical patients. *Am. J. Surg., 171*, 374–382.
18. Dixon, R. A., Dey, P. M., & Lamb, C. J., (1983). Phythoalexin: Enzymology and molecular biology. *Adv. Enzymol. Relat. Areas Mol. Biol., 55*, 1–69.
19. Dorman, H. J., & Deans, S. G., (2000). Antimicrobial agents from plants: Antibacterial activity of plant volatile oils. *J. Appl. Microbial., 88*(2), 308–316.

20. Falodun, A., Imieje, V., Erhiruiji, O., Ahomafor, J., Jacob, M. R., Khen, S. I., & Hamann, M. T., (2014). Evaluation of 3 medicinal plants extracts against *Plasmodium falciparum* and selected microorganisms. *Afr. J. Trad. Complent. Altern. Med.*, *11*(4), 142–146.
21. Farjana, A., Zerin, N., & Kabir, S., (2014). Antimicrobial activity of medicinal plant leaf extracts against pathogenic bacteria. *Asian Pacific J. of Tropical Diseases, 2*(2), 920–923.
22. Finch, R. D., Greenwood, D., Norly, S. R., & Whitley, R. J., (2003). *Antibiotic and Chemotheraphy: Anti-Infective Agent and Their Use in Therapy* (8th edn., pp. 45–49). Living Stone, Churchill, UK. ISBN-13–978–04430712.
23. Fujioka, T., & Kashiwada, Y., (1994). Anti-AIDS agents: 11 Betulinic acid and plantonic acid as anti-HIV principles from *Syzigiumclariflorum* and the anti-HIV activity of structurally related triterpenoids. *J. Nat. Prod, 57*, 243–247.
24. Geissman, T. A., (1963). *Flavonoids Compounds, Tannins, Lignins and Related Compounds in Pyrole Pigments, Isoprenoid Compounds and Phenolic Plant Constituents* (pp. 265–234). Elsevier, New York, N. Y.
25. Ghoshal, S., Krishna-Prasad, B. N., & Lakshmi, V., (1996). Anti-amoebic activity of *Piper langum* fruits against *Entamoeba histolytica in vitro* and *in vivo*. *J. Ethno Pharmacol., 50*, 167–170.
26. Gislene, G. G., Nascimento, F., Locatelli, J., Freitas, P. C., & Silva, G. L., (2000). Antibacterial activity of plant extracts and phytochemicals on antibiotic resistant bacteria. *Braz. J. Microbiol., 31*(4), 1678–4405.
27. Govaerts, R., (2001). How many species of seed plants are there? *Taxon,* 1085–1090.
28. Graybill, J. R., (1988). Systemic fungi infections: Diagnosis and treatment with therapeutic agents. *Infectious Disease Clinic of North America, 14,* 805–825.
29. Hostelman, K., & Marston, A., (2002). Twenty years of research into medicinal plants: Results and perspectives. *Phytochem. Res., 2*, 278–285.
30. Khan, U. A., Rahman, H., Niaz, Z., & Rehman, B., (2013). Antibacterial activity of some medicinal plant against selected human pathogenic bacteria. *European Journal of Microbiology and Immunology, 4,* 272–274.
31. Kirby, G. C., (1996). Medicinal plants and the control of parasites. *Trans. Roy. Soc. Trop. Med. Hyg., 90*, 605–609.
32. Manchandra, V., Sanharta, S., & Singh, N. P., (2010). Multidrug resistant acinetobacter. *Journal of Global Infectious Diseases, 2*(3), 291–304.
33. Marker, B., (2010). *Antibiotic Resistance in Salmonella, Marler Blog*. www.marlerblog.com/case-news/antibiotic-resistance-in-*salmonella/* (Accessed on 29 July 2019).
34. Martinez, M. J., Vasquez, S. M., Espinosa-Perez, C., Dias, M., & Hermera, S., (1994). Antimicrobial properties of Argentatine isolated from *Parthenium argentatum. Filoterapia, 65*, 371–372.
35. Martinez, M. J., Betancourt, J., Alanso-Gonzalez, N., & Jauregui, A., (1996). Screening of some Cuban medicinal plant for adults. *J. Eth. Pharmacol, 52*, 171–174.
36. Matthew, A., Gissel, R., & Liamthong, S., (2000). Antibiotic resistance in bacteria associated with food animals: United States perspective of livestock production. *Foodborne Path. and Dis., 4*(2), 115–133.
37. Narang, G. D., Nayar, S. M., & Cndiratta, D. K., (1962). Antibacteria activity of some indigenous drugs. *J. Vet. Animal Hus, 6*(1), 22–25.
38. Nasir, I. A., Babyo, A., Emeribe, A. U., & Sani, W. O., (2015). Surveillance for antibiotic resistance in Nigeria: Challenges and possible solutions. *Trends in Medical Research, 10,* 106–113.

39. Ndihokubwayo, J. B., Yahaya, A. A., Detsa, A. T., Ki-zerbo, G., & Odei, E. A., (2013). Antimicrobial resistance in the African region: Issues, challenges and actions proposed. *Afr. Health Min.*, *16*, 27–30.
40. Ng, P. C., (1994). Systemic fungi infections in Neonates. *Arch. Dis. of Childhood*, *71*, 130–135.
41. Nikado, H., (1994). Prevention of drug access to bacterial targets. Permeability barriers and active efflux. *Science*, *264*(5157), 382–388.
42. Orhan, D. D., Ozpeluk, B., Ozgen, S., & Ergun, F., (2010). Antibacterial, antifungal, and antiviral activities of some flavonoids. *Microbiological Research*, *165*(6), 496–504.
43. Ozcelik, B., Aslan, M., Orhan, I., & Kahaoglu, T., (2005). Antibacterial, antifungal and antiviral activities of the lipohylic extracts of *Pistaciavera. Microbiol. Res*, *160*(2), 159–164.
44. Pitout, J. J. D., (2008). Multi resistant enterobacteriaceae: New threat of an old problem. *Expert Review of Anti-Infective Therapy*, *6*(5), 657–669.
45. Ruberto, G., Baratta, M. T., Deans, S. G., & Dorman, H. J., (2000). Antioxidant and antimicrobial activity of *Foeniculum vulgare* and *Crithmummaritimum* essential oils. *Plant Med.*, *66*(8), 687–693.
46. Scalbert, A., (1992). Antimicrobial properties of tannins phytochemistry. In: *Proceedings of the 2nd North American Tannin Conference* (p. 8). Plenum Press, New York.
47. Seradzki, K., Wu, S., & Tomasz, A., (1999). Inactivation of the methicillin resistance gene mecA in Vancomycin-resistant *Staphylococcus aureus*. *Micro Drug Resist*, *5*(4), 253–257.
48. Smith, A., (2003). Bacteria resistance to Ab. In: Denyer, S. P., Hodges, N. A., & Arman, S. P., (eds.), *Hugo and Russel's Pharmaceutical Microbiology* (8th edn., pp. 220–224). Blackwell Science Ltd., Massachusetts, USA.
49. Starr, M. P., & Reynolds, D. M., (1951). Steptomycin resistance of coliform bacteria from turkey feed. *Am. J. Public Health*, *25*, 1375–1380.
50. Tahl, W. H., & Mahasneh, A. M., (2010). Antimicrobial, cytotoxicity and phytochemical screening of Jordamain plants used in traditional medicine. *Molecules*, *15*(3), 1811–1824.
51. Tenover, F. C., (2006). Mechanism of antimicrobial resistance in bacteria. *Am. J. Infect. Control*, *34*(5), 3–10.
52. Tringali, C., (2001). *Bioactive Compounds from Natural Sources: Isolation, Characteristics and Biological Properties* (p. 261). Taylor and Francis Group, London.
53. Webber, M. A., & Piddock, L. J., (2003). The importance of efflux pumps in bacterial antibiotic resistance. *The Journal of Antimicrobial Chemotherapy*, *51*, 9–11.
54. WHO., (2001). *Strategy for Containment of Antimicrobial Res. Report by WHO/CDS/CSR/DRS/2001* (pp. 1–105). Department of communicable disease surveillance and response, WHO, Geneva, Switzerland.
55. Wilke, M. S., Lovering, A. L., & Strynadka, N. C. J., (2005). β-Lactam antibiotic resistance: A current structural perspective. *Curr. Opin. Microbiol.*, *8*, 525–533.
56. Zhag, R., Eggleston, K., Rotimi, V., & Zeckhauser, R. J., (2006). Antibiotic resistance as a global threat: evidence from China, Kuwait and the United States. *Global Health*, *2*, 6–8.
57. Zhou, G., Shi, Q. S., Huang, X. M., & Xie, X. B., (2015). The three bacterial lines of defense against antimicrobial agents. *Int. J. Mol. Sci.*, *16*(9), 21711–21733.

# CHAPTER 7

# PLANT-BASED NATURAL PRODUCTS AGAINST HUNTINGTON'S DISEASE: PRECLINICAL AND CLINICAL STUDIES

BANADIPA NANDA, SAMAPIKA NANDY, ANURADHA MUKHERJEE, and ABHIJIT DEY

## ABSTRACT

Plant-based medicines have been in use for their proven benefits with least side effects against complex medical conditions, including neurological disorders. Botanicals have been used traditionally against neurological disorders in Japan, Korea, India, and China. Huntington's disease (HD) or Huntington's chorea is a hereditary neurological disorder responsible for brain cell death due to an autosomal dominant mutation in any one copy of the Huntingtin gene pair. Until now, there are only some available treatments for symptomatic relief. Hence, herb-based traditional uses of extracts and biocompounds are being investigated in a number of pre-clinical and clinical studies. The review in this chapter focuses on different types of plant extracts, active fractions, and natural biocompounds, monoherbal, and polyherbal formulations with their proven efficacy in cellular and animal models of HD. The present review utilizes a number of popular search engines to retrieve literature involving an array of anti-HD medicinal plants and natural compounds. However, further study is needed to evaluate the medical efficacy of plant-based medicines for human use.

## 7.1 INTRODUCTION

Huntington's disease (HD) is a chronic, progressive, neurodegenerative disorder, characterized by a combination of choreoathetotic movements and cognitive and psychiatric disturbances associated with neuronal death in corticostriatal circuits. Symptoms develop insidiously either as brief, jerky movements of the extremities, trunk, face, and neck (chorea) or as a change in personality and sometimes both [14]. Fine motor in-coordination and impairment of rapid eye movements are early features. Occasionally, choreic movements are less prominent than the predominance of bradykinesia and dystonia in the early onset of symptoms that occurs before age 20. With the progression of the disease, the severity of involuntary movements is higher with development of dysarthria and dysphagia, and patients exhibit a typical sporadic, rapid, involuntary limb movement, limb stiffness with impeded balance. The cognitive disorder manifests first as sluggish mental processing and difficulty in organizing complex tasks along with progressive dementia [16].

HD is characterized by prominent striatal neuronal loss of brain [78]. Atrophy of the structures proceeds in an orderly fashion, first affecting the caudate nucleus failure and then proceeding anteriorly from mediodorsal to ventrolateral. Other areas of the brain are also affected though to a lesser extent. The degree of atrophy is directly related to the severity and duration of the disease. In the late stages, the caudate nucleus takes on instead a flattened or concave appearance. As a result of the tissue loss, the ventricular system becomes correspondingly widened, especially the frontal horn. Along with these changes in the basal ganglia, there is characteristic diffuse gyral atrophy, most severe over the convex aspect of the brain.

Usually beginning in mid-adult life, HD is transmitted as an autosomal dominant disease, generally caused by a kind of genetic stutter, leading to unstable expansion of a trinucleotide (CAG) repeat in Huntington gene coding region, located on the terminal segment of the chromosome 4 short arm (4p16.3) [83]. The inherited mutation results in the production of an elongated poly-Q mutant Huntingtin protein (mHtt), widely expressed outside the CNS. The cellular functions of Htt protein have still not been fully elucidated; however, the expression of mHtt and the transcriptional dysregulation associated with the disturbance of histone-modifying complexes and changed interaction with chromatin related factors are interrelated. Moreover, altered energy production, altered neurotransmitter metabolism, receptors, and growth factors are also attributed to mHtt functionality. Peripheral signs of HD often include the weight loss and enhancement in pro-inflammatory signaling, but their role in HD pathophysiology is still unresolved. The occurrence of symptoms usually begins in middle age, although approximately 5% become symptomatic as juveniles; also, the pathophysiologic changes manifest in the brain years prior to the appearance of HD symptoms [90].

Because of its distressing and incapacitating nature and its implications for members of any family in which it appears (50% risk in all children of an affected parent), the disease has attracted attention in recent years. Rare and incurable, HD has a worldwide occurrence of 3 to 7 persons per 100,000 individuals with no gender predominance; and approximately 20 are reported as carriers per 100,000 people [38]. Moreover, a new mutation rate of 1–3% has also been observed. In the Caucasian population, the prevalence of HD is up to 5–10 per 100,000 individuals. Meta-regression revealed a significantly lower prevalence of HD in Asia, compared to the European, North American, and Australian population [1]. This alarmingly high occurrence of this disease in these regions than in Asia can be largely explained by the different mutation rates and geographic differences in the Huntington gene haplotypes [96, 137].

Now, HD is an incurable neurodegenerative disorder, for which present treatments are useful against some symptoms but do not target the underlying mechanism, mostly due to the lack of ample knowledge on its pathogenesis since the discovery of its genetics. Clinical symptoms of HD are mostly treated with conventional therapies as the number of treatments is limited to stop the progressive neuronal loss and behavioral and psychiatric disorders due to the selectiveness of the Blood-Brain-Barrier (BBB). The non-pharmacological treatments include genetic counseling and therapy along with palliative care [54]. Physical, cognitive, and behavioral dysfunction in HD, along with chorea, have a very limited number of treatments available. Patients treated with Tetrabenzene for their chorea are at risk from the drug interactions and potentially serious adverse side effects [113]. Antipsychotic agents like Olazopine and Aripiprazole and selective serotonin reuptake inhibitors (SSRIs) with adequate efficacy are used to regulate the early symptoms of HD.

Molecular therapeutic interventions against HD employ some new methods to selectively silence the gene and neutralization of its toxic protein product [81]. Number of preclinical studies indicated to possible efficacy of ribozymes, DNA enzymes, antisense oligonucleotides (ASOs), RNA interference (RNAi) and genome-editing methods to either silence or restore the mHtt gene [27]. In this context, the effective botanicals, functional on various biochemical targets with new mode of actions and low toxicity, are at full swing in the pharmaceutical industry. Plants in nature form the basis of sophisticated traditional medicines that have been effective to cure a wide range of diseases.

In this chapter, authors have discussed utility aspects of the biochemistry of natural products that include bioactive constituents of multiple varieties of plants, herbs, and crude extracts recommended by traditional practitioners to cure senile neurodegenerative disorders [25].

## 7.2 METHODOLOGY

Citations were retrieved from the PUBMED database by searching with keywords like "HD," "Huntington's medicinal plants," "Huntington's herb," "3-Nitropropionic acid," "Huntingtin," "Chinese herbal," "Ayurveda," "neurological disorder, medicinal plants," "HD, models" etc. Potential papers, reviews, books, and reports were assembled by cross-referencing the retrieved literature. After initial screening, herbal medicines (extracts or isolated compounds) are depicted in Tables 7.1 and 7.2, followed by summarization of major biological effects and possible molecular mode of action of the anti-HD herbal medicines on neurotoxic HD models. Total of 17 plant species belonging to 16 genera and 16 families, 27 bioactive phytochemicals and two herbal formulations have been described in this chapter. The nomenclature of the plants and the plant families and author's citations were verified from the Missouri Botanical Garden's electronic database (www.tropicos.org). In this chapter, the characteristics of herbs and herbal constituents, their biological targets, the underlying molecular mechanisms of action, applicability, and possible clinical trials have been discussed.

## 7.3 PROPERTIES OF MEDICINAL PLANTS: ANTI-HUNTINGTON'S DISEASE (HD)

Since ancient times, traditional formulation of plant-derived products has been a rich source of various bioactive natural products with neuroprotective potentials against various neurological disorders, thus revealing the nature as the inexhaustible resource of bioeffective chemicals and pharmacophores. Natural products present in plant and animal kingdom are results of series of biosynthetic processes that have been modulated over time. Natural products, including different plant-extracts, can provide complex molecules, otherwise inaccessible and exhibits profound neuroprotective abilities by treating chronic neurodegenerative disorders like HD.

HD, along with many other neuropsychiatric diseases, is practically on a rampage throughout the society due to the stressful lifestyle and number of other unknown reasons. Nootropic herbs (exhibiting neuroprotective functions by active phytochemicals such as, alkaloids, flavonoids, terpenoids, steroids, saponins, phenolics, etc.) have are alternative treatment protocols to avoid the devastating side-effects caused by prolonged administration of synthetic drugs, that are being used to treat HD.

Recent reports suggest that herbal extracts and medicines are ahead of the conventional therapies in the prevention of some CNS diseases. For example, *Bacopa monnieri or Brahmi* extracts was reported effective against ADHD, epilepsy, dementia, Alzheimer's disease (AD) [117, 118, 134] and an array of other neurological diseases. *Centella asiatica* or *Thankuni* has proven its therapeutic ability through various *in-vivo*/*in-vitro* models against Parkinson's Disease (PD), learning, and memory deficit and migraine; and its anti-HD potential in 3-NP treated brain mitochondria has been examined in recent studies [114–116]. The *Panax ginseng* possesses anti-anxiety, anti-depressant, and pro-cognitive properties. A partially purified extract (Rb extract) containing ginsenosides, Rb1, Rb3, and Rd significantly improved 3-NP induced motor-impairment and striatal cell loss [68]. Korean Red ginseng hot-water extract (@50, 100 and 250 mg/kg/day, per os (P. O)) reduces neurological impairment and loss and down-regulates the levels of nuclear factor-κB (NF-κB), tumor necrosis factor-α (TNF-α) and interleukin (IL)-1β, IL-6 in 3-NP in treated male mice.

Therefore, rather than designing drugs by extracting only a single molecule, rationally designed polyherbal formulations should also be under consideration, as an alternative in multi-targeted therapeutics and prophylaxis. Table 7.1 summarizes some neuroprotective medicinal herbs and their mode of action.

## 7.4 BIOACTIVITIES OF PHYTOCHEMICALS: ANTI-HUNTINGTON'S DISEASE (HD)

### *7.4.1 a-MANGOSTIN*

α-Mangostin is a natural xanthonoid, which is derived from the edible *Garcinia mangostana* L. (Clusiaceae) fruits [88]. Being a potential antioxidant, this compound showed modulatory effects on the GSH system in synaptosomes of rat brain treated with ferrous sulfate ($FeSO_4$) [76].

### *7.4.2 ASTRAGALAN*

Isolated from *Astragalus membranaceus* Moench (Fabaceae), Astragalan is a natural polysaccharide [143], which acts as an anti-apoptotic agent, and it showed neuroprotective effects in rats with ischemic brain injury [141]. This

**TABLE 7.1** Anti-Huntington's Disease Activity of Crude/Semi-Purified Plant Extracts/Fractions

| Plant Source | Family | Experimental Models | Mechanism of Action | | References |
|---|---|---|---|---|---|
| | | | Up-Regulation | Down-Regulation | |
| *Bacopa monnieri* (L.): leaf powder ethanolic extract | Plantaginaceae | 3-NP, PP mice | Antioxidant | Cytotoxicity, mitochondrial dysfunction | [134] |
| *Boerhaaviadiffusa* L.: polyphenol-rich ethanolic extract | Nyctaginaceae | 3-NP, rat brain homogenates | Antioxidant | - | [9] |
| *Centellaasiatica* (L.): leaf powder &aqueous extract | Apiaceae | 3-NP, PP male mice | Antioxidant | - | [115] |
| *C. asiatica:* aqueous extract | | 3-NP, male mice | Antioxidant, increase in GSH, thiols | Mitochondrial dysfunction, | [114] |
| *C. asiatica:* aqueous extract | | 3-NP, PP mice | antioxidant, increase in GSH, thiols | - | [116] |
| *Convolvulus pluricaulis:* ethyl acetate fraction of a methanol extract of the whole plant | Convolvulaceae | 3-NP, rats | Locomotor activity, MDA, nitrite, SOD, reduced GSH | Behavioral damage, body weight deficit, | [53] |
| *C. pluricaulis:* standardized hydro-methanol extract and fraction | | 3-NP, rats | Locomotor activity, memory, oxidative defense | Body weight deficit | [73] |
| *Gastrodiaelata:* Blume | Orchidaceae | Rats | Molecular chaperons, A(2A)-R, protein kinase A (PKA) activity, proteasome activity | - | [43] |
| *Lueheadivaricata:* aqueous extract | Malvaceae | 3-NP, rats | Locomotor activity, GSH/GSSG ratio, AChE activity | ROS production, lipid peroxidation | [24] |

**TABLE 7.1** *(Continued)*

| Plant Source | Family | Experimental Models | Mechanism of Action | | References |
|---|---|---|---|---|---|
| | | | Up-Regulation | Down-Regulation | |
| *Panax ginseng* (Korean Red Ginseng): extract | Araliaceae | 3-NP, mice | - | Neurological impairment, mortality, lesion formation, neuronal loss, microglial activation, MAPKs, NF-κB signal pathway, mRNA expression of TNF-α, IL-1β, IL-6, inducible NO synthase | [48] |
| *P. quinquefolius* L. (American ginseng): leaf and stem | Araliaceae | 3-NP, rodents | Behavioral score, | Striatal lesion volume | [68] |
| *P. quinquefolius* L. (Rb extract (with ginsenosides Rb1, Rb3, and Rd)) | | 3-NP, rodents | - | Motor impairment, striatum cell loss | [68] |
| *Psoraleacorylifolia:* water and 80% ethanol seed extract | Fabaceae | 3-NP, PC12 cells | ↑Mitochondrial respiration, | Apoptosis | [44] |
| *Punica Granatum* L. | Lythraceae | 3-NP, PC12 cells | Antioxidant, | ROS production, lipid peroxidation, extracellular NO, lactate/pyruvate ratio, lactase dehydrogenase | [5] |
| *Valeriana officinalis* L.: ethanolic extract | Caprifoliaceae | 3-NP, rat brain homogenates | - | TBARS activity, antioxidation | [123] |

compound was found to modulate the transcription factors like abnormal dauer formation-16/ forkhead box O (DAF-16/FOXO) [143] that reduces polyQ-exposed proteotoxicity in *C. elegans*.

### *7.4.3 BERBERINE*

Berberine is an ammonium salt from the protoberberine group of isoquinoline alkaloids. Isolated from *Berberis* sp. (Berberidaceae) and other plants, it attenuates motor dysfunction and increasing the survival duration in the transgenic HD (N171-82Q) mice by degrading the mHtt, and promoting autophagy [50]. Moreover, berberine prevented neuronal damage by inhibiting *in vitro* and *in vivo* glial-mediated inflammation in TBI [18].

### *7.4.4 CELASTROL*

It is a triterpenoid derived from the herb *Tripterygium wilfordii*, which protects against mitochondrial injury and prevents p38 mitogen-triggered protein kinase (p38 MAPK) in neurotoxic PD models [21]. Due to the presence of anti-inflammatory and antioxidant activities, celastrol showed neuroprotective effects in Drosophila model [30].

### *7.4.5 CURCUMIN*

Curcumin is a diarylheptanoid and the major curcuminoid of dietary turmeric (*Curcuma longa* L. (Zingiberaceae)) [84]. Adult neurogenesis in an AD model was induced by Curcumin-coated poly(lactic-co-glycolic) acid (PLGA) nanoparticles (Cur-PLGA-NPs), via canonical wnt/β-catenin pathway [131].

### *7.4.6 (-)-EPIGALLOCATECHIN-GALLATE (EGCG)*

(-)-Epigallocatechin-gallate (EGCG) are polyphenols (a type of catechin) extracted from *Camellia sinensis* (L.) [138]. Application of EGCG (@10, 20, and 40 mg/kg for 2weeks) shielded 3-NP-induced-HD rat models against oxidative stress, behavioral changes, mitochondrial dysfunction, and striatal injury possibly via NO regulation [58].

### 7.4.7 FERULIC ACID

It is the hydroxycinnamic acid, a phenolic compound obtained from plants [105]. Ferulic acid (with fish oil) showed pronounced neuroprotective activity compared to its individual components, in protecting rats from 3-NP toxicity by lowering the levels of MDA, hydroperoxides, and NO, restoring AChE and dopamine levels and by inhibiting mitochondrial dysfunctions [26].

### 7.4.8 FISETIN

Fisetin, a dietary bioflavonoid that naturally occurs in vegetables and fruits, has a neuroprotective function as an anti-inflammatory agent against aluminum chloride ($AlCl_3$)-induced neurotoxicity [94]. In PC12 cells and mutant Httex1 Drosophila model and R6/2 HD mouse, fisetin prevented mHtt-mediated damage via up-regulation of extracellular signal-regulated kinases (ERK) [72].

### 7.4.9 GALANTAMINE

Galantamine is an alkaloid derived from the bulbs and flowers of *Galanthus* sp. and some other genera [11]. In a 3-NP-insulted rat model of HD, galantamine showed neuroprotective activity through the modulation of nicotinic acetylcholine receptor (nAChR) [89].

### 7.4.10 GENISTEIN

Genistein is an isoflavone type phytoestrogen extracted from *Genista* species [132]. Genistein (@5, 10 and 20 mg/kg) prevented memory loss in 3-NP (20 mg/kg)-induced ovariectomized rats with its antioxidant, anti-inflammatory, and cholinesterase inhibitory properties [79].

### 7.4.11 GINSENOSIDES

Ginsenosides are plant saponins extracted from ginseng plant (*Panax quinquefolius* L.) [45]. In a rodent model of HD, ginsenosides (Rb1, Rb3, or Rd) improved mortality, motor functionality, and reduced toxin-induced striatal lesion volume [68].

### 7.4.12 HESPERIDIN

Hesperidin, a flavanone glycoside isolated from *Citrus* sp. (Rutaceae) [2], provides neuroprotection in rats as an antioxidant and an anti-apoptotic agent along with elevating the MDA levels and improving locomotor activity in the animals [80]. Hesperidin pretreatment protected neonatal rat from neuronal hypoxia-ischemic brain injury via antioxidation and phosphorylation of protein kinase B (Akt) activation [102].

### 7.4.13 KAEMPFEROL

This flavanol is prevalent among many plant foods [4]. Brain injury and neuroinflammation in rats were improved by kaempferol glycosides isolated from *Carthamustinctorius* via inhibition of signal transduction as well as transcription 3 (STAT3) and NF-κB activation [142]. In 3-NP-treated rats, kaempferol increased animal life span and protected against motor impairment and striatal lesions [62].

### 7.4.14 L-THEANINE

L-Theanine is an analog of amino acid found in green tea (*C. sinensis* L.) [10]. L-theanine (@100 and 200 mg/kg) protected 3-NP (10 mg/kg)-administered rats through the reduction of the oxidative stress and via restoring the levels of SOD, GSH, catalase (CAT) and succinate dehydrogenase (SDH) [129].

### 7.4.15 LUTEIN

Lutein, a xanthophyll, is considered as a carotenoid-based nutritional source derived from plant-based dietary sources [12]. Lutein protected 3-NP-insulted rats via antioxidative mechanisms [13].

### 7.4.16 LYCOPENE

Lycopene, a carotene, is found in *Lycopersicum* spp. (tomato) and other red-colored fruits and vegetables [23]. Mitochondrial dysfunctions in

3-NP-intoxicated rats were diminished by lycopene (@10 mg/kg, orally for 15 days) by its antioxidative ability [106].

### 7.4.17 MELATONIN

Melatonin (*N*-acetyl-5-methoxytryptamine) is found in many eukaryotes and bacteria. 3-NP-induced rats are characterized by behavioral damage, modulation of brain-derived neurotrophic factor (BDNF), glial cell-derived neurotrophic factor (GDNF) and neuronal loss, which were effectively reduced by melatonin (@1 mg/kg/bodyweight for 21 days) [128].

### 7.4.18 NARINGIN

Naringin is a dietary flavonoid extracted from *Citrus* sp. and other plants [145]. Naringin upregulated BDNF and vascular endothelial growth factor (VEGF) and prevented neuronal apoptosis caused by spinal cord injury [101].

### 7.4.19 NICOTINE

Nicotine is a stimulant and parasympathomimetic alkaloid isolated from *Nicotiana tabacum* L. Nicotine was found to prolong the lifespan and repair olfactory and motor deficits and reduces levodopa-induced dyskinesias in Drosophila and monkey PD models respectively [17, 98].

### 7.4.20 ONJISAPONIN B

Onjisaponin-B is isolated from *Polygala tenuifolia* Willd. In PC12 cell expressing mHtt, autophagy is triggered via 5'-adenosine monophosphate-activated protein kinase-mammalian target of rapamycin (AMPK-mTOR signaling) [139].

### 7.4.21 PROTOPANAXTRIOL

Protopanaxtriol is extracted from *Panax ginseng*. Protopanaxtriol (@5, 10, 20 mg/kg) functions as an antioxidant in 3-NP-intoxicated rats and upregulates

Hsp70 expression, increasing the body weight, preventing the changes in behavior, reducing ROS levels and enhancing nuclear factor (erythroid-derived 2)-like 2 [Nrf2] entering nucleus [33].

### 7.4.22 PUERARIN

Puerarin, an isoflavonoid and isolated from the root part of *Puerarialobata* (Willd.), prevents weight loss, oxidative stress, and hypothermia and can restore locomotor activity and neurotransmitters anomalies in 3-NP (20 mg/kg)-insulted rats [77].

### 7.4.23 QUERCETIN

Quercetin is a flavanol found in many plants. Quercetin protects against hippocampal neurodegeneration in rats triggered by hypobaric hypoxia [95], prevents oxidative damage and neuronal apoptosis in rat hippocampus [52], and inhibits diabetic neuropathy [31].

### 7.4.24 RESVERATROL

A stilbenoid and a phytoalexin, Resveratrol (@5 and 10 mg/kg, orally, once daily for 8 days) was found to block 3-NP-induced motor and cognitive damage in rats [61]. In R6/2 mouse model of HD, resveratrol showed neuroprotective function via ERK activation [72].

### 7.4.25 S-ALLYLCYSTEINE

*S*-Allylcysteine is garlic (*Allium sativum* L.) -derived organic compound and is a derivative of cysteine [32]. *S*-Allylcysteine blocked oxidative damage to prevent focal cerebral ischemia [7] and mitigated QA-induced neurotoxicity in rats [86].

### 7.4.26 (-) SCHISANDRIN B

(-)Schisandrin-Bis and antioxidant from *Schisandra chinensis* (Turcz.) Baill [34]. The compound showed anti-apoptotic and anti-necrotic properties via neutralization of 3-NP toxicity in PC12 cells [63].

### 7.4.27 SESAMOL

Sesamol is a lignin derivative extracted from *Sesamum indicum* L. The effect of PD in rats was reversed [6], and neuroinflammation in a rat model of diabetic neuropathy was suppressed [22] by Sesamol.

### 7.4.28 SPERMIDINE

Spermidine (@5 and 10 mg/kg), a potent polyamine with antioxidant and anti-inflammatory properties, attenuated 3-NP-induced striatal toxicity in rats through regulation of oxidative stress, motor coordination, and striatal neurotransmitters levels [46]. Polyamines (including spermidine) are active against age-induced memory impairment (AMI) [121].

### 7.4.29 SULFORAPHANE

Sulforaphane, an isothiocyanate isolated from *Brassica oleracea* L. (broccoli) or other cruciferous vegetables, protected mice from 3-NP--induced striatal toxicity via upregulation of Kelch-like ECH-associated protein 1 (Keap1)-Nrf2-ARE pathway and downregulation of MAPKs and NF-κB pathways [47]. Sulforaphane also exhibited anticonvulsant properties and improved mitochondrial function in mice [15].

### 7.4.30 TREHALOSE

Trehalose is a natural α-linked disaccharide that protected against spinal cord ischemia in rabbits [124]. When administered orally, it prevented polyQ aggregation in cerebrum and liver and increased polyQ generated motor function in transgenic mice [125].

### 7.4.31 VANILLIN

Vanillin (and agomelatine) prevented weight loss, enhanced locomotor activity and learning-memory and shielded against toxin-induced biochemical changes in 3-NP--induced rats [37] (Table 7.3).

**TABLE 7.2** Anti-HD Activity of Phytochemicals

| Compound | Source Plant | Plant Part | Family | In-Vitro/ In-Vivo Model | Mode of action | | | References |
|---|---|---|---|---|---|---|---|---|
| | | | | | Nature | Up-Regulation | Down-Regulation | |
| α-mangostin | *Garcinia mangostana* L. | Pericarp | Clusiaceae | 3-NP, CGNs | Antioxidant | - | ROS production | [91] |
| Astragalan | *Astragalus membrana-ceus*Moench | Root | Fabaceae | Mutant polyQ, *C. elegans* | - | Adult lifespan, daf-2, age-1lifespan | polyQ aggregation, DAF-16/ FOXOmodulation | [143] |
| Cannabidiol | *Cannabissativa* L. | Plant | Cannabaceae | 3-NP, rats | Antioxidant | - | Striatal atrophy | [104] |
| Celastrol | *Tripterygium wilfordii* | Hook root bark | Celastraceae | Mutant polyQ, HeLa, and PC12 cells | HSP modulator | - | - | [144] |
| Curcumin encapsulated solid lipid nanoparticles (C-SLNs) | *Curcuma longa* L. | root | Zingiberaceae | 3-NP, rats | - | Neuromotor coordination | Mitochondrial dysfunction | [108] |
| (-)-Epigallo-catechin-gallate | *Camellia sinensis* (L.) | Green tea leaves | Theaceae | polyQ-mediated htt protein, HD yeast; HD fly overexpressing Httex1 protein | - | Motor function | Mutant Httex1 protein aggregation, cytotoxicity, photoreceptor degeneration, | [29] |
| Fisetin | Many plants | Fruits, vegetables, and other plant parts | PC12 cells expressing | Mutant Httex1; *Drosophila* expressing mutant Httex1; R6/2 HD mouse | - | ERK activation | mHtt activation | [72] |

**TABLE 7.2** *(Continued)*

| Compound | Source Plant | Plant Part | Family | In-Vitro/ In-Vivo Model | Mode of action | | | References |
|---|---|---|---|---|---|---|---|---|
| | | | | | Nature | Up-Regulation | Down-Regulation | |
| Galantamine | *Galanthus* sp. and others | Flowers and other parts | Amaryllidaceae | 3-NP, rats | - | nAChR modulation, | Striatal lesion, anti-apoptotic | [89] |
| Ginsenosides (Rb1, Rc, and Rg5) | *Panax ginseng* | Root | Araliaceae | Striatal MSNs from YAC128 HD mouse | - | - | Glutamate-induced Ca($^{2+}$) responses | [140] |
| Hesperidin | Citrus fruits | Fruits | Rutaceae | 3-NP, rats | Prevented altered locomotor activity, prepulse inhibition (PPI) response, antioxidant, anti-inflammatory | Cortical, striatal, and hippocampal MDA | - | [80] |
| Kaempferol | Many plants | Fruits and vegetables | - | 3-NP, rats | - | - | Motor deficit, mortality, striatal lesions, antioxidant | [62] |
| Lutein | Many plants | Green leafy vegetables | - | 3-NP, rats | Antioxidant | Body weight, neurobehavioral improvement, activity of mitochondrial enzymes complex | - | [13] |
| Lycopene | Tomatoes, many plants | Red fruits and vegetables | Solanaceae | 3-NP, rats | - | Behavioral and biochemical activities, NO modulation | - | [59] |

**TABLE 7.2** *(Continued)*

| Compound | Source Plant | Plant Part | Family | In-Vitro/ In-Vivo Model | Mode of action | | | References |
|---|---|---|---|---|---|---|---|---|
| | | | | | Nature | Up-Regulation | Down-Regulation | |
| Melatonin | Many plants | Fruits and Vegetables, nuts, grains | - | 3-NP, rats | Antioxidant | - | - | [133] |
| Naringin | Citrus fruits and others | Fruits | Rutaceae | 3-NP, rats | Antioxidant | NO modulation | Behavioral alterations, mitochondrial enzymes complex dysfunction, | [57] |
| Nicotine | *Nicotiana tabacum* L. | Leaves | Solanaceae | 3-NP, rats | - | - | Striatal DA. GSH depletion | [126] |
| Onjisaponin B | Radix Polygala (Yuan Zhi) | Roots and other parts | Polygalaceae | mHtt, PC12 cells | - | Autophagy through AMPK-mTOR signaling | - | [139] |
| Quercetin | Many plants | Cherries, berries, vegetables | - | 3-NP, rats | | | Mitochondrial dysfunction, oxidative stress, neurobehavioral deficits | [106] |
| Quercetin with fish-oil | - | - | - | 3-NP, rats | Antioxidant | | Elevated Ach-E activity, mitochondrial dysfunctions | [26] |
| Resveratrol | Red grapes and others | Red grape skin, berry fruits, cocoa, nuts | - | PC12 cells expressing mutant Httex1; *Drosophila* expressing mutant Httex1 | Antifungal, Powerful antioxidant | ERK activation | - | [72] |
| | | | | *Drosophila* expressing Httex1p Q93 | - | Modulation of SIRT1, | Neuronal degeneration | [87] |

**TABLE 7.2** *(Continued)*

| Compound | Source Plant | Plant Part | Family | In-Vitro/ In-Vivo Model | Mode of action | | | References |
|---|---|---|---|---|---|---|---|---|
| | | | | | Nature | Up-Regulation | Down-Regulation | |
| S-allyl cysteine | *Allium sativum* L. | Garlic extract | Amaryllidaceae | 3-NP, rat brain | Chemopreventive, Antioxidant | - | Mitochondrial dysfunction, lipid peroxidation | [92] |
| (-) Schisandrin B | *Schisandra chinensis* (Turcz.) Baill. | Fruits | Schisandraceae | 3-NP, rat PC12 cells | Anti-necrotic, anti-apoptotic | - | - | [63] |
| Sesamol | *Sesamum indicum*, L. | Seeds and Oil | Pedaliaceae | 3-NP, rats | Antioxidant | Free radical scavenging activity | - | [60] |
| Trehalose | Sunflower, Moonwrt Selaginella, and Sea algae; Mushrooms | Sunflower seeds and other plants | - | COS-7 and PC12 cells expressing mHtt (EGFP-HDQ74) | Antioxidant | Autophagy against mHtt | - | [110] |
| Zeatin riboside | Plants | - | - | PC12 cells expressing mHtt | Cytokinin | A(2A)-R signaling modulation | mHtt-induced protein aggregations | [66] |

**TABLE 7.3** Anti-HD Activity of Herbal Formulations

| Formulations | Mode of Action | | References |
|---|---|---|---|
| | Up-Regulation | Down-Regulation | |
| Chaihu-Jia-Longgu-Muli Tan (CLMT) + Yi-Gan San (YGS) | _ | HD symptoms | [111] |
| Yi-Gan San (YGS)+ Chaihu-Jia-Longgu-Muli Tan (CLMT) | _ | HD symptoms | [111] |

## 7.5 DISCUSSION

Autophagy is the cellular process that directs proteins, macromolecules, and organelles to the lysosome for their degradation. This important protein-quality-control system becomes dysfunctional in HD, leading to the production and accumulation of misfolded huntingtin protein because of the expansion of the N-terminus poly-Q tract in the mutated huntingtin protein [70].

Therefore, the autophagy pathway becomes a primary target for the treatment. Onjisaponin B modified autophagy through AMPK-mTOR signaling in PC12 cells expressing mHtt [128] while sulforaphane thwarted mHtt cytotoxicity and facilitated mHtt degradation in HEK293 cells that expressed mHtt-94Q by enhancing proteasome and autophagy activities [63]. Again, oxidative stress and the resulting downregulation and degeneration of key proteins leading to the energy deficiency, protein folding impairments and last but not the least, molecular damage inside the cell can be held accountable for the prevalence of the disease [113].

An array of natural antioxidants has been reported in this review exhibiting anti-HD properties by antioxidation via modulating the biochemical and molecular parameters associated with HD pathogenesis. In addition, impaired mitochondrial and metabolic dysfunction by mHtt that degrades cytosolic and mitochondrial calcium homeostasis play major contributing roles in HD pathogenesis that leads towards neuronal loss and death [17, 92]. A number of natural compounds have also been documented to modulate mitochondrial function in order to exert neuroprotective efficacy against HD. Upregulation and increased level of the pro-apoptotic proteins by mHtt contribute to neuronal cell death in HD pathogenesis [37] and in the striatum and cortex, neuronal cell death caused by Caspase-2 [36].

Compounds such as naringin and puerarin have anti-apoptotic and anti-HD properties [68]. Furthermore, the brain tissue of HD patients after autopsy displayed lower acetylcholine level and a number of AChE inhibitors are diagnosed against the cognitive decline and dementia in HD [125].

Most of the plant-derived products have some quality issues that include product authentication, contamination issues, and the use of fillers. These problems have greatly affected their effectiveness and thus have weakened the confidence of the users. Authentication of herbs, advanced chemical and biological standardization methods, and herbal product's quality control are of significance and can be carried out by elucidating the major bioactive subset of phytochemicals by mass spectroscopy, HPLC, etc. Recently, the evaluation of bioactivity by studying the expression using mRNA micro-array method has joined this list. To uplift the quality, the purity of herbal preparations is achieved through the DNA barcoding in combination with High Resolution Melting analysis (Bar-HRM) [79].

Another important subject of consideration, while manufacturing CNS-active drugs, is the BBB that segregates the circulating blood from intestinal fluid to brain and prevents drug permeability into the brain via modulating uptake and efflux of drugs, thus protecting the brain from toxic metabolites and xenobiotics and maintaining normal homeostasis [29, 46, 78, 124].

Herbal products are of no exception. A diverse range of natural products is reported to ameliorate many CNS diseases by regulating the signal transductions associated with the break-down of BBB. Virgin olive oil reportedly protected against hypoxia-reoxygenation induced rat brain by reducing BBB permeability [75]. Naringin too decreases BBB dysfunction in 3-NP-insulted rats [33]. Thus, natural products are becoming new and compelling options in the designing of novel therapeutics that prevent BBB breakdown, especially in the molecular pathogenesis of neurological disorders.

A new approach to overcoming poor bioavailability of phytochemicals, which is considered as one of the limitations of herbal therapy, and systemic toxicity is the application of novel delivery systems exploiting the pharmacokinetic properties of existing drugs [6]. A modified curcumin, known as Theracurmin under repetitive systemic exposures, exhibited remarkably enhanced oral bioavailability. Again, piperine reportedly nullified antidiabetic and antioxidant potential of curcumin owing to its biotransformation and also improved absorption and bioavailability of curcumin among human and animals' subjects [111].

The absolute bioavailability of α-mangostin, another reported anti-HD phytochemical, was also increased in animals when administered orally as a soft capsule [135]. Similarly, unmatched biochemical properties of celastrol were nullified by nanoencapsulation, which also contributed to its enhanced bioactive efficacy [102]. It has also been reported that the self-micro emulsifying drug delivery system (SMEDDS) dispersible tablet can be the mode of oral administration for celastrol [90].

Chitosan and aspartic acid encapsulated nanoparticles (NPs), when administered, showed a significant enhancement of the effectiveness of EGCG [39]. Again, the combination of cyclosophoroase dimer and fisetin improved bioavailability and solubility of fisetin [44]. Liposomal encapsulation and nan-emulsion increased bioavailability and bioactivity of fisetin [105, 93]. On the other hand, the availability of EGCG was reportedly increased by Piperine, while per-acetylation increased the *in vitro* bioactivity and bioavailability of the compound [58, 59].

The CNS herbal drugs, although showed no adverse side effects, suffer from inadequacy and inconclusiveness owing to many methodological limitations, such as the small size of the samples, poor experimental designing and statistical analysis and the timespan of treatment offering mostly symptomatic relief within that short period [120]. Many of these herbal constituents are disease-modulators for pre-symptomatic people, and against early signs of neurodegenerative diseases, insufficiency in biomarker selection has been a big issue for the evaluation of pre-symptomatic efficiency of the herbal preparations [38].

To elucidate the bioefficacy of herbs, herbal products, and phytochemicals against complex syndromes such as neurological disorders including HD, extensive, and thorough preclinical and clinical trials are required [25]. In order to accept these traditional and alternative preparations against neurological disorders not only to give symptomatic relief but as long-lasting disease-modifying agents, standardization of herbal preparations to work on multiple targets and their stability, formulations of doses, study of potential side effects and development of tolerance are needed to be examined via rigorous pre and post-clinical trials [61].

## 7.6 SUMMARY

Traditional herbal therapeutics stand as an alternative and complementary solution and efficient aid to reduce HD compared to conventional medicine. Despite being used for the relief of neurological symptoms since ancient times, the mechanisms of action of herbs and phytochemicals have started to unveil. In this chapter, a comprehensive account of anti-HD properties of natural products has been discussed, with their proven efficacy against the biochemical and molecular modifications associated with HD pathophysiology in HD models and patients. However, most of the preclinical studies are not supported by the clinical trials, and thus there is a need of clinical evaluation for being included in mainstream medicine. Only the traditional

Chinese formulation Yi-Gan San (YGS) and Chaihu-Jia-Longgu-Muli Tan (CLMT) have been administered in HD patients in a cross over manner.

## KEYWORDS


- **3-nitropropionic acid**
- **alternative and complementary therapy**
- **autophagy**
- **botanicals**
- **Huntington's chorea**
- **Huntington's disease**
- **neurodegenerative disorder**


## REFERENCES


1. Agostinho, L. A., Dos Santos, S. R., Alvarenga, R. M. P., & Paiva, C. L. A., (2013).A systematic review of the intergenerational aspects and the diverse genetic profiles of Huntington's disease. *Genet Mol Res*, *12*, 1974–1981.
2. Ahmadi, A., Shadboorestan, A., Nabavi, S. F., Setzer, W. N., & Nabavi, S. M., (2015). The role of hesperidin in cell signal transduction pathway for the prevention or treatment of cancer. *Current Medicinal Chemistry*, *22*, 3462–3471.
3. Ajayi, S. A., Ofusori, D. A., Ojo, G. B., Ayoka, O. A., Abayomi, T. A., & Tijani, A. A., (2011). The microstructural effects of aqueous extract of Garcinia kola (Linn) on the hippocampus and cerebellum of malnourished mice. *Asian Pacific Journal of Tropical Biomedicine*, *1*, 261–265.
4. Alkhalidy, H., Moore, W., Zhang, Y., McMillan, R., Wang, A., Ali, M., & Hulver, M., (2015). Small molecule kaempferol promotes insulin sensitivity and preserved pancreatic β-cell mass in middle-aged obese diabetic mice. *Journal of Diabetes Research*.
5. Al-Sabahi, B. N., Fatope, M. O., Essa, M. M., Subash, S., Al-Busafi, S. N., Al-Kusaibi, F. S., & Manivasagam, T., (2017). Pomegranate seed oil: Effect on 3-nitropropionic acid-induced neurotoxicity in PC12 cells and elucidation of unsaturated fatty acids composition. *Nutritional Neuroscience*, *20*, 40–48.
6. Angeline, M. S., Sarkar, A., Anand, K., Ambasta, R. K., & Kumar, P., (2013). Sesamol and naringenin reverse the effect of rotenone-induced PD rat model. *Neuroscience*, *254*, 379–394.
7. Aqil, F., Munagala, R., Jeyabalan, J., & Vadhanam, M. V., (2013). Bioavailability of phytochemicals and its enhancement by drug delivery systems. *Cancer Letters*, *334*, 133–141.

8. Ashafaq, M., Khan, M. M., Raza, S. S., Ahmad, A., Khuwaja, G., Javed, H., & Islam, F., (2012). S-allyl cysteine mitigates oxidative damage and improves neurologic deficit in a rat model of focal cerebral ischemia. *Nutrition Research, 32*, 133–143.
9. Ayyappan, P., Palayyan, S. R., & Kozhiparambil, G. R., (2016). Attenuation of oxidative damage by Boerhaaviadiffusa L. against different neurotoxic agents in rat brain homogenate. *Journal of Dietary Supplements, 13*, 300–312.
10. Ben, P., Zhang, Z., Xuan, C., Sun, S., Shen, L., Gao, Y., & Luo, L., (2015). Protective effect of L-theanine on cadmium-induced apoptosis in PC12 cells by inhibiting the mitochondria-mediated pathway. *Neurochemical Research, 40*, 1661–1670.
11. Berkov, S., Bastida, J., Sidjimova, B., Viladomat, F., & Codina, C., (2011). Alkaloid diversity in Galanthus elwesii and Galanthus nivalis. *Chemistry & Biodiversity, 8*, 115–130.
12. Bernstein, P. S., Li, B., Vachali, P. P., Gorusupudi, A., Shyam, R., Henriksen, B. S., & Nolan, J. M., (2016). Lutein, zeaxanthin, and meso-zeaxanthin: The basic and clinical science underlying carotenoid-based nutritional interventions against ocular disease. *Progress in Retinal and Eye Research, 50*, 34–66.
13. Binawade, Y., & Jagtap, A., (2013). Neuroprotective effect of lutein against 3-nitropropionic acid–induced Huntington's disease–like symptoms: Possible behavioral, biochemical, and cellular alterations. *Journal of Medicinal Food, 16*, 934–943.
14. Bruyn, G. W., Bots, G. T. A. M., & Dom, R., (1979). Huntington's chorea: Current neuropathological status. *Advances in Neurology, 23*, 83–93.
15. Carrasco-Pozo, C., Tan, K. N., & Borges, K., (2015). Sulforaphane is anticonvulsant and improves mitochondrial function. *Journal of Neurochemistry, 135*, 932–942.
16. Carroll, J. B., Bates, G. P., Steffan, J., Saft, C., & Tabrizi, S. J., (2015). Treating the whole body in Huntington's disease. *The Lancet Neurology, 14*, 1135–1142.
17. Chambers, R. P., Call, G. B., Meyer, D., Smith, J., Techau, J. A., Pearman, K., & Buhlman, L. M., (2013). Nicotine increases lifespan and rescues olfactory and motor deficits in a Drosophila model of Parkinson's disease. *Behavioral Brain Research, 253*, 95–102.
18. Chen, C. C., Hung, T. H., Lee, C. Y., Wang, L. F., Wu, C. H., Ke, C. H., & Chen, S. F., (2014). Berberine protects against neuronal damage via suppression of glia-mediated inflammation in traumatic brain injury. *PloS One, 9*, e115694.
19. Chen, C. M., (2011). Mitochondrial dysfunction, metabolic deficits, and increased oxidative stress in Huntington's disease. *Chang Gung Med. J., 34*(2), 135–152.
20. Chen, Z., Jalabi, W., Shpargel, K. B., Farabaugh, K. T., Dutta, R., Yin, X., & Trapp, B. D., (2012). Lipopolysaccharide-induced microglial activation and neuroprotection against experimental brain injury is independent of hematogenous TLR4. *Journal of Neuroscience, 32*, 11706–11715.
21. Choi, B. S., Kim, H., Lee, H. J., Sapkota, K., Park, S. E., Kim, S., & Kim, S. J., (2014). Celastrol from 'Thunder God Vine'protects SH-SY5Y cells through the preservation of mitochondrial function and inhibition of p38 MAPK in a rotenone model of Parkinson's disease. *Neurochemical Research, 39*, 84–96.
22. Chopra, K., Tiwari, V., Arora, V., & Kuhad, A., (2010). Sesamol suppresses neuro-inflammatory cascade in experimental model of diabetic neuropathy. *The Journal of Pain, 11*, 950–957.
23. Cooperstone, J. L., Ralston, R. A., Riedl, K. M., Haufe, T. C., Schweiggert, R. M., King, S. A., et al., (2015). Enhanced bioavailability of lycopene when consumed as

cis-isomers from tangerine compared to red tomato juice, a randomized, cross-over clinical trial. *Molecular Nutrition & Food Research*, *59*, 658–669.
24. Courtes, A. A., Arantes, L. P., Barcelos, R. P., Da Silva, I. K., Boligon, A. A., Athayde, M. L., Puntel, R. L., & Soares, F. A. A., (2015). Protective effects of aqueous extract of Lueheadivaricata against behavioral and oxidative changes induced by 3-nitropropionic acid in rats. *Evidence-Based Complementary and Alternative Medicine.* E-article ID: 723431. http://dx.doi.org/10.1155/2015/723431
25. Cristina, C. H., Divino, D. R. M., Pereira, D. V. F., Carolina, N. P., Calve, F. P., Alberto, M. F. C., Barreiro, E., & Viegas, C., (2011). The role of natural products in the discovery of new drug candidates for the treatment of neurodegenerative disorders I: Parkinson's disease. *CNS & Neurological Disorders-Drug Targets (Formerly Current Drug Targets-CNS & Neurological Disorders)*, *10*, 239–250.
26. Denny, J. K. M., (2014). Neuroprotective efficacy of a combination of fish oil and ferulic acid against 3-nitropropionic acid-induced oxidative stress and neurotoxicity in rats: Behavioral and biochemical evidence. *Applied Physiology, Nutrition, and Metabolism*, *39*, 487–496.
27. Dey, A., (2017). Natural products against Huntington's disease (HD): Implications of neurotoxic animal models and transgenics in preclinical studies. *Neuroprotective Natural Products: Clinical Aspects and Mode of Action*, 185–246.
28. Dey, A., & De, J. N., (2015). Neuroprotective therapeutics from botanicals and phytochemicals against Huntington's disease and related neurodegenerative disorders. *Journal of Herbal Medicine*, *5*, 1–19.
29. Ehrnhoefer, D. E., Duennwald, M., Markovic, P., Wacker, J. L., Engemann, S., Roark, M., et al., (2006). Green tea (−)-epigallocatechin-gallate modulates early events in huntingtin misfolding and reduces toxicity in Huntington's disease models. *Human Molecular Genetics*, *15*, 2743–2751.
30. Faust, K., Gehrke, S., Yang, Y., Yang, L., Beal, M. F., & Lu, B., (2009). Neuroprotective effects of compounds with antioxidant and anti-inflammatory properties in a Drosophila model of Parkinson's disease. *BMC Neuroscience*, *10*, 109.
31. Ferreira, P. E. B., Lopes, C. R. P., Alves, A. M. P., Alves, É. P. B., Linden, D. R., Zanoni, J. N., & Buttow, N. C., (2013). Diabetic neuropathy: An evaluation of the use of quercetin in the cecum of rats. *World Journal of Gastroenterology: WJG*, *19*, 6416–6426.
32. Fricker, G., (2008). Drug interactions with natural products at the blood brain barrier. *Current Drug Metabolism*, *9*, 1019–1026.
33. Gao, Y., Chu, S. F., Li, J. P., Zhang, Z., Yan, J. Q., Wen, Z. L., Xia, C. Y., Mou, Z., Wang, Z. Z., He, W. B., & Guo, X. F., (2015). Protopanaxtriol protects against 3-nitropropionic acid-induced oxidative stress in a rat model of Huntington's disease. *Acta Pharmacologica Sinica*, *36*, 311.
34. Giridharan, V. V., Thandavarayan, R. A., Arumugam, S., Mizuno, M., Nawa, H., Suzuki, K., & Konishi, T., (2015). Schisandrin B ameliorates ICV-infused amyloid β induced oxidative stress and neuronal dysfunction through inhibiting RAGE/NF-κB/MAPK and up-regulating HSP/Beclin expression. *PLoS One*, *10*, e0142483.
35. Gómez-Sierra, T., Molina-Jijón, E., Tapia, E., Hernández-Pando, R., García-Niño, W. R., Maldonado, P. D., & Pedraza-Chaverri, J., (2014). S-allylcysteine prevents cisplatin-induced nephrotoxicity and oxidative stress. *Journal of Pharmacy and Pharmacology*, *66*, 1271–1281.

36. Gopinath, K., & Sudhandiran, G., (2015). Protective effect of naringin on 3-nitropropionic acid-induced neurodegeneration through the modulation of matrix metalloproteinases and glial fibrillary acidic protein. *Canadian Journal of Physiology and Pharmacology, 94*, 65–71.
37. Gupta, S., & Sharma, B., (2014). Pharmacological benefits of agomelatine and vanillin in experimental model of Huntington's disease. *Pharmacology Biochemistry and Behavior*, *122*, 122–135.
38. Harper, P. S., (1992). The epidemiology of Huntington's disease. *Human Genetics, 89*, 365–376.
39. Hermel, E., Gafni, J., Propp, S. S., Leavitt, B. R., Wellington, C. L., Young, J. E., et al., (2004). Specific caspase interactions and amplification are involved in selective neuronal vulnerability in Huntington's disease. *Cell Death and Differentiation, 11*, 424.
40. Hickey, M. A., & Chesselet, M. F., (2003). Apoptosis in Huntington's disease. *Progress in Neuro-Psychopharmacology and Biological Psychiatry, 27*, 255–265.
41. Ho, Y. S., So, K. F., & Chang, R. C. C., (2010). Anti-aging herbal medicine—how and why can they be used in aging-associated neurodegenerative diseases? *Ageing Research Reviews*, *9*, 354–362.
42. Hong, Z., Xu, Y., Yin, J. F., Jin, J., Jiang, Y., & Du, Q., (2014). Improving the effectiveness of (−)-epigallocatechin gallate (EGCG) against rabbit atherosclerosis by EGCG-loaded nanoparticles prepared from chitosan and polyaspartic acid. *Journal of Agricultural and Food Chemistry*, *62*, 12603–12609.
43. Huang, C. L., Yang, J. M., Wang, K. C., Lee, Y. C., Lin, Y. L., Yang, Y. C., & Huang, N. K., (2011). Gastrodiaelata prevents huntingtin aggregations through activation of the adenosine A2A receptor and ubiquitin proteasome system. *Journal of Ethnopharmacology*, *138*, 162–168.
44. Im, A. R., Chae, S. W., Jun, Z. G., & Lee, M. Y., (2014). Neuroprotective effects of psoraleacorylifolia Linn seed extracts on mitochondrial dysfunction induced by 3-nitropropionic acid. *BMC Complementary and Alternative Medicine*, *14*, 370.
45. Ivanov, D. A., Georgakopoulos, J. R., & Bernards, M. A., (2016). The chemoattractant potential of ginsenosides in the ginseng–Pythium irregular pathosystem. *Phytochemistry*, *122*, 56–64.
46. Jamwal, S., & Kumar, P., (2016). Spermidine ameliorates 3-nitropropionic acid (3-NP)-induced striatal toxicity: Possible role of oxidative stress, neuroinflammation, and neurotransmitters. *Physiology & Behavior*, *155*, 180–187.
47. Jang, M., & Cho, I. H., (2016). Sulforaphane ameliorates 3-nitropropionic acid-induced striatal toxicity by activating the Keap1-Nrf2-ARE pathway and inhibiting the MAPKs and NF-κB pathways. *Molecular Neurobiology*, *53*, 2619–2635.
48. Jang, M., Lee, M. J., Kim, C. S., & Cho, I. H., (2013). Korean red ginseng extract attenuates 3-Nitropropionic acid-induced huntington's-like symptoms. *Evidence-Based Complementary and Alternative Medicine,* E-article ID: 237207. doi: 10.1155/2013/237207.
49. Jeong, D., Choi, J. M., Choi, Y., Jeong, K., Cho, E., & Jung, S., (2013). Complexation of fisetin with novel cyclosophoroase dimer to improve solubility and bioavailability. *Carbohydrate Polymers*, *97*, 196–202.
50. Jiang, W., Wei, W., Gaertig, M. A., Li, S., & Li, X. J., (2015). Therapeutic effect of berberine on Huntington's disease transgenic mouse model. *PLoS One*, *10*, e0134142.

51. Kam, A., M Li, K., Razmovski-Naumovski, V., Nammi, S., Chan, K., Li, Y. Q., & Li, G., (2012). The protective effects of natural products on blood-brain barrier breakdown. *Current Medicinal Chemistry, 19*, 1830–1845.
52. Kanter, M., Unsal, C., Aktas, C., & Erboga, M., (2016). Neuroprotective effect of quercetin against oxidative damage and neuronal apoptosis caused by cadmium in hippocampus. *Toxicology and Industrial Health, 32*, 541–550.
53. Kaur, M., Prakash, A., & Kalia, A. N., (2016). Neuroprotective potential of antioxidant potent fractions from convolvulus *pluricaulis Chois* in 3-nitropropionic acid challenged rats. *Nutritional Neuroscience, 19*, 70–78.
54. Killoran, A., & Biglan, K. M., (2014). Current therapeutic options for Huntington's disease: Good clinical practice versus evidence-based approaches? *Movement Disorders, 29*, 1404–1413.
55. Kim, J. H., Kim, S., Yoon, I. S., Lee, J. H., Jang, B. J., Jeong, S. M., et al., (2005). Protective effects of ginseng saponins on 3-nitropropionic acid-induced striatal degeneration in rats. *Neuropharmacology, 48*, 743–756.
56. Kumar, P., & Kumar, A., (2009). Possible neuroprotective effect of withaniasomnifera root extract against 3-nitropropionic acid-induced behavioral, biochemical, and mitochondrial dysfunction in an animal model of Huntington's disease. *Journal of Medicinal Food, 12*, 591–600.
57. Kumar, P., & Kumar, A., (2010). Protective effect of hesperidin and naringin against 3-nitropropionic acid induced Huntington's like symptoms in rats: possible role of nitric oxide. *Behavioural Brain Research, 206*, 38–46.
58. Kumar, P., & Kumar, A., (2009). Protective effects of epigallocatechin gallate following 3-nitropropionic acid-induced brain damage: Possible nitric oxide mechanisms. *Psychopharmacology, 207*, 257–270.
59. Kumar, P., Kalonia, H., & Kumar, A., (2009). Lycopene modulates nitric oxide pathways against 3-nitropropionic acid-induced neurotoxicity. *Life Sciences, 85*, 711–718.
60. Kumar, P., Kalonia, H., & Kumar, A., (2010). Protective effect of sesamol against 3-nitropropionic acid-induced cognitive dysfunction and altered glutathione redox balance in rats. *Basic & Clinical Pharmacology & Toxicology, 107*, 577–582.
61. Kumar, P., Padi, S. S. V., Naidu, P. S., & Kumar, A., (2006). Effect of resveratrol on 3-nitropropionic acid-induced biochemical and behavioral changes: possible neuroprotective mechanisms. *Behavioral Pharmacology, 17*, 485–492.
62. Lagoa, R., Lopez-Sanchez, C., Samhan-Arias, A. K., Gañan, C. M., Garcia-Martinez, V., & Gutierrez-Merino, C., (2009). Kaempferol protects against rat striatal degeneration induced by 3-nitropropionic acid. *Journal of Neurochemistry, 111*, 473–487.
63. Lam, P. Y., & Ko, K. M., (2012). Beneficial effect of (–) Schisandrin B against 3-nitropropionic acid-induced cell death in PC12 cells. *Biofactors, 38*, 219–225.
64. Lambert, J. D., Hong, J., Kim, D. H., Mishin, V. M., & Yang, C. S., (2004). Piperine enhances the bioavailability of the tea polyphenol (–)-epigallocatechin-3-gallate in mice. *The Journal of Nutrition, 134*, 1948–1952.
65. Lambert, J. D., Sang, S., Hong, J., Kwon, S. J., Lee, M. J., Ho, C. T., & Yang, C. S., (2006). Peracetylation as a means of enhancing in vitro bioactivity and bioavailability of epigallocatechin-3-gallate. *Drug Metabolism and Disposition, 34*, 2111–2116.
66. Lee, Y. C., Yang, Y. C., Huang, C. L., Kuo, T. Y., Lin, J. H., Yang, D. M., & Huang, N. K., (2012). When cytokinin, a plant hormone, meets the adenosine A2A receptor: A

novel neuroprotectant and lead for treating neurodegenerative disorders? *PloS One*, *7*, e38865.

67. Li, X. Z., Zhang, S. N., Liu, S. M., & Lu, F., (2013). Recent advances in herbal medicines treating Parkinson's disease. *Fitoterapia*, *84*, 273–285.
68. Lian, X. Y., Zhang, Z., & Stringer, J. L., (2005). Protective effects of ginseng components in a rodent model of neurodegeneration. *Annals of Neurology*, *57*, 642–648.
69. Liu, Y., Hettinger, C. L., Zhang, D., Rezvani, K., Wang, X., & Wang, H., (2014). Sulforaphane enhances proteasomal and autophagic activities in mice and is a potential therapeutic reagent for Huntington's disease. *Journal of Neurochemistry*, *129*, 539–547.
70. Mahdy, H. M., Mohamed, M. R., Emam, M. A., Karim, A. M., Abdel-Naim, A. B., & Khalifa, A. E., (2014). Puerarin ameliorates 3-nitropropionic acid-induced neurotoxicity in rats: Possible neuromodulation and antioxidant mechanisms. *Neurochemical Research*, *39*, 321–332.
71. Mahdy, H. M., Tadros, M. G., Mohamed, M. R., Karim, A. M., & Khalifa, A. E., (2011). The effect of Ginkgo biloba extract on 3-nitropropionic acid-induced neurotoxicity in rats. *Neurochemistry International*, *59*, 770–778.
72. Maher, P., Dargusch, R., Bodai, L., Gerard, P. E., Purcell, J. M., & Marsh, J. L., (2010). ERK activation by the polyphenols fisetin and resveratrol provides neuroprotection in multiple models of Huntington's disease. *Human Molecular Genetics*, *20*, 261–270.
73. Malik, J., Choudhary, S., & Kumar, P., (2015). Protective effect of Convolvulus pluricaulis standardized extract and its fractions against 3-nitropropionic acid-induced neurotoxicity in rats. *Pharmaceutical Biology*, *53*, 1448–1457.
74. Mani, R. B., (2004). The evaluation of disease modifying therapies in Alzheimer's disease: A regulatory viewpoint. *Statistics in Medicine*, *23*, 305–314.
75. Manyam, B. V., Giacobini, E., & Colliver, J. A., (1990). Cerebrospinal fluid acetylcholinesterase and choline measurements in Huntington's disease. *Journal of Neurology*, *237*, 281–284.
76. Márquez-Valadez, B., Maldonado, P. D., Galván-Arzate, S., Méndez-Cuesta, L. A., Pérez-De La Cruz, V., Pedraza-Chaverrí, J., & Santamaría, A., (2012). Alpha-mangostin induces changes in glutathione levels associated with glutathione peroxidase activity in rat brain synaptosomes. *Nutritional Neuroscience*, *15*, 13–19.
77. Martin, D. D., Ladha, S., Ehrnhoefer, D. E., & Hayden, M. R., (2015). Autophagy in Huntington disease and huntingtin in autophagy. *Trends in Neurosciences*, *38*, 26–35.
78. McBride, J. L., & Clark, R. L., (2016). Stereotaxic surgical targeting of the nonhuman primate caudate and putamen: Gene therapy for Huntington's disease. In: *Gene Therapy for Neurological Disorders* (pp. 409–428). Humana Press, New York, NY.
79. Menze, E. T., Esmat, A., Tadros, M. G., Abdel-Naim, A. B., & Khalifa, A. E., (2015). Genistein improves 3-NPA-induced memory impairment in ovariectomized rats: impact of its antioxidant, anti-inflammatory and acetylcholinesterase modulatory properties. *PLoS One*, *10*, e0117223.
80. Menze, E. T., Tadros, M. G., Abdel-Tawab, A. M., & Khalifa, A. E., (2012). Potential neuroprotective effects of hesperidin on 3-nitropropionic acid-induced neurotoxicity in rats. *Neurotoxicology*, *33*, 1265–1275.
81. Merienne, N., & Déglon, N., (2015). Approches de gene silencing pour le traitement de la maladie de Huntington. *Médecine /Sciences*, *31*, 159–167.
82. Mohagheghi, F., Bigdeli, M. R., Rasoulian, B., Zeinanloo, A. A., & Khoshbaten, A., (2010). Dietary virgin olive oil reduces blood brain barrier permeability, brain edema, and brain injury in rats subjected to ischemia-reperfusion. *The Scientific World Journal*, *10*, 1180–1191.

83. Myers, R. H., MacDonald, M. E., Koroshetz, W. J., Duyao, M. P., Ambrose, C. M., Taylor, S. A. M., et al., (1993). De novo expansion of a (CAG) n repeat in sporadic Huntington's disease. *Nature Genetics*, *5*, 168.
84. Naz, R. K., Lough, M. L., & Barthelmess, E. K., (2016). Curcumin: A novel non-steroidal contraceptive with antimicrobial properties. *Frontiers in Bioscience (Elite Edition)*, *8*, 113–128.
85. Ohtsuki, S., (2004). Physiological function of blood-brain barrier transporters as the CNS supporting and protecting system. *Yakugakuzasshi: Journal of the Pharmaceutical Society of Japan*, *124*, 791–802.
86. Osathanunkul, M., Madesis, P., & De Boer, H., (2015). Bar-HRM for authentication of plant-based medicines: Evaluation of three medicinal products derived from Acanthaceae species. *Plos One*, *10*, e0128476.
87. Pallos, J., Bodai, L., Lukacsovich, T., Purcell, J. M., Steffan, J. S., Thompson, L. M., & Marsh, J. L., (2008). Inhibition of specific HDACs and sirtuins suppresses pathogenesis in a Drosophila model of Huntington's disease. *Human Molecular Genetics*, *17*, 3767–3775.
88. Pan-In, P., Wongsomboon, A., Kokpol, C., Chaichanawongsaroj, N., & Wanichwecharungruang, S., (2015). Depositing α-mangostin nanoparticles to sebaceous gland area for acne treatment. *Journal of Pharmacological Sciences*, *129*, 226–232.
89. Park, J. E., Lee, S. T., Im, W. S., Chu, K., & Kim, M., (2008). Galantamine reduces striatal degeneration in 3-nitropropionic acid model of Huntington's disease. *Neuroscience Letters*, *448*, 143–147.
90. Paulsen, J. S., Zhao, H., Stout, J. C., Brinkman, R. R., Guttman, M., & Ross, C. A., (2001). Huntington study group, clinical markers of early disease in persons near onset of Huntington's disease. *Neurology*, *57*, 658–662.
91. Pedraza-Chaverrí, J., Reyes-Fermín, L. M., Nolasco-Amaya, E. G., Orozco-Ibarra, M., Medina-Campos, O. N., González-Cuahutencos, O., & Mata, R., (2009). ROS scavenging capacity and neuroprotective effect of α-mangostin against 3-nitropropionic acid in cerebellar granule neurons. *Experimental and Toxicological Pathology*, *61*, 491–501.
92. Pérez-De La Cruz, V., González-Cortés, C., Pedraza-Chaverrí, J., Maldonado, P. D., Andrés-Martínez, L., & Santamaría, A., (2006). Protective effect of S-allylcysteine on 3-nitropropionic acid-induced lipid peroxidation and mitochondrial dysfunction in rat brain synaptosomes. *Brain Research Bulletin*, *68*, 379–383.
93. Pérez-Severiano, F., Rodríguez-Pérez, M., Pedraza-Chaverrí, J., Maldonado, P. D., Medina-Campos, O. N., Ortíz-Plata, A., & Santamaría, A., (2004).S-allylcysteine, a garlic-derived antioxidant, ameliorates quinolinic acid-induced neurotoxicity and oxidative damage in rats. *Neurochemistry International*, *45*, 1175–1183.
94. Prakash, D., Gopinath, K., & Sudhandiran, G., (2013). Fisetin enhances behavioral performances and attenuates reactive gliosis and inflammation during aluminum chloride-induced neurotoxicity. *Neuromolecular Medicine*, *15*, 192–208.
95. Prasad, J., Baitharu, I., Sharma, A. K., Dutta, R., Prasad, D., & Singh, S. B., (2013). Quercetin reverses hypobaric hypoxia-induced hippocampal neurodegeneration and improves memory function in the rat. *High Altitude Medicine & Biology*, *14*, 383–394.
96. Pringsheim, T., Wiltshire, K., Day, L., Dykeman, J., Steeves, T., & Jette, N., (2012). The incidence and prevalence of Huntington's disease: A systematic review and meta-analysis. *Movement Disorders*, *27*, 1083–1091.

97. Qi, X., Qin, J., Ma, N., Chou, X., & Wu, Z., (2014). Solid self-microemulsifying dispersible tablets of celastrol: Formulation development, characterization and bioavailability evaluation. *International Journal of Pharmaceutics, 472*, 40–47.
98. Quik, M., Mallela, A., Ly, J., & Zhang, D., (2013). Nicotine reduces established levodopa-induced dyskinesias in a monkey model of Parkinson's disease. *Movement Disorders, 28*, 1398–1406.
99. Quintanilla, R. A., & Johnson, G. V., (2009). Role of mitochondrial dysfunction in the pathogenesis of Huntington's disease. *Brain Research Bulletin, 80*, 242–247.
100. Ragelle, H., Crauste-Manciet, S., Seguin, J., Brossard, D., Scherman, D., Arnaud, P., & Chabot, G. G., (2012). Nanoemulsion formulation of fisetin improves bioavailability and antitumor activity in mice. *International Journal of Pharmaceutics, 427*, 452–459.
101. Rong, W., Wang, J., Liu, X., Jiang, L., Wei, F., Hu, X., Han, X., & Liu, Z., (2012). Naringin treatment improves functional recovery by increasing BDNF and VEGF expression, inhibiting neuronal apoptosis after spinal cord injury. *Neurochemical Research, 37*, 1615–1623.
102. Rong, Z., Pan, R., Xu, Y., Zhang, C., Cao, Y., & Liu, D., (2013). Hesperidin pretreatment protects hypoxia–ischemic brain injury in neonatal rat. *Neuroscience, 255*, 292–299.
103. Sagredo, O., Pazos, M. R., Satta, V., Ramos, J. A., Pertwee, R. G., & Fernández-Ruiz, J., (2011). Neuroprotective effects of phytocannabinoid-based medicines in experimental models of Huntington's disease. *Journal of Neuroscience Research, 89*, 1509–1518.
104. Sagredo, O., Ramos, J. A., Decio, A., Mechoulam, R., & Fernández-Ruiz, J., (2007). Cannabidiol reduced the striatal atrophy caused 3-nitropropionic acid in vivo by mechanisms independent of the activation of cannabinoid, vanilloid TRPV1 and adenosine A2A receptors. *European Journal of Neuroscience, 26*, 843–851.
105. Sajjadi, S. E., Shokoohinia, Y., & Moayedi, N. S., (2012). Isolation and identification of ferulic acid from aerial parts of Kelussiaodoratissima Mozaff. *Jundishapur Journal of Natural Pharmaceutical Products, 7*, 159.
106. Sandhir, R., & Mehrotra, A., (2013). Quercetin supplementation is effective in improving mitochondrial dysfunctions induced by 3-nitropropionic acid: Implications in Huntington's disease. *Biochimica et Biophysica Acta (BBA)-Molecular Basis of Disease, 1832*, 421–430.
107. Sandhir, R., Mehrotra, A., & Kamboj, S. S., (2010). Lycopene prevents 3-nitropropionic acid-induced mitochondrial oxidative stress and dysfunctions in nervous system. *Neurochemistry International, 57*, 579–587.
108. Sandhir, R., Yadav, A., Mehrotra, A., Sunkaria, A., Singh, A., & Sharma, S., (2014). Curcumin nanoparticles attenuate neurochemical and neurobehavioral deficits in experimental model of Huntington's disease. *Neuromolecular Medicine, 16*, 106–118.
109. Sanna, V., Chamcheu, J. C., Pala, N., Mukhtar, H., Sechi, M., & Siddiqui, I. A., (2015). Nanoencapsulation of natural triterpenoid celastrol for prostate cancer treatment. *International Journal of Nanomedicine, 10*, 6835.
110. Sarkar, S., Davies, J. E., Huang, Z., Tunnacliffe, A., & Rubinsztein, D. C., (2007). Trehalose, a novel mTOR-independent autophagy enhancer, accelerates the clearance of mutant huntingtin and α-synuclein. *Journal of Biological Chemistry, 282*, 5641–5652.
111. Satoh, T., Takahashi, T., Iwasaki, K., Tago, H., Seki, T., Yaegashi, N., & Arai, H., (2009). Traditional Chinese medicine on four patients with Huntington's disease. *Movement Disorders, 24*, 453–455.

112. Seguin, J., Brullé, L., Boyer, R., Lu, Y. M., Romano, M. R., Touil, Y. S., & Chabot, G. G., (2013). Liposomal encapsulation of the natural flavonoid fisetin improves bioavailability and antitumor efficacy. *International Journal of Pharmaceutics, 444*, 146–154.
113. Setter, S., Neumiller, J., Dobbins, E., Wood, L., Clark, J., DuVall, C., & Santiago, A., (2009). Treatment of chorea associated with Huntington's disease: Focus on tetrabenazine. *The Consultant Pharmacist®, 24*, 524–537.
114. Shinomol, G. K., (2008). Effect of *Centellaasiatica* leaf powder on oxidative markers in brain regions of prepubertal mice in vivo and it's in vitro efficacy to ameliorate 3-NPA-induced oxidative stress in mitochondria. *Phytomedicine, 15*, 971–984.
115. Shinomol, G. K., (2008). Prophylactic neuroprotective property of Centellaasiatica against 3-nitropropionic acid induced oxidative stress and mitochondrial dysfunctions in brain regions of prepubertal mice. *Neurotoxicology, 29*, 948–957.
116. Shinomol, G. K., & Ravikumar, H., (2010). Prophylaxis with *Centellaasiatica* confers protection to prepubertal mice against 3-nitropropionic-acid-induced oxidative stress in brain. *Phytotherapy Research, 24*, 885–892.
117. Shinomol, G. K., & Bharath, M. S., (2012). Pretreatment with *Bacopa monnieri* extract offsets 3-nitropropionic acid induced mitochondrial oxidative stress and dysfunctions in the striatum of prepubertal mouse brain. *Canadian Journal of Physiology and Pharmacology, 90*, 595–606.
118. Shinomol, G. K., & Bharath, M. S., (2012). Neuromodulatory propensity of Bacopa monnieri leaf extract against 3-nitropropionic acid-induced oxidative stress: *In vitro* and *in vivo* evidences. *Neurotoxicity Research, 22*, 102–114.
119. Shivasharan, B. D., Nagakannan, P., Thippeswamy, B. S., Veerapur, V. P., Bansal, P., & Unnikrishnan, M. K., (2013). Protective effect of calendula officinalis Linn. Flowers against 3-nitropropionic acid induced experimental Huntington's disease in rats. *Drug and Chemical Toxicology, 36*, 466–473.
120. Shoba, G., Joy, D., Joseph, T., Majeed, M., Rajendran, R., & Srinivas, P. S. S. R., (1998). Influence of piperine on the pharmacokinetics of curcumin in animals and human volunteers. *Planta Medica, 64*, 353–356.
121. Sigrist, S. J., Carmona-Gutierrez, D., Gupta, V. K., Bhukel, A., Mertel, S., Eisenberg, T., & Madeo, F., (2014). Spermidine-triggered autophagy ameliorates memory during aging. *Autophagy, 10*, 178–179.
122. Sorolla, M. A., Rodríguez-Colman, M. J., Vall-llaura, N., Tamarit, J., Ros, J., & Cabiscol, E., (2012). Protein oxidation in Huntington disease. *Biofactors, 38*, 173–185.
123. Sudati, J. H., Fachinetto, R., Pereira, R. P., Boligon, A. A., Athayde, M. L., Soares, F. A., De Vargas Barbosa, N. B., & Rocha, J. B. T., (2009). *In vitro* antioxidant activity of *Valeriana officinalis* against different neurotoxic agents. *Neurochemical Research, 34*, 1372.
124. Takahashi, S., Isaka, M., Hamaishi, M., Imai, K., Orihashi, K., & Sueda, T., (2014). Trehalose protects against spinal cord ischemia in rabbits. *Journal of Vascular Surgery, 60*, 490–496.
125. Tanaka, M., Machida, Y., Niu, S., Ikeda, T., Jana, N. R., Doi, H., & Nukina, N., (2004). Trehalose alleviates polyglutamine-mediated pathology in a mouse model of Huntington disease. *Nature Medicine, 10*, 148.

126. Tariq, M., Khan, H. A., Elfaki, I., Al Deeb, S., & Al Moutaery, K., (2005). Neuroprotective effect of nicotine against 3-nitropropionic acid (3-NP)-induced experimental Huntington's disease in rats. *Brain Research Bulletin, 67*, 161–168.
127. Tasset, I., Agüera, E., Olmo-Camacho, R., Escribano, B., Sánchez-López, F., Delgado, M. J., Cruz, A. H., Gascón, F., Luque, E., Peña, J., Jimena, I. M., & Túnez, I., (2011). Melatonin improves 3-nitropropionic acid induced behavioral alterations and neurotrophic factors levels. *Prog. Neuropsychopharmacol Biol. Psychiatry, 35*, 1944–1949.
128. Tasset, I., Pontes, A. J., Hinojosa, A. J., De la Torre, R., & Túnez, I., (2011). Olive oil reduces oxidative damage in a 3-nitropropionic acid-induced Huntington's disease-like rat model. *Nutritional Neuroscience, 14*, 106–111.
129. Thangarajan, S., Deivasigamani, A., Natarajan, S. S., Krishnan, P., & Mohanan, S. K., (2014). Neuroprotective activity of L-theanine on 3-nitropropionic acid-induced neurotoxicity in rat striatum. *International Journal of Neuroscience, 124*, 673–684.
130. Tian, J., Shi, J., Zhang, X., & Wang, Y., (2010). Herbal therapy: a new pathway for the treatment of Alzheimer's disease. *Alzheimer's Research & Therapy, 2*, 30.
131. Tiwari, S. K., Agarwal, S., Seth, B., Yadav, A., Nair, S., Bhatnagar, P., & Srivastava, V., (2013). Curcumin-loaded nanoparticles potently induce adult neurogenesis and reverse cognitive deficits in Alzheimer's disease model via canonical Wnt/β-catenin pathway. *ACS Nano, 8*, 76–103.
132. Tosun, F., Erdem, C. K., & Eroğlu, Y., (2003). Determination of genistein in the Turkish Genista L. species by LC-MS. *Die Pharmazie-An International Journal of Pharmaceutical Sciences, 58*, 549–550.
133. Túnez, I., Montilla, P., Del Carmen, M. M., Feijóo, M., & Salcedo, M., (2004). Protective effect of melatonin on 3-nitropropionic acid-induced oxidative stress in synaptosomes in an animal model of Huntington's disease. *Journal of Pineal Research, 37*, 252–256.
134. Uabundit, N., Wattanathorn, J., Mucimapura, S., & Ingkaninan, K., (2010). Cognitive enhancement and neuroprotective effects of *Bacopa monnieri* in Alzheimer's disease model. *Journal of Ethnopharmacology, 127*, 26–31.
135. Urquhart, B. L., & Kim, R. B., (2009). Blood– brain barrier transporters and response to CNS-active drugs. *European Journal of Clinical Pharmacology, 65*, 1063.
136. Vattakatuchery, J. J., & Kurian, R., (2013). Acetylcholinesterase inhibitors in cognitive impairment in Huntington's disease: A brief review. *World Journal of Psychiatry, 3*, 62.
137. Warby, S. C., Visscher, H., Collins, J. A., Doty, C. N., Carter, C., Butland, S. L., Kanazawa, I., Ross, C. J., & Hayden, M. R., (2011). HTT haplotypes contribute to differences in Huntington disease prevalence between Europe and East Asia. *European Journal of Human Genetics, 19*, 561.
138. Wei, J. C. C., Huang, H. C., Chen, W. J., Huang, C. N., Peng, C. H., & Lin, C. L., (2016). Epigallocatechin gallate attenuates amyloid β-induced inflammation and neurotoxicity in EOC 13.31 microglia. *European Journal of Pharmacology, 770*, 16–24.
139. Wu, A. G., Wong, V. K. W., Xu, S. W., Chan, W. K., Ng, C. I., Liu, L., & Law, B. Y. K., (2013). Onjisaponin B derived from radix polygalae enhances autophagy and accelerates the degradation of mutant α-synuclein and huntingtin in PC-12 cells. *International Journal of Molecular Sciences, 14*, 22618–22641.
140. Wu, J., Jeong, H. K., Bulin, S. E., Kwon, S. W., Park, J. H., & Bezprozvanny, I., (2009). Ginsenosides protect striatal neurons in a cellular model of Huntington's disease. *Journal of Neuroscience Research, 87*, 1904–1912.

141. Yan, L., & Zhou, Q. H., (2012). Study on neuroprotective effects of astragalan in rats with ischemic brain injury and its mechanisms. *Zhongguoyingyong sheng li xue za zhi=Zhongguoyingyongshenglixuezazhi (Chinese Journal of Applied Physiology)*, *28*, 373–377.
142. Yu, L., Chen, C., Wang, L. F., Kuang, X., Liu, K., Zhang, H., & Du, J. R., (2013). Neuroprotective effect of kaempferol glycosides against brain injury and neuroinflammation by inhibiting the activation of NF-κB and STAT3 in transient focal stroke. *PloS One*, *8*, e55839.
143. Zhang, H., Pan, N., Xiong, S., Zou, S., Li, H., Xiao, L., & Huang, Z., (2012). Inhibition of polyglutamine-mediated proteotoxicity by Astragalus membranaceus polysaccharide through the DAF-16/FOXO transcription factor in Caenorhabditis elegans. *Biochemical Journal*, *441*, 417–424.
144. Zhang, Y. Q., & Sarge, K. D., (2007). Celastrol inhibits polyglutamine aggregation and toxicity though induction of the heat shock response. *Journal of Molecular Medicine*, *85*, 1421–1428.
145. Zhang, Y. S., Li, Y., Wang, Y., Sun, S. Y., Jiang, T., Li, C., Cui, S. X., & Qu, X. J., (2016). Naringin, a natural dietary compound, prevents intestinal tumorigenesis in Apc Min/+ mouse model. *Journal of Cancer Research and Clinical Oncology*, *142*, 913–925.
146. Zhao, Y., Tang, G., Tang, Q., Zhang, J., Hou, Y., Cai, E., & Wang, S., (2016). A method of effectively improved α-mangostin bioavailability. *European Journal of Drug Metabolism and Pharmacokinetics*, *41*, 605–613.

**FIGURE 2.1** *C. edulis* (sour fig).

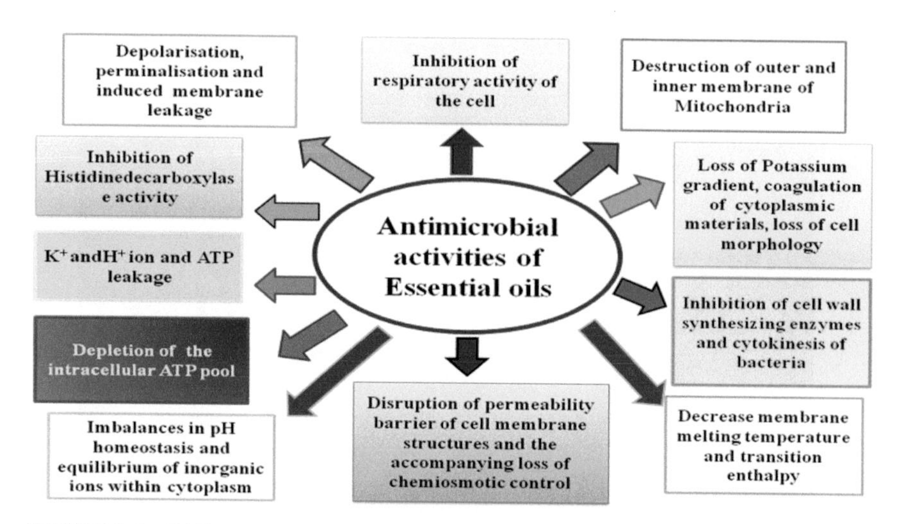


**FIGURE 3.2** Different modes of action of essential oils as antimicrobial agent.

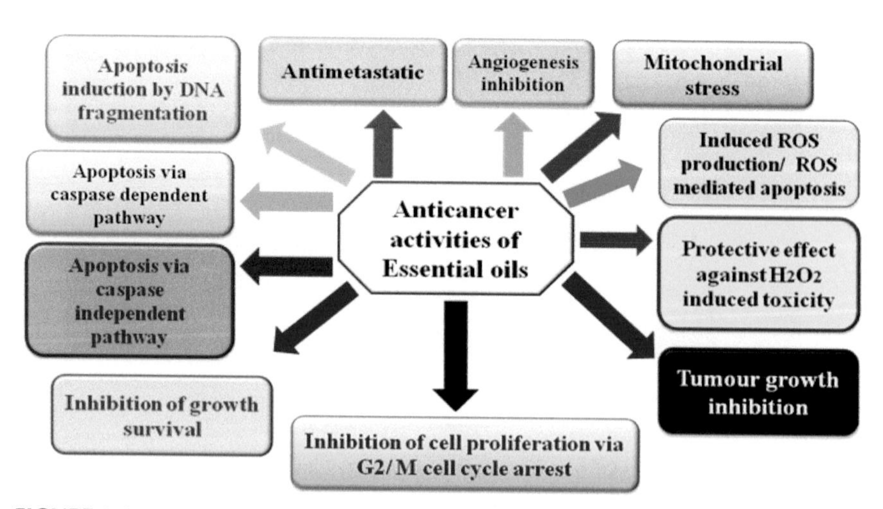


**FIGURE 3.3** Essential oils exhibiting different modes of anticancer action.

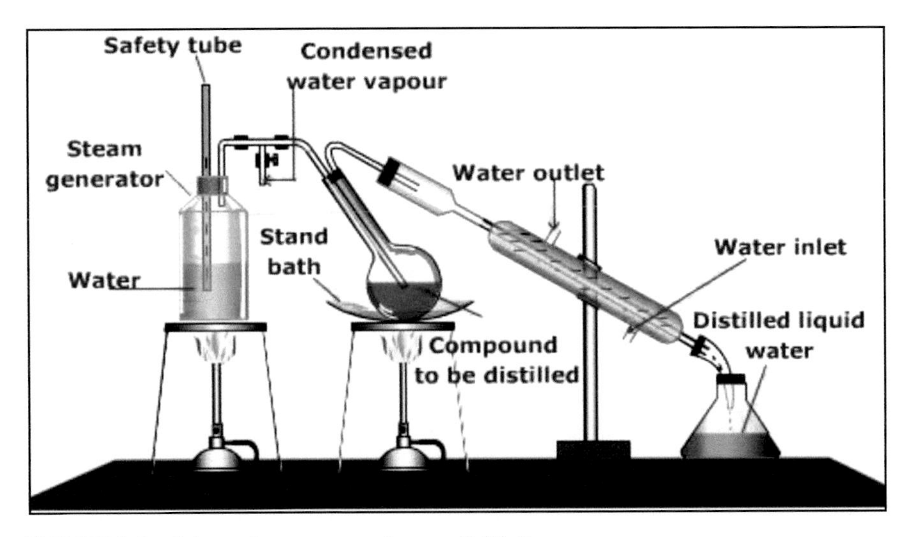


**FIGURE 4.3** Schematic apparatus of steam distillation.

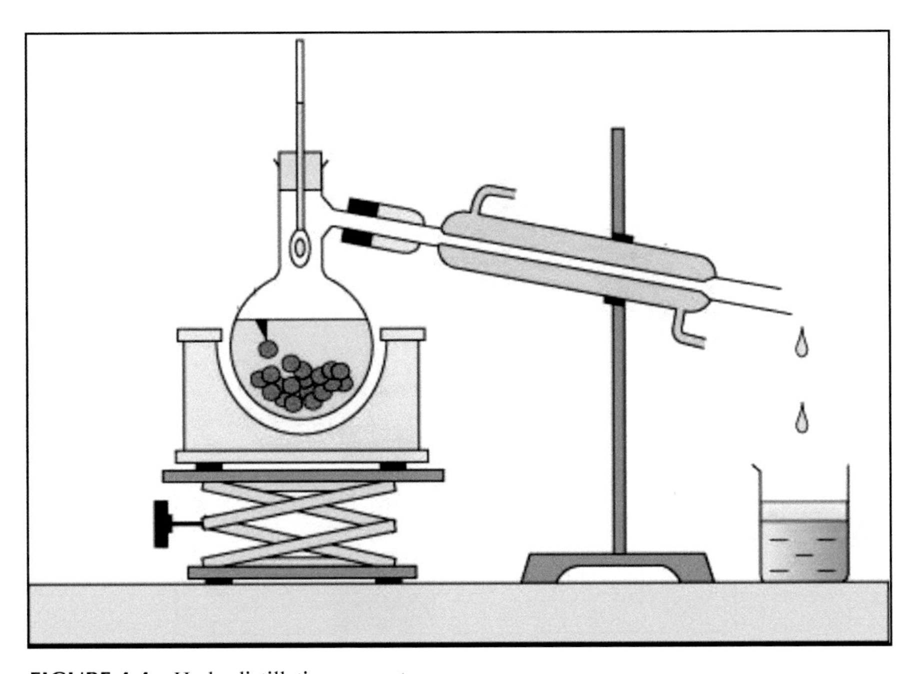

**FIGURE 4.4** Hydrodistillation apparatus.

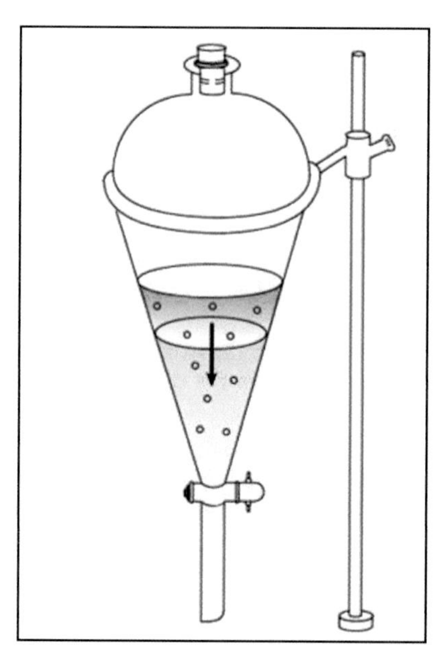

**FIGURE 4.5** Organic solvent extraction.

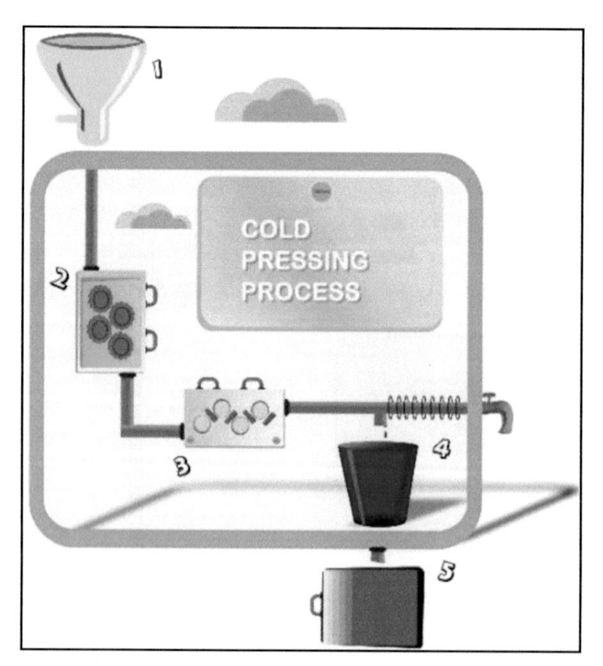


**FIGURE 4.6** Expression methods (*Legends:* 1–Filtering; 2–Milling; 3–Pressing; 4–filtering; 5–decantation).

**FIGURE 4.7** Enfleurage method.

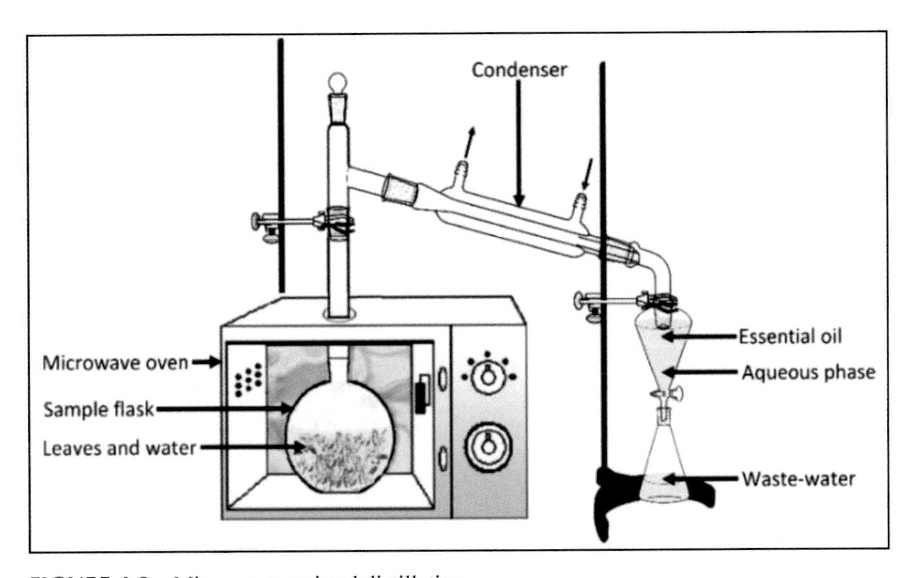


**FIGURE 4.8** Microwave-assisted distillation.

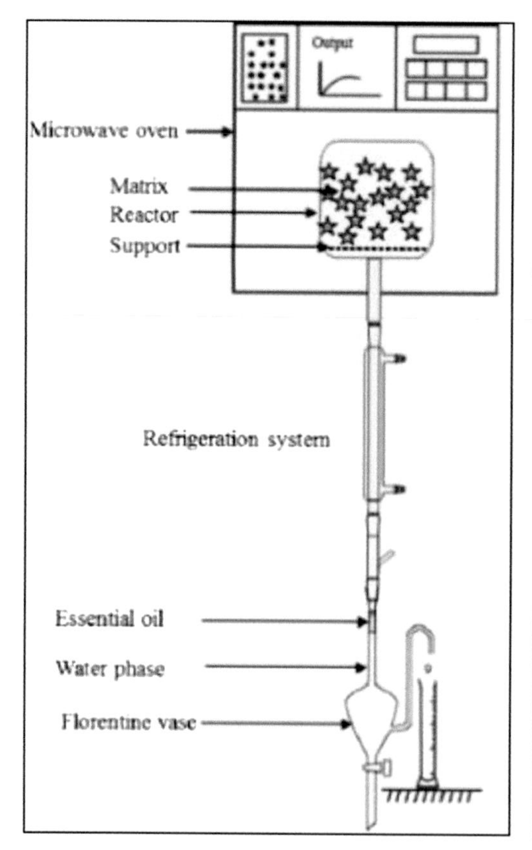

**FIGURE 4.9** Microwave hydrodiffusion and gravity (MHG).

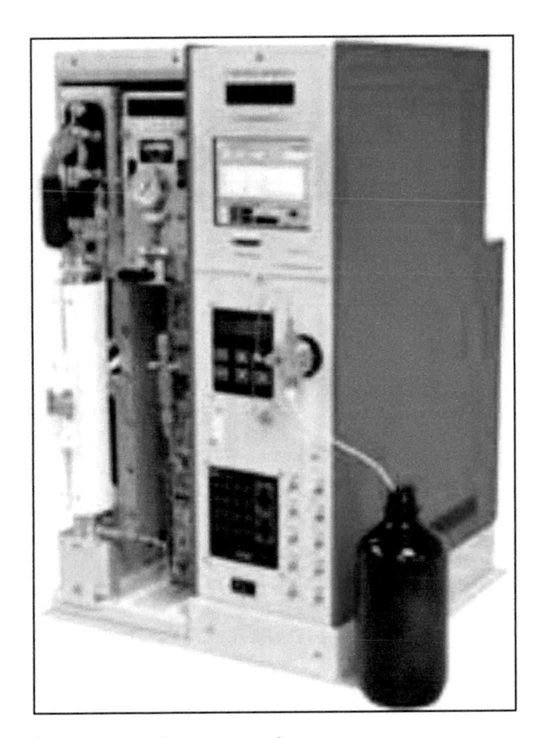

**FIGURE 4.10** High-pressure solvent extraction.

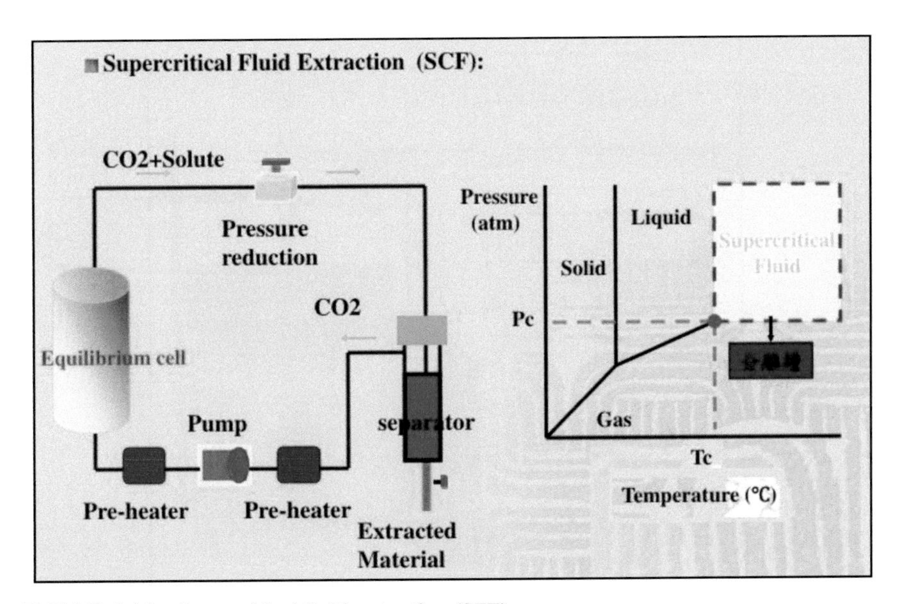


**FIGURE 4.11** Supercritical fluid extraction (SFE).

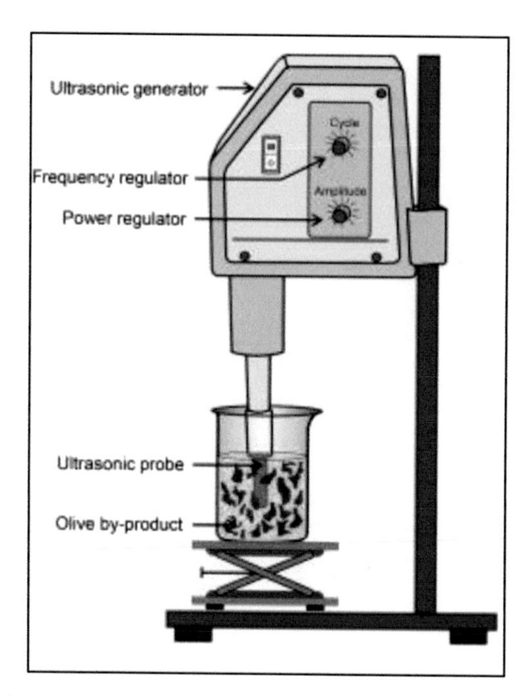


FIGURE 4.12 Ultrasonic extraction.

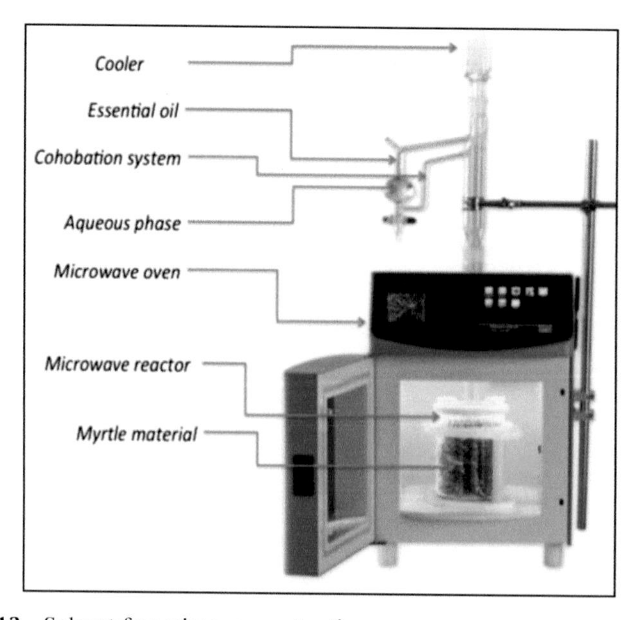


FIGURE 4.13 Solvent-free microwave extraction.

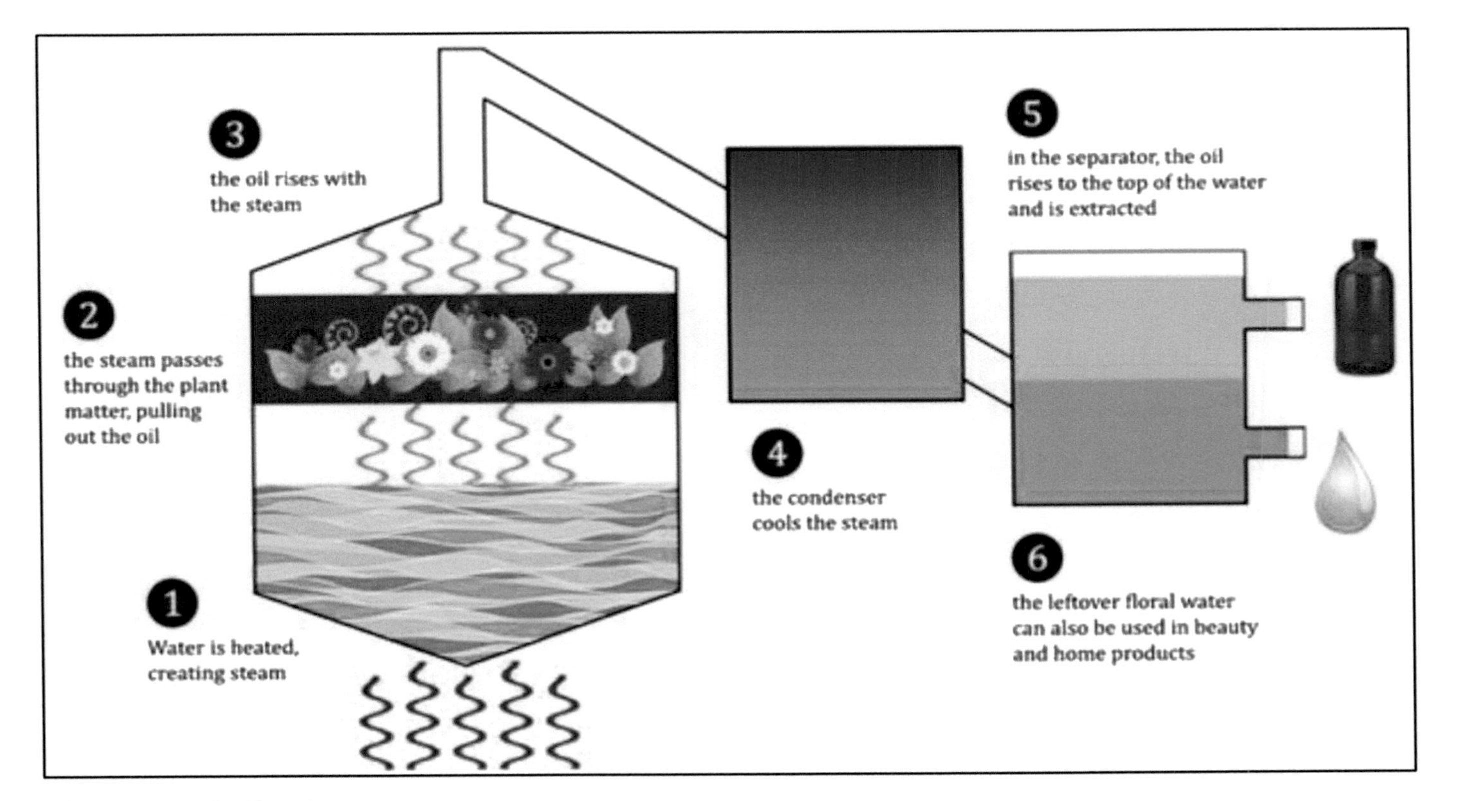


**FIGURE 4.14** The Phytonic process.

**FIGURE 5.1** A typical *tulsi* plant with flowers and seeds.

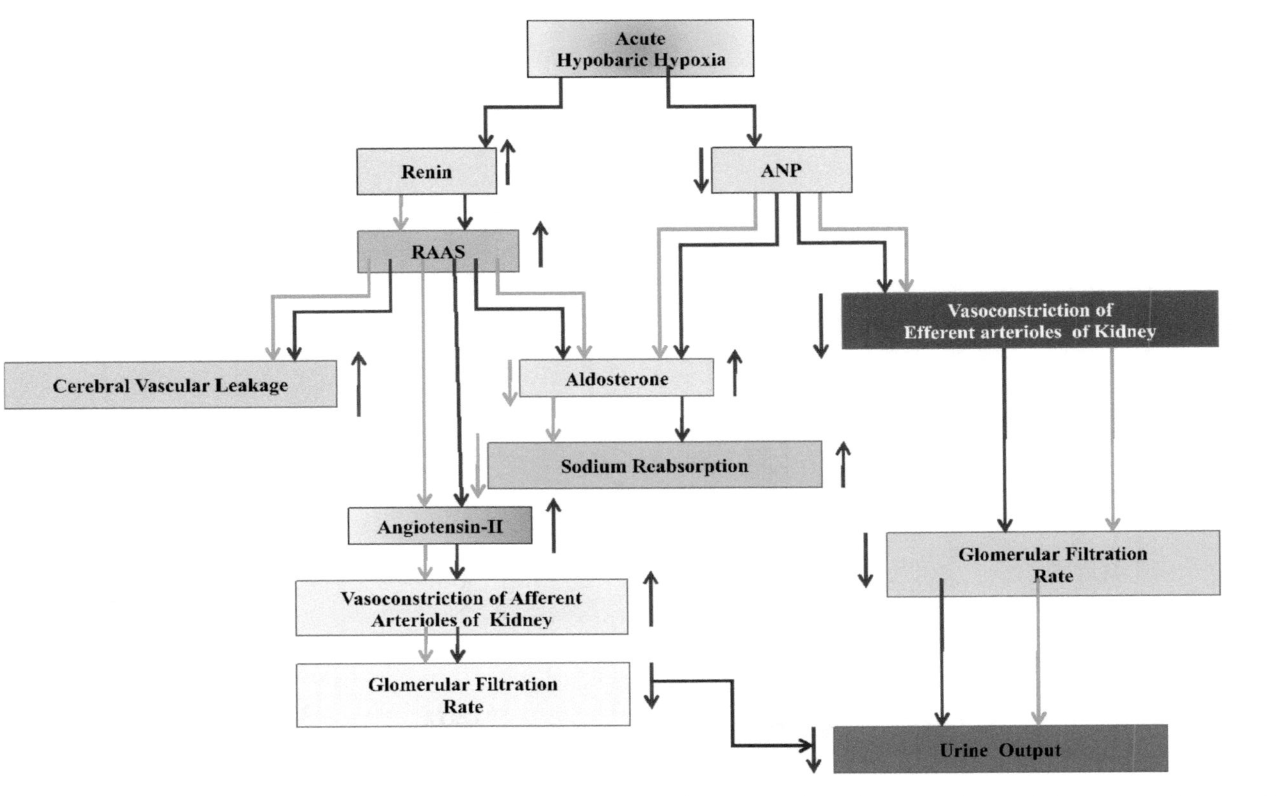


**FIGURE 8.1** Possible molecular mechanism of hypobaric hypoxia-induced altered dieresis, sodium retention, and fluid accumulation in AMS positive patients, where ANP: Arterial natriuretic peptide; RAAS: Renin-angiotensin-aldosterone system.

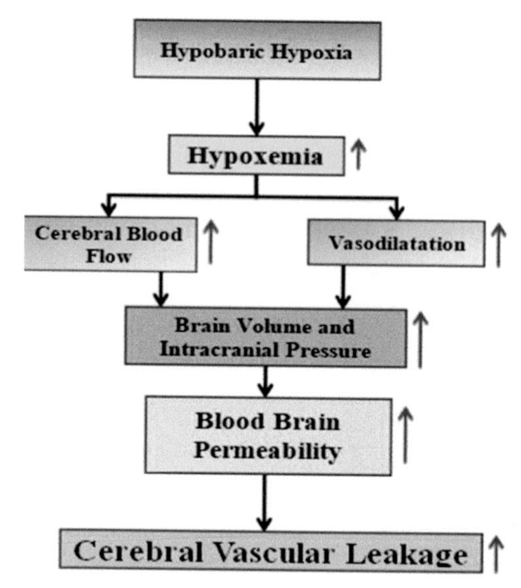


**FIGURE 8.2** Acute hypobaric hypoxia-induced altered blood-brain permeability and CVL.

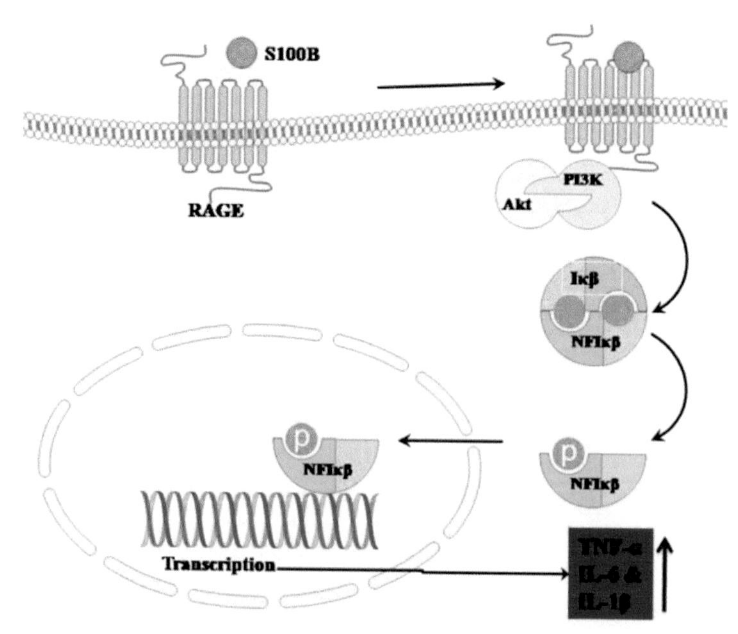


**FIGURE 8.3** Role of S100B protein in neuro-inflammation (Legends: S100B: S100 calcium-binding protein B; RAGE: Receptor for the advanced glycation end products; PKB: Protein kinase B; PI3K: Phosphatidylinositol-4,5-bisphosphate 3-kinase; Iκβ: Inhibitor of kappa B; Nuclear factor kappa-light-chain-enhancer of activated B cells; TNF-α: Tumor necrosis factor α; IL-1: Interleukin 6; IL-1β: Interleukin 1β).

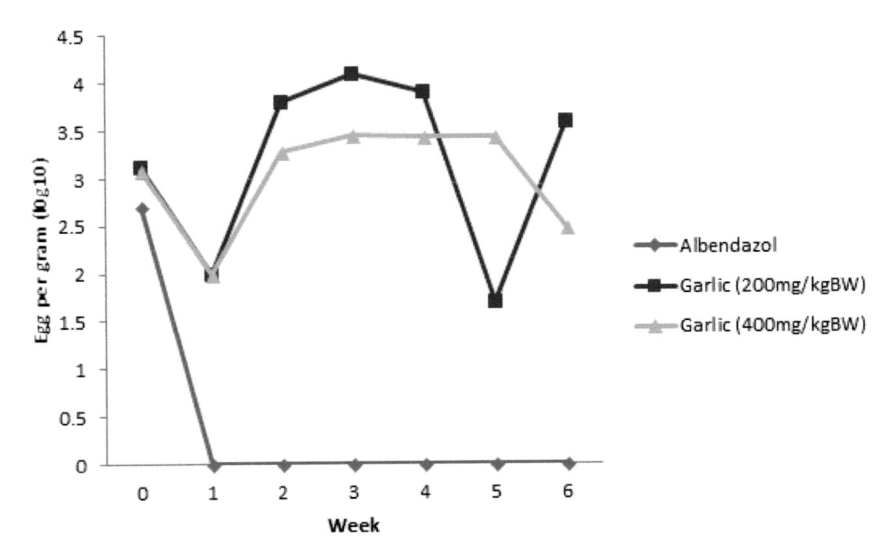


**FIGURE 9.1** Line-graph showing egg count per gram of fecal.

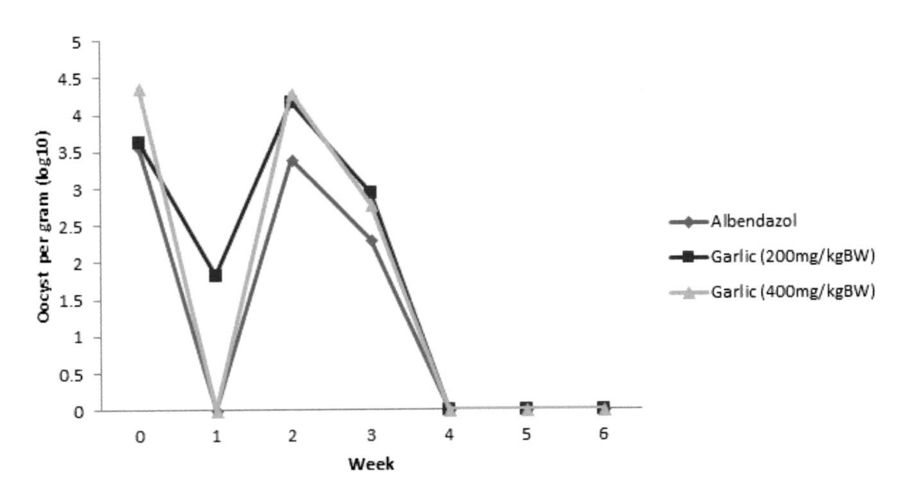


**FIGURE 9.2** Line-graph showing oocyst count per gram of fecal.

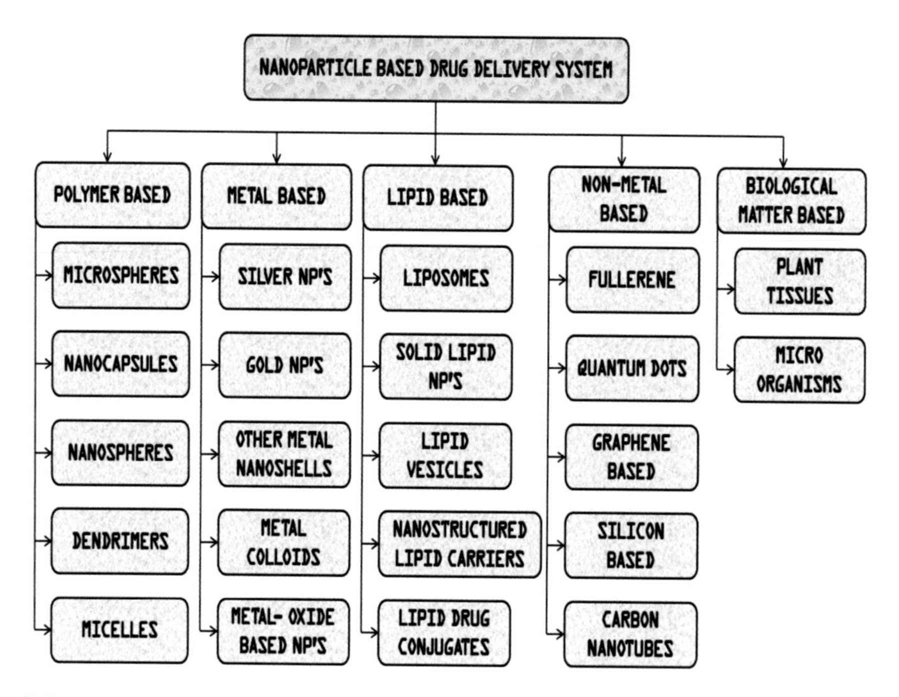


**FIGURE 15.3** Summary of nanoparticle-based drug delivery systems.

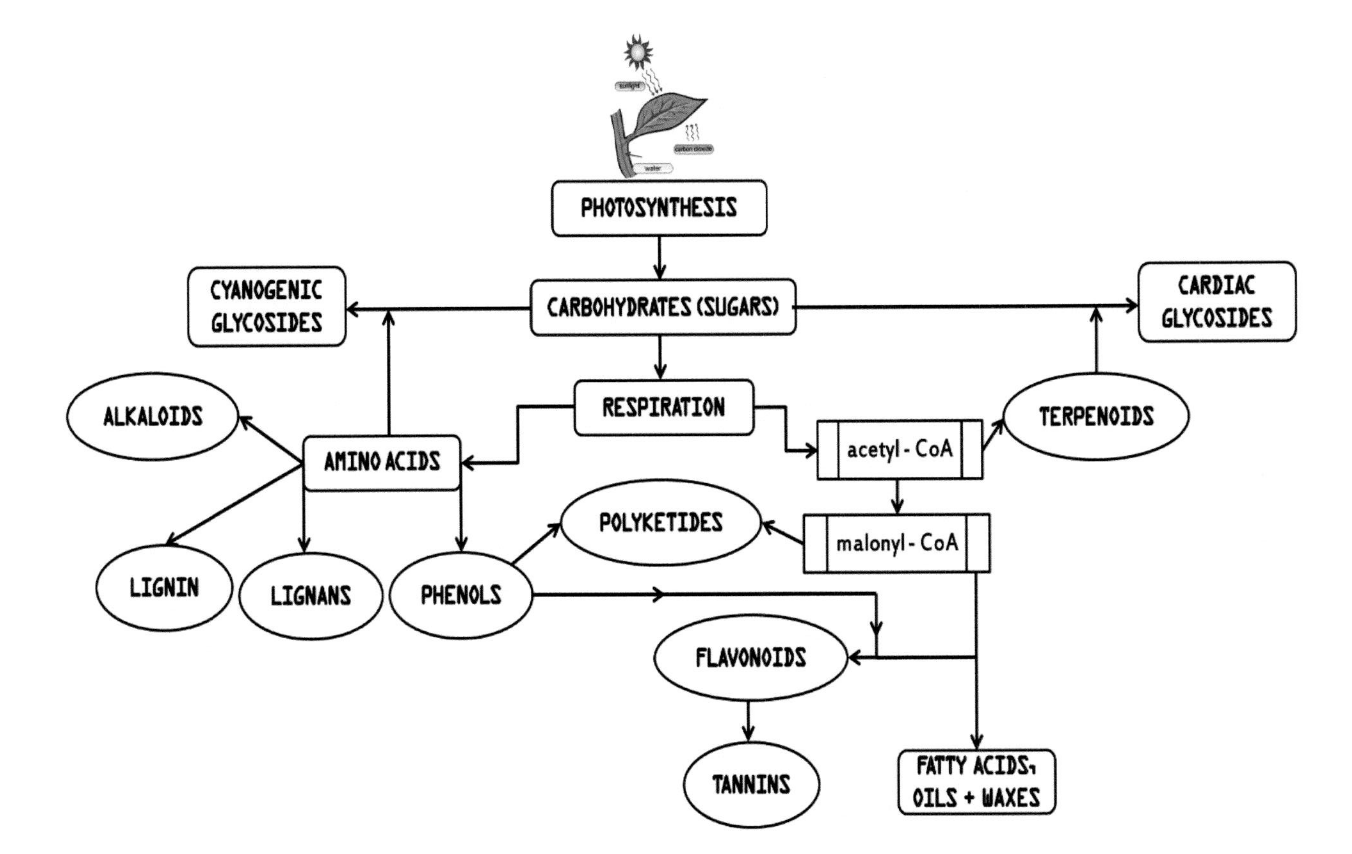


**FIGURE 15.4** Pathway to produce some secondary plant metabolites.

# CHAPTER 8

# PATHOPHYSIOLOGY AND MANAGEMENT OF ACUTE MOUNTAIN SICKNESS (AMS)

KUSHAL KUMAR, KALPANA KUMARI BARWAL, and
SUNIL KUMAR HOTA

## ABSTRACT

Rapid ascend to high altitude in non-acclimatized person may lead to acute mountain sickness (AMS). The symptoms associated with AMS are headache, nausea, vomiting, insomnia, fatigue, and dizziness. The occurrence and the severity depend on the rate of ascend, the altitude attained, altitude at which dwellers sleep and individual vulnerability to the cardinal manifestations associated with AMS. The only questionnaire for accessing the severity of symptoms associated with AMS is *Lake Louise Score (LLS)*. The exact pathophysiology of AMS is still not clear. The excessive accumulation of fluid, neuro-inflammation, and hypovolemia at high altitude could be the factors behind the pathophysiology of AMS. The symptoms associated with AMS could be managed by taking pharmacological and non-pharmacological measures. This chapter discusses the history, symptoms, diagnosis, pathophysiology, and management of symptoms associated with AMS.

## 8.1 INTRODUCTION

Rapid ascend to high altitude in non-acclimatized individuals may lead to acute mountain sickness (AMS), usually begins within the few hours of ascend and cardinal manifestations consist of headache, nausea, vomiting, insomnia, fatigue, and dizziness [40]. The occurrence and severity depend on the rate of ascend, the altitude attained, altitude at which dwellers sleep and individual vulnerability to the cardinal manifestations associated with AMS [4]. Depending on the severity and treatment, the symptoms coupled with AMS may last from a few days to a week [18]. AMS may be graded from grade 1 to grade 4 depending upon the severity of symptoms. However, patient have the minimal symptoms of AMS at grade 1 and grade 2 and can continue to trek with some precautionary measures; patient has the severe manifestations at grade 3 and cannot continue to trek and requires rest; and the patient cannot ambulate at grade 4, symptoms are progressive, and disturbance of consciousness or gait may be present and required prompt medical treatment with oxygen and descent. At Grade 3 and 4, patients with severity of symptoms may be incapacitating and could be associated with HAPE [32].

This chapter discusses history, pathophysiology, and non-pharmacological, and pharmacological management of AMS.

## 8.2 HISTORY OF ACUTE MOUNTAIN SICKNESS (AMS)

- **37–32 BC:** The symptoms of high altitude ailments were reported by a Chinese official of the Western Han Dynasty, Too Kin during the reign of Chung Li [24].
- **1590 A.D:** Priest Jose de Acosta also had given portrayal about the symptoms allied with an ascent to the Andean range in Peru (4800 m) [54, 84].
- **1772 A.D:** Interestingly with the discovery of the existence of oxygen gas by a Swedish chemist, first scientific expedition to Mont Blanc was accomplished by Horace-Benedict de Saussure, and documented pulse and respiration at various altitudes made the clear possible role of oxygen underlying the symptoms of AMS [9, 25].
- **1862 A.D:** Two balloonists died, while ascend to 29, 000 ft. and death was due to extreme altitude sickness [19].
- **1891 A.D:** H. Kronecker, had started working on a decompression chamber taking two people at a time to a pressure equivalent to 13,000 feet [31].

  With the end of the nineteenth century, Paul Bert made a remarkable logical achievement by finding the mechanism of the oxygen-carrying capacity of hemoglobin and its relationship with the partial pressure of oxygen by using decompression chamber as a model of high altitude and AMS symptoms, and he was the father of high altitude physiology [87]. At the end of the nineteenth century, researchers now had started to understand the AMS as well as the association of its symptoms with fall in the partial pressure of oxygen. Joseph Barcroft, a British physiologist, one another pioneer in the field of high altitude physiology, has started doing experiment with him own as a subject in a sealed room for 6 days simulating the altitude of 18,000 ft.

- **1922 A.D:** a first systematic study on AMS was conducted by the International High Altitude Expedition and expanded the knowledge about the high altitude physiology and concluded impairment of both physical and mental abilities during high altitude ascend [87].

  Alexander M. Kellas, a British physiologist, attracted attention of researchers towards the Himalayas, by his conclusive review of Mt. Everest, "*Mount Everest could be ascended by a man of excellent physical and mental constitution in first-rate training, without adventitious aids if the physical difficulties of the mountain are not too great*" [80].
- **1966 A.D:** A clinical study in the New England Journal of Medicine gained popularity for providing the direct correlation between hypoxia and the symptoms of AMS [70].

HAPE and HACE, both pathologically and clinically, are considered being an extension of AMS [28]. Mounting evidences advocating the progression of AMS to HAPE and HAPE with 36 hours and symptoms HAPE and HACE occur concomitantly or may progress individually [32]. Historically, the theory of alteration in neurological functions during rapid ascend to high altitude was first proposed in 1898, by, Mosso, and further confirmed in 1965, by Singh et al., in 1969 [27, 72]. While, 1911, post-mortem studies of a doctor died during an expedition to Mont Blanc and subsequently in 1913, T.H. Ravenhill, high altitude researcher, provided with a diagnostic framework of HAPE [64].

## 8.3 SYMPTOMS ASSOCIATED WITH ACUTE MOUNTAIN SICKNESS (AMS)

The most leading complication of AMS is a headache, and it is estimated that 25, 80 and 100% non-acclimatized dwellers experience a headache at an altitude of 2750, 3000 and 4500 m, respectively [8]. Additional non-specific cardinal manifestations include lassitude, fatigue, malaise, dizziness, and nausea [45]. In addition to this, insomnia has been found to be prevalent in individuals at high altitude and has been considered as major symptoms of AMS [66].

Rapid ascend to high altitude has been considered as the main risk factor of headache associated with high altitude. Headache has been found to appear as isolated symptoms within a few hours of high altitude exposure in non-acclimatized individuals; and can be constitutively present with other non-specific symptoms of AMS [33, 47]. Headache associated with AMS has commonly found to be diffusive and steady, and non-acclimatized sojourns may experience pain in the frontal, frontoparietal, or holo-cranial regions [69, 71]. Headache at high altitude may be due to the activation of pain receptors of large blood vessels of trigeminal ganglia that projects to the cortex and innervates meninges [67]. The activation of pain receptors may be due to the increased intracranial pressure, brain swelling, or due to the release of nociceptive substances [45].

Loss of appetite has been observed in non-acclimatized dwellers due to hypobaric hypoxia [83]. Reductions in appetite result in the reduction of caloric and protein intake to 30 and 40%, respectively [82]. The possible molecular mechanism is the increased activity of hypoxia-inducible factor 1 in the hypothalamus resulting in increased expression of leptin, a responsible protein for control of appetite [58]. Nausea and vomiting have been reported to cause by rhythmic labyrinthine stimulation of afferent neurons during motion sickness [32].

Data emerges from studies provide an evidence of altered sleep cycle and deep sleep that can cause insomnia [56]. In addition to this, insomnia at high altitude has been positively correlated with symptoms of anxiety [20]. Hypobaric hypoxia-induced can decrease plasma volume, and excessive ventilation-induced can decrease cerebral blood flow that could be the reason for central cardinal manifestations of AMS (e.g., dizziness, faintness, mental confusion, and ataxia) [32]. Thus, AMS can be owed to more of cerebral and above-mentioned neurological symptoms along with a deficit in learning and memory, focusing, and impaired finger tapping speed [18, 30]. Magnetic resonance imaging reveals edema in *globus pallidus* and cortical dysfunctions [36].

## 8.4 LAKE LOUISE QUESTIONNAIRE FOR ACUTE MOUNTAIN SICKNESS (AMS)

In 1991, a simplified AMS questionnaire, Lake Louise Score (LLS), was proposed and was named after the venue of International Hypoxia Conference at Lake Lousie, Canada [17]. The LLS has been validated clinically, and nowadays, this questionnaire has been widely used for assessment of incidence and symptoms associated with AMS [38, 48]. LLS has five questions with 4-point scale; and a total score of a patient with 3–5 has mild symptoms while score more than 6 has severe symptoms of AMS [17] (Table 8.1).

**TABLE 8.1** Lake Louise Score (LLS) for AMS

| Symptoms | Score | Interpretation |
|---|---|---|
| **Headache** | 0 | No |
| | 1 | Mild |
| | 2 | Severe |
| | 3 | Incapacitating |
| **Gastrointestinal** | 0 | No |
| | 1 | Poor appetite or nausea |
| | 2 | Moderate appetite or nausea |
| | 3 | Severe nausea or vomiting |
| **Fatigue and/or weakness** | 0 | No tired or weak |
| | 1 | Mild fatigue or weakness |
| | 2 | Moderate fatigue or weakness |
| | 3 | Severe fatigue or weakness |
| **Dizziness or light-headedness** | 0 | No dizziness |
| | 1 | Mild dizziness |
| | 2 | Dizziness or light-headedness |
| | 3 | Severe dizziness |
| **Difficulty in sleeping** | 0 | Sleep as well as usual |
| | 1 | Difficulty in sleeping |
| | 2 | Woke up many times, poor sleep |
| | 3 | Could not sleep et al. |

## 8.5 PATHOPHYSIOLOGY OF ACUTE MOUNTAIN SICKNESS (AMS)

Indeed, the exact pathophysiological processes that cause symptoms of AMS in dwellers are still not clear [4]. There are many proposed hypothesis describing multiple factors correlated with the cardinal manifestations of AMS [45]. More specifically, symptoms associated with AMS have been found to be cumulative of various hypothesis viz., fluid, and electrolyte imbalance, cerebral swelling induced by vasodilatation and cellular edema, the release of local mediators and increased blood-brain-barrier (BBB) permeability and hypovolemia hypothesis of AMS [11].

### 8.5.1 FLUID AND ELECTROLYTE IMBALANCE HYPOTHESIS

Fluid and electrolyte balance plays a dominant role in optimal health maintenance [40, 65]. Approximately, 70% of the bodyweight of an individual is the total body water present in the human body and is distributed between intra and extracellular compartments. Intracellular fluid is $2/3^{rd}$ of total body water consisting of potassium, magnesium, and phosphates (ATP, ADP, and AMP) as major ions while extracellular fluid constitutes $1/3^{rd}$ of total body water consisting of sodium, bicarbonate, and chloride as major ions in the system [26].

Though the exact mechanism has not been put forth, numerous hypotheses have been proposed regarding the pathophysiological mechanisms leading to AMS symptoms. One such hypothesis, which indicates the role of hyper-activated adrenergic nervous system medicated imbalance of circulatory renin-angiotensin-aldosterone system (RAAS) in the pathophysiological mechanisms associated with AMS, is known as the fluid and electrolyte imbalance hypothesis of AMS [40, 43, 77].

Interestingly, there is an excessive accumulation of fluid has been found to observed in dwellers with positive symptoms of AMS, supporting hyper-activated RAAS mediated fluid imbalance hypothesis of AMS [72]. Under pathophysiological conditions, angiotensin-II, a peptide hormone, and a key component of RAAS amplifies sodium re-absorption and causes a significant decrease in GFR [78]. In addition to this, angiotensin-II also has been found to control the release of aldosterone, another key component of RAAS [40]. This angiotensin II release is regulated by Atrial natriuretic peptide (ANP), a peptide hormone, secreted from the atria, which also controls the action of several hormones viz., angiotensin-II, aldosterone, and vasopressin and various other physiological parameters of fluid balance in the body like vasodilation, fluid, and electrolyte balance and GFR [5, 41, 90]. Excessive accumulation of fluid and electrolyte could be the reason for edema of lung and brain, pointing an arrow towards the advancement of AMS towards HAPE and HACE if AMS remains untreated [40] (Figure 8.1).

### 8.5.2 VASOGENIC HYPOTHESIS OF ACUTE MOUNTAIN SICKNESS (AMS)

High metabolic rate and restricted substrate storage capacity of the brain has led the brain to accurately auto-regulate blood flow to maintain a constant supply of nutrients and oxygen [85]. In has been reported that, altered oxygen

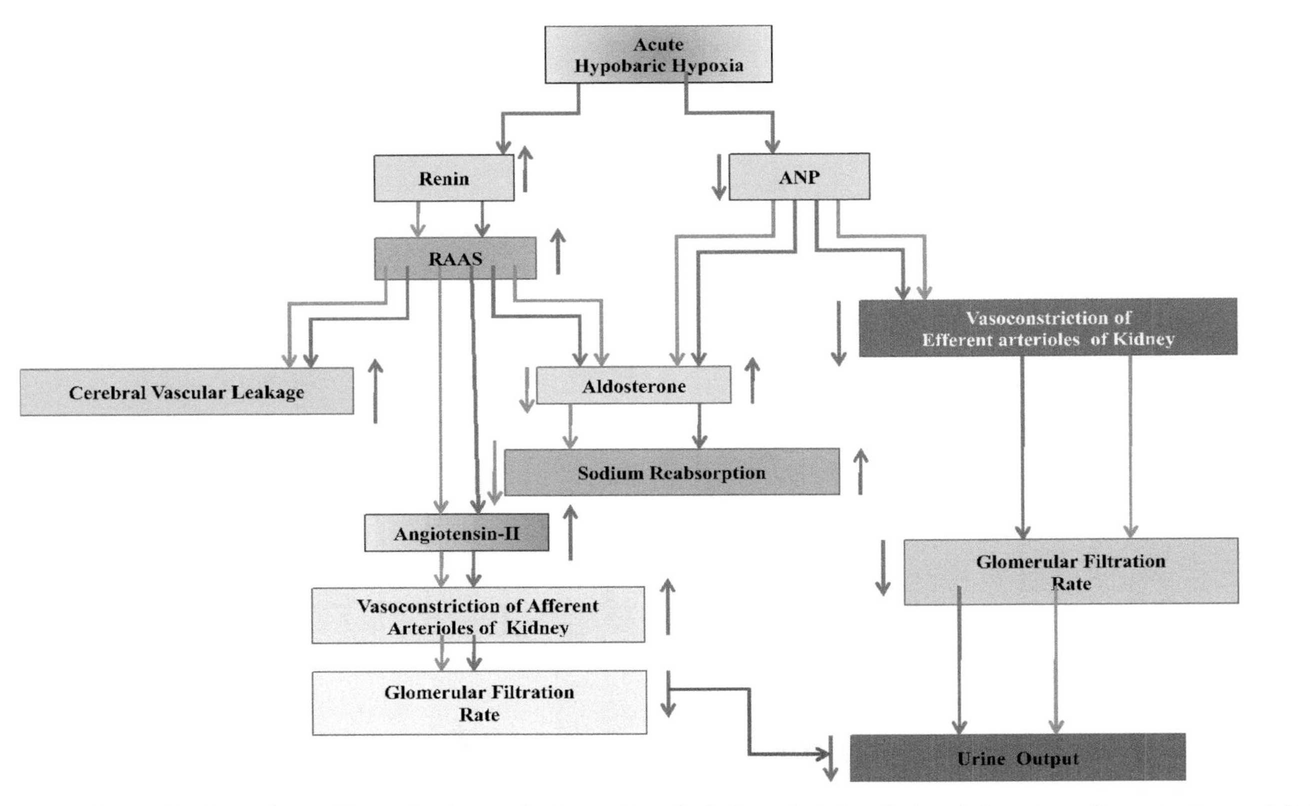


**FIGURE 8.1 (See color insert.)** Possible molecular mechanism of hypobaric hypoxia-induced altered dieresis, sodium retention, and fluid accumulation in AMS positive patients, where ANP: Arterial natriuretic peptide; RAAS: Renin-angiotensin-aldosterone system.

supply to brain, partial pressure of carbon dioxide, mean arterial pressure, and autonomic nervous system majorly regulate cerebral blood flow [3, 86].

Acute exposure to hypobaric hypoxia results in decreased blood supply to the brain, and hyperventilation induced excessive accumulation of carbon dioxide could be working synergistically in controlling cerebral blood flow at high altitude [89]. To compensate with decreased cerebral blood flow, vascular resistance has been found to significantly reduce in small arteries and arterioles and increase in cerebral blood flow [57]. Increase cerebral blood flow and vasodilatation in arteries and arterioles cause increased brain volume and intracranial pressure in the brain [12] (Figure 8.2).

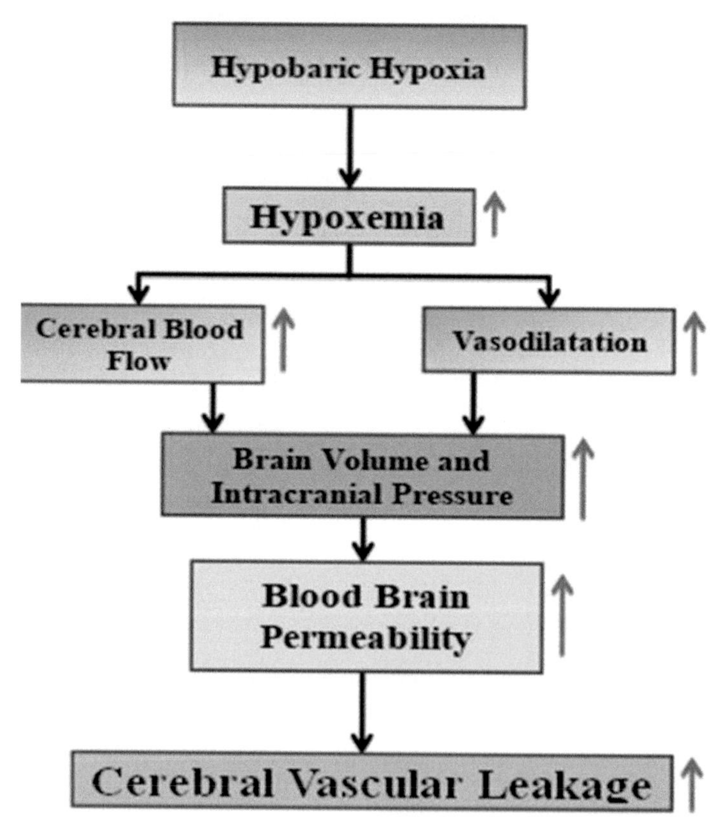


**FIGURE 8.2 (See color insert.)** Acute hypobaric hypoxia-induced altered blood-brain permeability and CVL.

Augmented cerebral blood flow and vasodilatation may overcome capillary vasoconstriction, increase blood-brain permeability that finally resulting in CVL [6, 37, 69]. Increased brain volume and elevated intracranial pressure result in amplified BBB permeability that finally leads to augmented CVL referred to as 'tight fit hypotheses of symptoms associated with AMS [86].

BBB, formed by the brain endothelial cells, plays a dominant role in maintaining fluid homeostasis by regulating the flux of fluid and substances between the systemic circulation and brain microenvironment and protecting the brain from harmful xenobiotics that may cause damage to neuronal and non-neuronal cells present in brain microenvironment [1]. Acute exposure to hypobaric hypoxia has been well reported to cause disruption of BBB permeability and is associated with increased CVL [73]. Hypobaric hypoxia-mediated increased neuroinflammation has been reported to play a role in acute hypobaric hypoxia-induced altered BBB permeability [37]. Under physiological conditions, homeostasis between pro- and anti-inflammatory cytokines plays a pivotal role in maintaining the immune responses in the body [35]. However, under pathophysiological conditions, dysregulation of homeostasis between pro- and anti-inflammatory cytokines leads to neuroinflammation and disruption BBB [79] (Figure 8.2).

The S100 classes of proteins are low molecular weight proteins that have been characterized by their two calcium-binding sites with helix-loop-helix conformation [49]. The 21 different types of S100 class of proteins have been reported in the literature and considered as DAMPs [50]. The major function of S100 class of proteins is their function as $Ca^{2+}$ sensing, and once activated, they may interact with other proteins resulting in regulation of their activity. S100B firstly identified S100 class of protein has been reported to secrete by astrocytes [74]. Secreted S100B protein may exert regulatory activities intra- and extracellular signals and has been considered as a specific marker of brain injury [34, 75]. More specifically, elevated serum S100B has been considered as a marker of neuroinflammation associated with acute and chronic injury [51].

High serum S100B protein levels have been found in subjects exposed to acute hypobaric hypoxia with symptoms of AMS [88]. However, the exact molecular of S100B dependent regulation of inflammatory pathways is not clear during hypobaric hypoxia. The receptor the advanced glycation endproducts (RAGE), could work as a signal traducer receptor and causes microglia activation result in neuro-inflammation through NF-κB dependent mechanisms [2, 63]. Interestingly, dissociated NF-κB has been found in the literature to controls hundreds of genes emerged to be a mediator of inflammatory processes [2] (Figure 8.3).

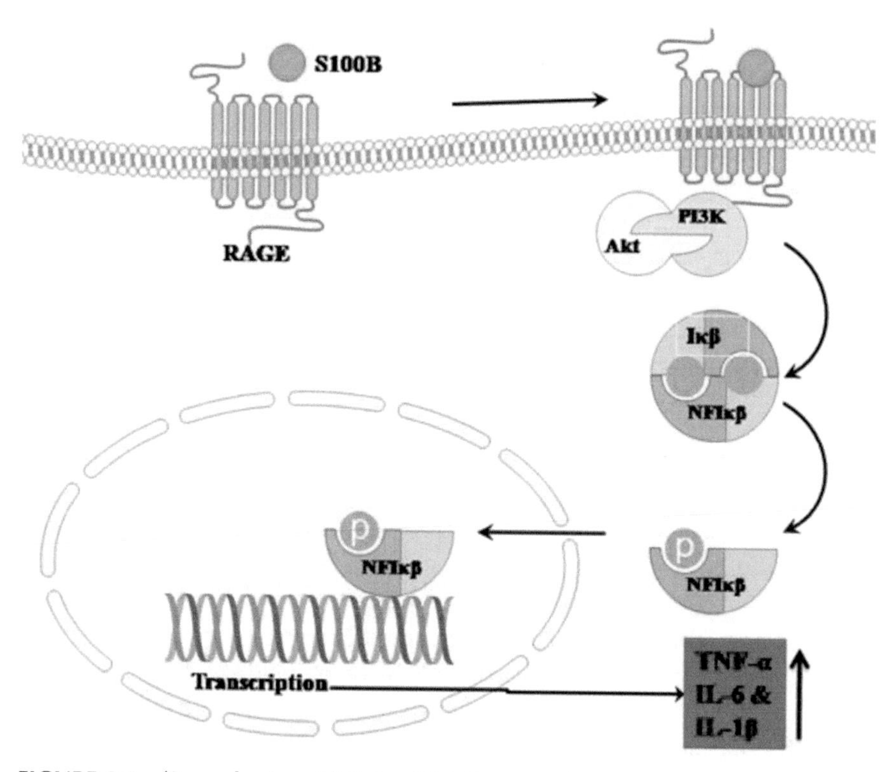


**FIGURE 8.3** **(See color insert.)** Role of S100B protein in neuro-inflammation (Legends: S100B: S100 calcium-binding protein B; RAGE: Receptor for the advanced glycation end products; PKB: Protein kinase B; PI3K: Phosphatidylinositol-4,5-bisphosphate 3-kinase; Iκβ: Inhibitor of kappa B; Nuclear factor kappa-light-chain-enhancer of activated B cells; TNF-α: Tumor necrosis factor α; IL-1: Interleukin 6; IL-1β: Interleukin 1β).

### *8.5.3 HYPOVOLEMIA HYPOTHESIS OF ACUTE MOUNTAIN SICKNESS (AMS)*

Exposure to hypobaric hypoxia has been found to decrease the plasma volume and has been positively correlated with the symptoms of AMS [55]. The mechanism behind high altitude-induced dehydration and hypovolemia is not clear. It could be due to the cold environment at high altitude that leads to excessive diuresis and poor availability of water [29].

Also, this dehydration at high altitude has been found to compromise physical performance, and this may contribute to the positive correlation of AMS symptoms with dehydration [11]. However, further studies are required to propose molecular mechanisms to justify the correlation between dehydration and manifestations of AMS.

Magnesium has been reported to play a dominant role in maintaining cellular functions, and deficiency of magnesium has been associated with many of neurological and cardiac dysfunction [76]. In addition to this, there is a direct link of magnesium with endurance and performance. It has been reported that animals deficient in magnesium perform at a snail's pace in trade mill test [39]. Magnesium also has a role in chronic fatigue syndrome. The patients with chronic fatigue syndrome have low RBC magnesium, further supporting the link of magnesium with performance and endurance [15]. Exposure to dehydration has been previously reported to decrease serum magnesium concentration [7]. Previous studies on shifting of magnesium ions from serum into RBC during marathon provide a plausible explanation for the decrease in serum Magnesium concentration [10, 44]. Franceschi et al., [16] indicated that supplementation of magnesium reduces RBC dehydration in sickle cell disease [16].

## 8.6 NON-PHARMACOLOGICAL AND PHARMACOLOGICAL MEASURES TO PREVENT ACUTE MOUNTAIN SICKNESS (AMS)

No specific drug molecule has been identified to treat symptoms pertaining to AMS. This may be because of non-selective symptoms associated with the pathophysiology of AMS. However, the symptoms could be managed by non-pharmacologically and pharmacologically measures [45].

### *8.6.1 NON-PHARMACOLOGICAL MEASURES FOR ACUTE MOUNTAIN SICKNESS (AMS)*

Majorly there are three factors (Table 8.2) that describe the incidence and severity of AMS, e.g., speed of ascent, altitude attained, and the previous acclimatization [32]. Rapid ascent to high altitude has been found to be a major contributor for the symptoms of AMS. It has been reported that about 25% of dwellers experience symptoms of AMS following rapid ascend to high altitude [32, 68]. Therefore, by controlling the speed of ascent in terms of meters gain per day has been found to be the best method to control symptoms associated with AMS [45, 46]. In addition to this, sleep at altitude gain in a day has been additionally considered for preventing symptoms [46].

**TABLE 8.2** Description About Non-Pharmacological Measures of AMS

| Measures | Description |
|---|---|
| Speed of ascend | 300–500 m/day ascend with 1 day rest every 3–4 days |
| Altitude attained | The percentage prevalence of AMS is above 90% at the altitude more than 4000 m |
| Pre-acclimatization | 5 or more days above the altitude of 3000 m in the last 2 months |

Prevalence and severity of cardinal manifestations associated with AMS have been found to increase with altitude attained [32]. Montgomery et al., have found that 25% of individual experience symptoms of AMS at an altitude of 2000 m [52, 53]. Subsequently, 79% of subjects were found AMS positive at the altitude of 3660 m and approximately 96% of subjects had symptoms of AMS at the altitude of 4,232 m [32, 60, 61].

### *8.6.2 PHARMACOLOGICAL MEASURES FOR ACUTE MOUNTAIN SICKNESS*

Acetazolamide, a carbonic anhydrase inhibitor, @ 125 mg per day, has been suggested sufficient for preventing symptomatic AMS [32, 91]. Acetazolamide has been reported to cause bicarbonate dieresis and metabolic acidosis resulting in stimulation of ventilator responses to hypobaric hypoxia at high altitude exposure [33, 81]. On the other hand, intravenous therapy of acetazolamide has also been found to increase cerebral blood flow and respiration by causing carbonic acidosis in brain tissue [42]. Dexamethasone, a synthetic glucocorticoid, has been reported to be the best alternative to acetazolamide [45]. The molecular mechanism of dexamethasone was found to be acting on a reduction of capillary bed permeability by inhibiting the production of prostaglandins [13, 14, 91]. The use of dexamethasone over acetazolamide is still debatable [62]. However, dexamethasone has been found highly effective as compare to acetazolamide in cases, where rapid ascend for a short period is required [21–23]. Non-steroidal anti-inflammatory drug (viz., ibuprofen) has also been found to be effective for preventing symptoms associated with AMS [45, 59].

## 8.7 SUMMARY

Rapid ascend to high altitude in un-acclimatized individuals could be a responsible factor for the symptoms associated with AMS. Exact pathophysiology associated with AMS is still not clear. Acute hypobaric hypoxia-induced fluid and electrolyte imbalance, neuro-inflammation, and decreased plasma volume could be synergistically playing a role in the cardinal manifestations associated with AMS. AMS could be managed non-pharmacologically pharmacologically depending on the severity of the symptoms associated with AMS.

## KEYWORDS


- **acute mountain sickness**
- **atrial natriuretic peptide**
- **fluid and electrolyte balance**
- **high altitude cerebral edema**
- **high altitude pulmonary edema**
- **hypobaric hypoxia**
- **neuroinflammation**


## REFERENCES


1. Abbott, N. J., (2002). Astrocyte-endothelial interactions and blood-brain barrier permeability. *J. Anat., 200*(6), 629–638.
2. Aggarwal, B. B., Prasad, S., Reuter, S., Kannappan, R., & Yadev, V. R., (2011). Identification of novel anti-inflammatory agents from Ayurvedic medicine for prevention of chronic diseases: "Reverse pharmacology" and "bedside to bench" approach. *Curr. Drug Targets, 12*(11), 1595–1653.
3. Balaban, D. Y., Machina, M. A., Han, J. S., Katznelson, R., & Minkovich, L. L., (2012). The interaction of carbon dioxide and hypoxia in the control of cerebral blood flow. *Pflugers Arch., 464*, 345–351.
4. Barry, J. W., & Pollard, A. J., (2003). Altitude illness. *BMJ, 326*, 915–919.
5. Baxter, J. D., Lewicki, J. A., & Gardner, D. G., (1988). Atrial natriuretic peptide. *Nature Biotechnology, 6*, 529–546.
6. Brown, R. C., Mark, K. S., & Egleton, R. D., (2003). Protection against hypoxia-induced increase in blood-brain barrier permeability: Role of tight junction proteins and NFkappaB. *J Cell Sci., 116*, 693–700.
7. Buchman. A. L., Keen, C., Commisso, J., Killip, D., Ou, C. N., & Rognerud, C. L., (1998). The effect of a marathon run on plasma and urine mineral and metal concentrations. *J. Am. Coll. Nutr., 17*(2), 124–127.
8. Carod-Artal, F. J., (2014) High-altitude headache and acute mountain sickness. *Neurologia, 29*(9), 533–540.

9. Carozzi Albert, V., Crettaz, Bernard., & Ripoll, D., (1999). *Les plis du temps, Mythe* (The folds of time, myth). In: De Saussure, H. B., (ed.), *Science* (p. 368). Ethnographic Museum, Conches Annex. Geneva.
10. Casoni, I., Guglielmini, C., Graziano, L., Reali, M. G., & Mazzotta, D., (1990). Changes of magnesium concentrations in endurance athletes. *Int. J. Sports Med., 11*, 234–237.
11. Castellani, J. W., Muza, S. R., Cheuvront, S. N., Sils, I. V., & Fulco, C. S., (2010). Effect of hypohydration and altitude exposure on aerobicexercise performance and acute mountain sickness. *J. Appl. Physiol., 109*(6), 1792–1800.
12. Cipolla, M. J., (2010). *The Cerebral Circulation* (p. 70). Morgan & Claypool Life Sciences, San Rafael (CA).
13. Coote, J. H., (1991). Pharmacological control of altitude sickness. *Trends in Pharmacological Sciences, 12*, 450–455.
14. Coote, J. H., (1995). Medicine and mechanisms in altitude sickness. *Recommendations in Sports Med., 20*(3), 148–159.
15. Cox, I. M., Campbell, M. J., & Dowson, D., (1991). Red blood cell magnesium and chronic fatigue syndrome. *Lancet, 337*(8744), 757–760.
16. De Franceschi, L., Bachir, D., Galacteros, F., et al., (1997). Oral magnesium supplements reduce erythrocyte dehydration in patients with sickle cell disease. *Journal of Clinical Investigation, 100*(7), 1847–1852.
17. Dellasanta, P., Gaillard, S., Loutan, L., & Kayser, B., (2007). Comparing questionnaires for the assessment of acute mountain sickness. *High Alt. Med. Biol., 8*(3), 184–191.
18. Deweber, K., & Scorza, K., (2010). Return to activity at altitude after high-altitude illness. *Sports Health, 2*(4), 291–300.
19. Doherty, M. J., (2003). James Glaisher's 1862 account of balloon sickness: Altitude, decompression injury, and hypoxemia. *Neurology, 60*(6), 1016–1018.
20. Dong, J. Q., Zhang, J. H., Qin, J., & Li, Q. N., (2013). Anxiety correlates with somatic symptoms and sleep status at high altitudes. *Physiol. Behav., 112, 113*, 23–31.
21. Ellsworth, A. J., Larson, E. B., & Strickland, D., (1987). A randomized trial of dexamethasone and acetazolamide for acute mountain sickness prophylaxis. *Am. J. Med., 83*(6), 1024–1030.
22. Ellsworth, A. J., (1989). Preventing and treating acute mountain sickness. *West J. Med., 150*(2), 196.
23. Ellsworth, A. J., Meyer, E. F., & Larson, E. B., (1991). Acetazolamide or dexamethasone use versus placebo to prevent acute mountain sickness on Mount Rainier. *West J. Med., 154*(3), 289–293.
24. Gilbert, D. L., (1983). The first documented report of mountain sickness: The China or Headache Mountain story. *Respir Physiol., 52*(3), 315–326.
25. Doherty, C., (2004). Breath of life: the evolution of oxygen therapy. *Journal of the Royal Society of Medicine, 97*(10), 489–493.
26. Hackett, P., & Mangione, M. P., (2015). Fluid and electrolyte balance. In: Sikka, P., Beaman, S., & Street, J., (eds.). *Basic Clinical Anesthesia* (p. 715). Springer, New York, NY.
27. Hackett, P. H., & Hornbein, T., (1988). Disorder of high altitude. In: Murray, J. F., & Nadel, J. A., (ed.), *Textbook of Respiratory Medicine* (pp. 1646–1663). Saunders, Philadelphia.
28. Hackett, P. H., & Roach, R. C., (2001). High-altitude illness. *N. Engl. J. Med., 345*, 107–114.

29. Hansen, J. E., & Evans, W. O., (1970). A hypothesis regarding the pathophysiology of acute mountain sickness. *Arch. Environ. Health*, *21*(5), 666–669.
30. Hornbein, T. F., (1992). Long term effects of high altitude on brain function. *Int. J. Sports Med.*, *13*, S43–S45.
31. Houston, C. S., (1980). *Going High* (p. 87). Queen City Printers, Burlington, VT.
32. Hultgren, H., (1997). *High Altitude Medicine* (p. 550). Hultgren Publications, Stanford –CA.
33. Imray, C., Wright, A., Subudhi, A., & Roach, R., (2010). Acute mountain sickness: Pathophysiology, prevention, and treatment. *Prog. Cardiovasc. Dis.*, *52*, 467–484.
34. James, N. T., Forough, M., & Thomas, G. P., (2010). Intracellular and extracellular effects of S100B in the cardiovascular response to disease. *Cardiovascular Psychiatry and Neurology.* 2010; E-article ID: 206073. doi: 10.1155/2010/206073.
35. Jeon, S. W., & Kim, Y. K., (2016). Neuroinflammation and cytokine abnormality in major depression: Cause or consequence in that illness? *World Journal of Psychiatry*, *6*(3), 283–293.
36. Jeong, J. H., Kwon, J. C., Chin, J., Yoon, S. J., & Na, D. L., (2002). Globus pallidus lesions associated with high mountain climbing. *J. Korean Med. Sci.*, *17*(6), 861–863.
37. Julian, C. G., Subudhi, A. W., Wilson, M. J., Dimmen, A. C., Pecha, T., & Roach, R. C., (1985). Acute mountain sickness, inflammation, and permeability: New insights from a blood biomarker study. *Journal of Applied Physiology*, *111*(2), 392–399.
38. Kayser, B., Aliverti, A., Pellegrino, R., & Dellaca, R., (2010). Comparison of a visual analogue scale and Lake Louise symptom scores for acute mountain sickness. *High Alt. Med. Biol.*, *11*(1), 69–72.
39. Keen, C. L., Lowney, P., Gershwin, M. E., Hurley, L. S., & Stern, J. S., (1987). Dietary magnesium intake influences exercise capacity and hematologic parameters in rats. *Metabolism*, *36*(8), 788–793.
40. Kumar, K., Sharma, S., Vashishtha, V., & Bhardwaj, P., (2016). *Terminaliaarjuna* bark extract improves diuresis and attenuates acute hypobaric hypoxia induced cerebral vascular leakage. *J. Ethnopharmacol.*, *180*, 43–53.
41. Laragh, J. H., (1330). Atrial natriuretic hormone, the renin-aldosterone axis, and blood pressure-electrolyte homeostasis. *N. Engl. J. Med.*, *313*, 1330–1340.
42. Leaf, D. E., & Goldfarb, D. S., (1313). Mechanisms of action of acetazolamide in the prophylaxis and treatment of acute mountain sickness. *J. Appl. Physiol.*, *102*(4), 1313–1322.
43. Loeppky, J. A., Icenogle, M. V., Maes, D., Riboni, K., Scotto, P., & Roach, R. C., (2003). Body temperature, autonomic responses, and acute mountain sickness. *High Alt. Med. Biol.*, *4*(3), 367–373.
44. Lukaski, H. C., Bolonchuk, W. W., Siders, W. A., & Milne, D. B., (1990). Chromium supplementation and resistance training: Effects on body composition, strength, and trace element status of men. *Am. J. Clin. Nutr.*, *63*, 954–965.
45. Luks, A. M., McIntosh, S. E., Grissom, C. K., & Auerbach, P. S., (2014). Wilderness medical society practice guidelines for the prevention and treatment of acute altitude illness, update. *Wilderness Environ Med.*, *25*, S4–14.
46. Luks, A. M., McIntosh, S. E., Grissom, C. K., & Auerbach, P. S., (2010). Wilderness medical society consensus guidelines for the prevention and treatment of acute altitude illness. *Wilderness Environ Med.*, *21*(2), 146–155.

47. Lyons, T. P., Muza, S. R., Rock, P. B., & Cymerman, A., (1995). The effect of altitude pre-acclimatization on acute mountain sickness during reexposure. *Aviat. Space Environ Med.*, *66*(10), 957–962.
48. Maggiorini, M., Buhler, B., Walter, M., & Oelz, O., (1990). Prevalence of acute mountain sickness in the Swiss Alps. *BMJ*, *301*, 853–854.
49. Marenholz, I., Heizmann, C. W., & Fritz, G., (2004). S100 proteins in mouse and man: from evolution to function and pathology (including an update of the nomenclature)." *Biochemical and Biophysical Research Communications*, *322*(4), 1111–1122.
50. Memari, B., Bouttier, M., Dimitrov, V., Ouellette, M., Behr, M. A., Fritz, J. H., & White, J. H., (2015). Engagement of the aryl hydrocarbon receptor in mycobacterium tuberculosis-infected macrophages has pleiotropic effects on innate immune signaling. *J. Immunol.*, *195*(9), 4479-4491.
51. Michetti, F., Corvino, V., Geloso, M. C., & Lattanzi, W., (2012). The S100B protein in biological fluids: More than a lifelong biomarker of brain distress. *J. Neurochem.*, *120*(5), 644–659.
52. Montgomery, A. B., Mills, J., & Luce, J. M., (1989). Incidence of acute mountain sickness at intermediate altitude. *JAMA*, *261*(5), 732–734.
53. Moraga, F. A., Osorio, J. D., & Vargas, M. E., (2002). Acute mountain sickness in tourists with children at Lake Chungará (4400 m) in northern Chile. *Wilderness Environ Med.*, *13*(1), 31–35.
54. Moore, L. G., Jahnigen, D., Rounds, S. S., & Reeves, J. T., (1982). Maternal hyperventilation helps preserve arterial oxygenation during high altitude pregnancy. *J. Appl. Physiol.*, *52*, 690–694.
55. Nerin, M. A., Palop, J., & Montano, J. A., (2006). Acute mountain sickness: Influence of fluid intake. *Wilderness Environ Med.*, *17*(4), 215–220.
56. Nussbaumer-Ochsner, Y., Ursprung, J., Siebenmann, C., Maggiorini, M., & Bloch, K. E., (2012). Effect of short-term acclimatization to high altitude on sleep and nocturnal breathing. *Sleep*, *35*(3), 419–423.
57. Otis, S. M., Rossman, M. E., & Schneider, P. A., (1988). Relationship of cerebral blood flow regulation to acute mountain sickness. *J. Ultrasound Med.*, *8*(3), 143–148.
58. Palmer, B. F., & Clegg, D. J., (2014). Ascent to altitude as a weight loss method: The good and bad of hypoxia inducible factor activation. *Obesity (Silver Spring, Md.).*, *22*(2), 311–317.
59. Pandit, A., Karmacharya, P., & Pathak, R., (2014). Efficacy of NSAIDs for the prevention of acute mountain sickness: A systematic review and meta-analysis. *Journal of Community Hospital Internal Medicine Perspectives*, *4*(4), 10.
60. Pigman, E. C., & Karakla, D. W., (1990). Acute mountain sickness at intermediate altitude: Military mountainous training. *Am. J. Emerg. Med.*, *8*(1), 7–10.
61. Pigman, E. C., (1991). Acute mountain sickness. Effects and implications for exercise at intermediate altitudes. *Sports Med.*, *12*(2), 71–79.
62. Rabold, M. B., (1992). Dexamethasone for prophylaxis and treatment of acute mountain sickness. *Journal of Wilderness Medicine*, *3*(1), 54–60.
63. Ramasamy, R., Yan, S. F., & Schmidt, A. M., (2011). Receptor for AGE (RAGE): Signaling mechanisms in the pathogenesis of diabetes and its complications. *Annals of the New York Academy of Sciences*, *1243*, 88–102.
64. Ravenhill, T., (1913). Some experiences of mountain sickness in the Andes. *J. Trop. Med. & Hyg.*, *20*, 313–320.

65. Rehrer, N. J., (2001). Fluid and electrolyte balance in ultra-endurance sport. *Sports Med., 31*(10), 701–715.
66. Sakamoto, R., Okumiya, K., Norboo, T., & Tsering, N., (2017). Sleep quality among elderly high-altitude dwellers in Ladakh. *Psychiatry Res., 249*, 51–57.
67. Sanchez, D. R. M., & Moskowitz, M. A., (1999). High altitude headache – lessons from headaches at sea level. In: Roach, R. C., Wagner, P. D., & Hackett, P. H., (eds.), *Hypoxia: Into the Next Millennium* (pp. 145–153). Kluwer Academic/Plenum Publishers, New York.
68. Schneider, M., & Bernasch, D., (1886). Acute mountain sickness: Influence of susceptibility, preexposure, and ascent rate. *Med. Sci. Sports Exerc., 34*(12), 1886–1891.
69. Schoch, H. J., Fischer, S., & Marti, H. H., (2002). Hypoxia-induced vascular endothelial growth factor expression causes vascular leakage in the brain. *Brain, 125*, 2549–2557.
70. Serrano, D. M., (2005). High altitude headache: Prospective study of its clinical characteristics. *Cephalalgia, 25*, 1110–1116.
71. Silber, E., Sonnenberg, P., & Collier, D. J., (2003). Clinical feature of headache at altitude: A prospective study. *Neurology, 60*, 1167–1671.
72. Singh, I., Khanna, P. K., & Srivastava, M. C., (1969). Acute mountain sickness. *N. Engl. J. Med., 280*(4), 175–184.
73. Song, T. T., Bi, Y. H., & Gao, Y. Q., (2016). Systemic pro-inflammatory response facilitates the development of cerebral edema during short hypoxia. *Journal of Neuroinflammation, 13*(1), 63.
74. Sorci, G., & Bianchi, R., (2010). S100B Protein, a damage-associated molecular pattern protein in the brain and heart, and beyond. *Cardiovascular Psychiatry and Neurology*, 656481.
75. Sun, B., Liu, H., & Nie, S., (2013). S100B protein in serum is elevated after global cerebral ischemic injury. *World J. Emerg. Med., 4*(3), 165–168.
76. Swaminathan, R., (2003). Magnesium metabolism and its disorders. *The Clinical Biochemist Reviews, 24*(2), 47–66.
77. Swenson, E. R., (1997). High altitude diuresis: Fact or fancy. In: Houston, C. S., & Coates, G., (eds.), *Hypoxia: Women at Altitude* (pp. 272–283). Queen City Printers, Burlington – VT.
78. Vos, P. F., Koomans, H. A., & Boer, P., (1992). Effects of angiotensin II on renal sodium handling and diluting capacity in man pretreated with high-salt diet and enalapril. *Nephrol Dial Transplant, 7*(10), 991–996.
79. Wang, W. Y., Tan, M. S., Yu, J. T., & Tan, L., (2015). Role of pro-inflammatory cytokines released from microglia in Alzheimer's disease. *Annals of Translational Medicine, 3*(10), 136–142.
80. West, J. B., & Alexander, M., (1985). Physiological challenge of Mt. Everest. *J. Appl. Physiol., 63*(1), 3–11.
81. West, J. B., (2013). Joseph Barcroft's studies of high-altitude physiology. *Am. J. Physiol. Lung. Cell Mol. Physiol., 305*(8), L523–L529.
82. Westerterp, K. R., & Kayser, B., (2006). Body mass regulation at altitude. *Eur. J. Gastroenterol. Hepatol., 18*, 1–3.
83. Westerterp-Plantenga, M. S., Westerterp, K. R., & Rubbens, M., (1999). Appetite at "high altitude" [Operation Everest III (Comex-'97)]: A simulated ascent of Mount Everest. *J. Appl. Physiol., 87*(1), 391–399.
84. White, A. P., (2001). High life: A history of high-altitude physiology and medicine. *The Yale Journal of Biology and Medicine, 74*(2), 139–141.

85. Willie, C. K., Tzeng, Y. C., Fisher, J. A., & Ainslie, P. N., (2014). Integrative regulation of human brain blood flow. *The Journal of Physiology*, *592*, 841–859.
86. Wilson, M. H., & Milledge, J., (2008). Direct measurement of intracranial pressure at high altitude and correlation of ventricular size with acute mountain sickness: Brian Cummins' results from the 1985 Kishtwar expedition. *Neurosurgery*, *63*(5), 970–975.
87. Windsor, J. S., & Rodway, G. W., (2007). Heights and hematology: The story of hemoglobin at altitude. *Postgraduate Medical Journal*, *83*(977), 148–151.
88. Winter, C. D., Whyte, T. R., Cardinal, J., & Rose, S. E., (2014). Elevated plasma S100B levels in high altitude hypobaric hypoxia do not correlate with acute mountain sickness. *Neurol Res.*, *36*(9), 779–785.
89. Xu, F., Liu, P., Pascual, J. M., Xiao, G., & Lu, H., (2012). Effect of hypoxia and hyperoxia on cerebral blood flow, blood oxygenation, and oxidative metabolism. *Journal of Cerebral Blood Flow & Metabolism*, *32*(10), 1909–1918.
90. Yasue, H., Obata, K., Okumura, K., & Kurose, M., (1989). Increased secretion of atrial natriuretic polypeptide from the left ventricle in patients with dilated cardiomyopathy. *Journal of Clinical Investigation*, *83*(1), 46–51.
91. Zell, S. C., & Goodman, P. H., (1988). Acetazolamide and dexamethasone in the prevention of acute mountain sickness. *West J. Med.*, *148*(5), 541–545.

CHAPTER 9

# ANTIHELMINTIC PROPERTIES OF *ALLIUM SATIVUM* L.: EFFECT ON FECAL EGG COUNT OF WEST AFRICAN DWARF (WAD) RAMS

AZEEZ OLANREWAJU YUSUF and
GANIYAT ABIOLA OLADUNMOYE

## ABSTRACT

This study was carried out to determine the effect of *Allium sativum* (garlic) as an alternative anti-helminthic on West African Dwarf (WAD) sheep. Twelve rams between 8–13kg were used in a 42-day experimental period. The rams were housed individually in pen fed, and the sheep were fed *Panicum maximum*, substituted with isoenergetic and isocaloric concentrate diet at 3% of their body weight for each animal, water was available at all time. The effects of the treatments were accessed by the extent of the decline in the slopes of the fecal egg count (FEC) of the experimental animals. Albendazole reduced the egg per gram of the fecal, while administered garlic showed no definite reduction but increase in the egg per gram of the fecal. However, for oocyst count, all administered treatment followed the same trend. It is concluded that drenching reconstituted *Alium sativum* at the rate of 200 and 400 mg/kgBW was not effective against gastrointestinal nematodes.

## 9.1 INTRODUCTION

In ruminant production, gastrointestinal parasites have been a major constraint that hinders the sustainability and profitability of livestock production, especially in sheep production [6]. An unmanageable load of endoparasites is associated with lowered outputs of animal products and by-products, thus contributing to production and productivity loss [14]. Helminthiasis caused by helminths is the most common and severest form of all gastrointestinal diseases with nematodes being the greatest challenge for conventional sheep producers. Sheep graze closer to the ground and directly over their manure, which is in pellet form. This increases their susceptibility to the nematodes more than other classes of livestock [7].

Since the 1960s, worm control has relied heavily on the use of synthetic anti-helminths, such as Invermectins, Levamisoles, and Thiabendazole [7]. The continuous and indiscriminate use of antihelmintic drugs has caused growing resistance of the parasite to conventional treatments [12, 13]. Also, the long-term regular uses of these anti-helminths have led to the loss of an animal's natural resistance and immunity. It was opined that the increasing resistance of parasites to antihelmintic drugs is a largely known phenomenon, this necessitates the search for several alternatives (nutrition manipulation, use of medicinal plants and rotational grazing) rather than the common chemotherapeutic means, because of the recent consumers concern for safe food [13].

The use of medicinal plants against parasite has been practiced since time immemorial till the present time. This can be referred as trado-medicine, which has its origin in ethnoveterinary medicine [1]. In ethnoveterinary practices, different parts of plant or extracts are employed in combating varying internal or external parasites in livestock. One of such plants is garlic (*Allium sativum*). Garlic is employed locally to cure different kinds of diseases like infectious bursal diseases, ulcers, diarrhea, and dermal infections [8]. Koch and Lawson [15] reported that garlic has antibacterial and antioxidants effects. It has anti-cancer, immunomodulatory, anti-inflammatory, hypoglycemic effects with cardiovascular protecting properties. Garlic is effective in treating gastrointestinal parasites in livestock animals [11]. Studies from organic sheep producers in the US reported the use of garlic as a viable alternative to commercial anti-helminth [16]. An antihelmintic effect of garlic in mice has been patented [2] with its efficiency on coccidia infections reported in rabbits [20].

Garlic is rich in aromatic oils, which aid in digestion and positively influence respiratory system when inhaled into air sacs and lungs of birds. Also, it was found that garlic has strong anti-oxidative effects [9]. Garlic was reported [5] to have parasiticide, amebicide, acarifuge, vermifuge, larvicide, fungicide, and immuno-stimulant properties. It was stated that garlic can be used to prevent influenza, toxicities, and kill internal parasites [3]. The oil has a broad-antimicrobial spectrum, which can serve as anti-bacterial, anti-fungal, antiviral, and anti-parasitic agent.

This study investigates the effect of garlic on gastro-intestinal parasites of West African Dwarf (WAD) sheep and hypothesizes that garlic extract will have no effect on the gastrointestinal parasites of WAD sheep.

## 9.2 MATERIALS AND METHODS

### 9.2.1 STUDY SITE

The study was carried out at the Directorate of University Farms (DUFARMS), Federal University of Agriculture, Abeokuta, Ogun State, Nigeria. The site is located within the derived savannah agro-ecological zone of South-Western Nigeria, with a mean annual temperature of 34°C with a relative humidity of 82%. It is within 7013'49.46"N and longitude 3026'11.98" E. The altitude is 76 m above sea level.

### 9.2.2 SOURCE OF EXPERIMENTAL ANIMALS

Twelve WAD sheep between 8–13 kg were purchased from villages around the University of Agriculture, Abeokuta, Nigeria. They were isolated for twenty-one days for proper acclimatization. They were administered Penstrep as anti-stress to treat any incidence of bacterial, viral, or other diseases. The initial parasite load of the animals was measured before the start of the experiment. Thereafter, the sheep were housed individually in pens made of wood and corrugated iron sheets and managed intensively. Throughout the period of the experimental, the sheep were fed *Panicummaximum*, substituted with isoenergetic and isocaloric concentrate diet at 3% of their body weight for each animal. The concentrate diet was composed of wheat offal, palm kernel cake, cassava peel, bone meal, and salt.

### 9.2.3 PLANT MATERIAL PREPARATION AND RECONSTITUTION

Garlic (*Allium sativum*) cloves were bought from the local market, sliced, dried, and pulverized into its powdered form. The powdered garlic was further processed and reconstituted to the required dosages for drenching;

using water as the solvent. Approximately 100 ml of water was used to reconstitute 20 grams (g) of garlic powder (treatment 2), and 40 g was reconstituted the same amount of water for (treatment 3) to give 200 mg/ml and 400 mg/ml garlic for treatments 2 and 3, respectively.

### 9.2.4 DOSAGE

The dosages of the treatments were: Albendazole 5 mg/kg BW of the animal, while the reconstituted *Allium satium* was administered at the rate of 200 mg and 400 mg/kg BW of the animal for treatments two and three, respectively. The dosages were administered for three days in a week throughout the period of the experiment.

### 9.2.5 EXPERIMENTAL DESIGN

The twelve experimental animals were divided into three (3) treatments, containing four animals per treatment. The details of the treatments are as follows:

- T1 = (control), 5 mg/kg BW of Albendazole
- T2 = 200 mg/kg (body weight) Garlic
- T3 = 400 mg/kg BW Garlic

### 9.2.6 DATA COLLECTION

#### 9.2.6.1 FECAL EGG COUNTS (FECS)

Samples of the fecal were collected weekly via the rectum of the experimental animals and analyzed using floatation technique (McMaster counting technique). The McMaster counting is a quantitative technique usually employed to quantify the amount of eggs and oocytes in each gram of fecal sample (e.p.g. and o.p.g). The eggs were separated from the said sample using a floatation liquid in a McMaster counting chamber. This method will identify 50 plus e.p.g. and o.p.g of the fecal sample [22].

All data obtained were analyzed using descriptive statistics using graph feature in SPSS 2006 software.

## 9.3 RESULTS

The composition of *Allium sativum* (garlic) powder consisted of: 13.03% moisture and 86.97% dry matter; 4.11% fat, 6.78% ash, 10.89% crude fiber, 18.97% crude protein (CP) and 59.26% carbohydrate.

The parasite eggs identified and counted were: Moniezaspp, Strongylespp, and Trichuris spp. The oocyst identified and counted was Eimeria spp. The effects of the Albendazole, 200 mg/kgBW, and 400 mg/kgBW *Allium sativum* administration on the egg count per gram (EPG) of fecal (EPG) is presented in Figure 9.1.

At the onset of the experiment, the experimental animals had between 500–1300 EPG of fecal (approximately 2.8–3.2 e.p.g; 10log10). From week zero of the experiment to week one, the EPG of rams administered Albendazole at 5mg/kgBW was reduced from 2.80 (10log) to zero and remained at this level throughout the experimental period. However, the EPG of rams administered 200mg/kgBW and 400mg/kgBW garlic was reduced from the onset of the experiment to week one and afterward was increased and decreased intermittently, with no definite reduction in egg count throughout the experimental period (Figure 9.2).

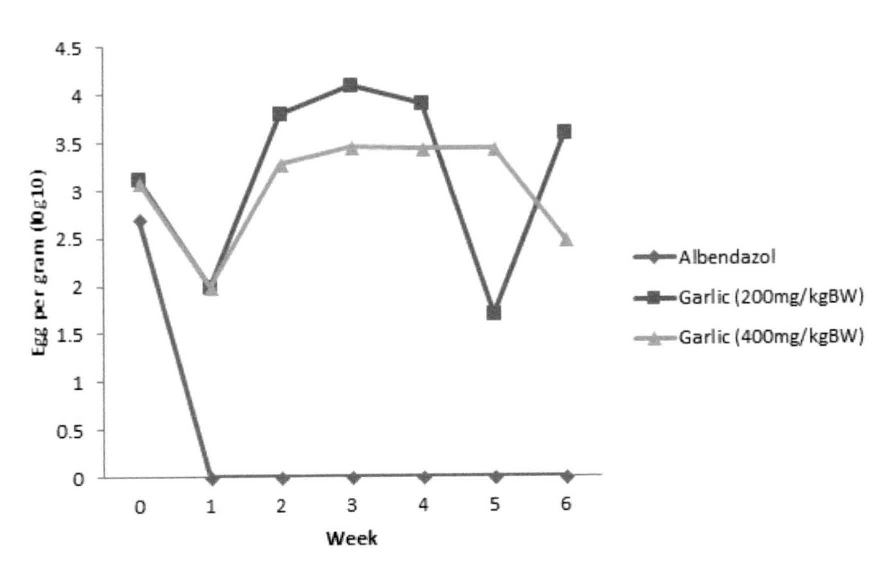


**FIGURE 9.1 (See color insert.)** Line-graph showing egg count per gram of fecal.

The effects of the Albendazole, 200 mg/kgBW *Allium sativum*, and 400 mg/kg BW *Allium sativum* administration on the oocyst count per gram of fecal (OPG) is presented in Figure 9.2. At the onset of the experiment, the experimental animals had between 3533–23000 oocyst per gram of fecal (approximately 3.7–4.3 10log10). The oocyst counts of animals administered Albendazole, 200 mg/kgBW, and 400 mg/kgBW garlic followed the same trend throughout the experiment. The OPG of the rams administered Albendazole, and those administered 400 mg/kg garlic were lower than that of animals administered 200 mg/kg garlic at week one.

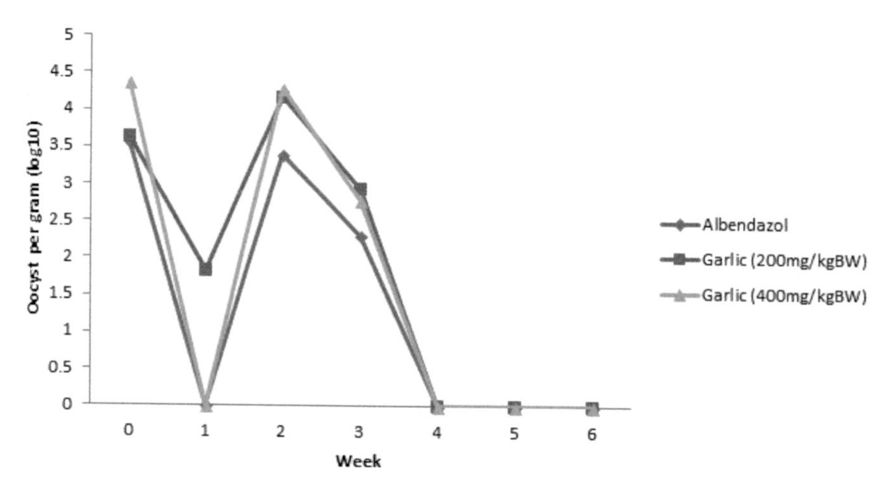


**FIGURE 9.2** **(See color insert.)** Line-graph showing oocyst count per gram of fecal.

From week one to week two, the OPG for all three treatments was increased with the slope of those administered Albendazole slightly lower than those drenched 200 mg/kg and 400 mg/kg garlic. From week two to week four, the OPG was reduced with the slope of animals given Albendazole slightly lower than those of animals given 200 mg/kg and 400 mg/kg garlic. At week four, Albendazole, and 200 mg/kgBW and 400 mg/kgBW garlic dropped to zero and remained at this level till the end of the experimental period.

## 9.4 DISCUSSION

The results of the chemical composition in this chapter disagreed with Nwinuka et al., [17], who reported lower values for moisture content (6.67%), fat content (0.68%) and crude fiber (5.53%). The ash (6.40%)

and CP (13.35%) from this author were comparable to the results of this study. However, the carbohydrate content obtained by Nwinuka [17] was higher than that obtained in this study (72.28%). The results by authors also opposed the report of Okolo et al., [18], who reported increased moisture content (68.09%) and lower fat (0.76%), ash (1.44%), crude fiber (0.69%), CP (8.54%) and carbohydrate (19.48%) content.

The results in this chapter further disagreed with Bhowmik et al., [4], who reported lower values of moisture content (6.57%), fat content (3.28%), ash content (3.31%) and crude fiber (0.73%). The carbohydrate content reported was higher than that obtained in this study (68.86%). The CP content obtained in this study (18.97%) is higher than that obtained by Bhowmik et al., [4] (17.55%), which is similar to that stated in Encyclopedia of Chemical Technology, 1980 (17.55%). The wide differences in the chemical composition of garlic can be hinged on variety differences, location, varying soil chemical composition, and processing techniques [4].

Results obtained for egg per gram (e.p.g) of the fecal showed that administration of Albendazole reduced the e.p.g in WAD sheep. The study revealed that synthetic anti-helminth remained as effective in the control of gastrointestinal parasite in WAD sheep in terms of the egg per gram of fecal than *Allium sativum*. This result agrees with the work of Gradé and his coworkers [10], who reported a lower egg count in the fecal of sheep treated with synthetic anti-helminth than those treated with garlic. Similar results were also reported by Worku et al., [21], who noticed no significant effect of garlic on fecal egg count among experimental treatment groups. The results of this study also corroborated with those of Gradé and his coworkers [10], who reported that animals treated with synthetic anthelmintic showed the greatest decline in the shedding of nematode eggs. The results of this study also agree with that of Sutton et al., [19], who reported that synthetic anti-helminth was more effective in reducing egg count in experimental animals than garlic. However, the results of this study disagreed with the findings of Fadzirayi et al., [7], who opined that the effect of *Allium sativum* and synthetic anthelmintics egg count in sheep is the same; also disagreed with that of Ayaz et al., [2], who reported that *Allium sativum* was more effective in controlling gastrointestinal parasites in infected mice than synthetic anti-helminth.

The results obtained for oocyst count showed that the control, synthetic anti-helminth, 200 mg/kgBW and 400 mg/kgBW garlic followed a similar trend for oocyst population; and hence, both synthetic anti-helminth and *Allium sativum* produced the same effect on oocyst count in sheep. This can be attributed to the fact that garlic can be used in the treatment of parasitic protozoa in sheep [2]. The results in this chapter, however, disagreed with the result of Gradé and his coworkers [10], who reported no difference in the decline of oocyst with garlic treatments and that of synthetic anti-helminths.

In this study, administration of synthetic anti-helminth showed superior effects on the reduction of fecal egg count (FEC) than *Allium sativum* at 200 mg/kgBW and 400 mg/kgBW garlic. A similar trend was observed in oocyst count for both synthetic anti-helminth and garlic.

## 9.5 SUMMARY

This study compared synthetic anti-helminth and *Allium sativum* on egg count of rams. Based on this research study, it is concluded that synthetic anti-helminth is more effective in controlling gastrointestinal nematode than *Allium sativum*. However, *Allium sativum* is as effective as synthetic anti-helminth in combating parasitic protozoa.

## KEYWORDS

- **albendazole**
- ***Allium sativum***
- **antibiotic**
- **ethnoveterinary medicine**
- **fecal**
- **gastro-intestine**
- **rams**

## REFERENCES

1. Athanasiadou, S., Githiori, J., & Kyriazakis, I., (1392). Medicinal plants for helminth parasite control: Facts and fiction, *1*(9), 1392–1400.
2. Ayaz, E., Turel, I., Gul, A., & Yilmaz, O., (2008). Evaluation of the anthelmintic activity of garlic (Allium sativum) in mice naturally infected with *Aspiculuristetraptera*. *Recent Patents on Anti-Infective Drug Discovery*, *3*(2), 149–152.

3. Bensky, D., A., Gamble, A., & Kaptchuk, T. J., (1993). *Chinese Herbal Medicine: Materiamedica* (3rd edn., p. 1325). Eastland Press, New York.
4. Bhowmik, S., Chowdhury, S. D., Kabir, M. H., & Ali, M. A., (2008). Chemical composition of some medicinal plant products of indigenous origin. *Bangladesh Veterinarian, 25*(1), 32–39.
5. Duke, J. A., (2002). *Handbook of Medicinal Herbs* (p. 896). CRC Press, Boca Raton–FL, 2002.
6. Eysker, M., & Ploeger, H., (2000). Value of present diagnostic methods for gastrointestinal nematode infections in ruminants. *Parasitology, 120*(7), 109–119.
7. Fadzirayi, C., Masamha, B., & Mukutirwa, I., (2010). Efficacy of Allium sativum (garlic) in controlling nematode parasites in sheep. *International Journal of Research, Veterinary Medicine, 8*(3), 161–169.
8. Fenwick, G. R., Hanley, A. B., & Whitaker, J. R., (1985). The genus Allium—part 1. *Critical Reviews in Food Science & Nutrition, 22*(3), 199–271.
9. Gardzielewska, J., Pudyszak, K., Majewska, T., Jakubowska, M., & Pomianowski, J., (2003). Effect of plant-supplemented feeding on fresh and frozen storage quality of broiler chicken meat. *Electronic Journal of Polish Agricultural University (Series Animal Husbandry), 6*(2), Open Access, http://yadda.icm.edu.pl/yadda/element/bwmeta1.element.agro-article-6155736c-7827–4bd7-a039-f4d40a4cc8d3 (Accessed on 29 July 2019).
10. Gradé, J. T., Arble, B. L., Weladji, R. B., & Van Damme, P., (2008). Anthelmintic efficacy and dose determination of Albiziaanthelmintica against gastrointestinal nematodes in naturally infected Ugandan sheep. *Veterinary Parasitology, 157*(3), 267–274.
11. Guarrera, P. M., (1999). Traditional antihelmintic, antiparasitic and repellent uses of plants in Central Italy. *Journal of Ethnopharmacology, 68*(1), 183–192.
12. Hoste, H., Gaillard, L., & Le Frileux, Y., (2005). Consequences of the regular distribution of sainfoin hay on gastrointestinal parasitism with nematodes and milk production in dairy goats. *Small Ruminant Research, 59*(2), 265–271.
13. Jackson, F., & Coop, R., (2000). The development of anthelmintic resistance in sheep nematodes. *Parasitology, 120*(7), 95–107.
14. Knox, M., (2000). Nutritional approaches to nematode parasite control in sheep. *Feed Mix, 8*(4), 12–15.
15. Koch, H. P., & Lawson, L. D., (1996). *Garlic: The Science and Therapeutic Application of Allium Sativum L. and Related Species* (Vol. XV, p. 329). Williams & Wilkins, Baltimore, Maryland.
16. Noon, J., (2003). *A Controlled Experiment to Measure the Effectiveness on Lambs of Wormers That Conform to the New Organic Standards*. FNE03–482 Farmer/Grower Northeast Sare Grant Report, Springvale, Maine, E-Report, http://www.garlicbarrier.com/2003_SARE_Report.html (Accessed on 29 July 2019).
17. Nwinuka, N., Ibeh, G., & Ekeke, G., (2005). Proximate composition and levels of some toxicants in four commonly consumed spices. *J. Appl. Sci. Environ. Mgt., 9*(1), 150–155.
18. Okolo, S. C., Olajide, O. O., Idowu, D. I., Adebiyi, A. B., Ikokoh, P. P., & Orishadipe, A. T., (2012). Comparative proximate studies on some Nigerian food supplements. *Annals of Biological Research, 3*(2), 773–779.
19. Sutton, G., & Haik, R., (1999). Efficacy of garlic as an anthelmintic in donkeys. *Israel Journal of Veterinary Medicine, 54,* 23–27.

20. Toulah, F., & Al-Rawi, M., (2007). Efficacy of garlic extract on hepatic coccidiosis in infected rabbits (Oryctolagus cuniculus): Histological and biochemical studies. *Journal of the Egyptian Society of Parasitology, 37*(3), 957–968.
21. Worku, M., Franco, R., & Baldwin, K., (2009). Efficacy of garlic as an anthelmintic in adult goats. *Archives of Biological Science Belgrade, 61*(1), 135–140.
22. Zajac, A. M., & Conboy, G. A., (2012). *Veterinary Clinical Parasitology* (8th edn., p. 368). Wiley-Blackwell, New York.

CHAPTER 10

# BURITI (*MAURITIA FLEXUOSA* L.) OIL SUPPLEMENTATION: EFFECTS ON OXIDATIVE STRESS AND HORMONAL CONCENTRATIONS IN MALE WISTAR RATS

BOITUMELO ROSEMARY MOSITO, NICOLE LISA BROOKS, and YAPO GUILLAUME ABOUA

## ABSTRACT

This chapter explores how Buriti oil may modulate oxidative stress biomarkers and hormonal function in an *in vitro* animal model using male Wistar rats. Results indicate that control groups experienced normal physiological oxidative stress; therefore, no oxidative stress was induced. Dietary supplementation of Buriti oil conferred protective effects against physiological oxidative stress by significantly decreasing lipid peroxidation (LPO) damages, increasing epidydimal and testicular weights and antioxidant enzyme activities (SOD, CAT, and GSH); and increasing the testosterone concertation in the plasma without having an effect on estradiol concentration.

## 10.1 INTRODUCTION

Male fertility can be affected by lifestyle, hormonal control, age, chronic diseases, and oxidative stress. Antioxidants have been shown to treat diseases caused by oxidative stress. Several researchers have demonstrated the successful use of antioxidant-rich foods such as nuts, green tea, and oils in the improvement of male fertility [7, 8, 10]. Buriti oil is a palm oil from the Buriti fruit known as *Mauritia flexuosa* (known as moriche palm) [39].

This palm tree is abundant in Amazonian parts of Southern America. Buriti oil is rich in antioxidants mainly carotenoids, tocopherols, and tocotrienols. Carotenoids are precursors of vitamin A, while vitamin E consists of tocopherols and tocotrienols.

Testosterone is the key hormone for spermatogenesis regulation. Normal concentration of testosterone in the body influences sexual behavior, muscle mass, energy, and bone integrity [46]. Studies have demonstrated that vitamin E supplementation increases testosterone concentration. Estradiol is a natural estrogen found in high concentrations in females and very low concentrations in males [49]. Research studies on the effects of vitamin E supplementation on testosterone and estradiol have concluded that it increases their concentration [15, 48]. Jargar concluded that supplementation of rats with vitamin E protected testicular tissue and was able to increase testosterone levels [21].

Damage due to oxidative stress can decrease body and organ weights. The consumption of antioxidant-rich foods has shown to reduce oxidative stress damage hence protecting tissues from lipid peroxidation (LPO). Yuce [50] demonstrated that supplementation with cinnamon oil causes an increase in reproductive organ weights of male rats. Jargar [21] reported that diabetic ratson vitamin E supplementation gained both body and organ weights because their tissues were protected against LPO.

LPO disrupts lipid membranes in tissues such as the heart, testis, and lung tissue. Antioxidant-rich foods have been reported to lower LPO caused by oxidative stress. Carotenoids remove free radicals before they interact with the cell membrane by transferring a hydrogen atom with a single electron to a free radical. Carotenoid derived radicals may further undergo bimolecular decay to generate non-radical products. Soluble polyphenols in green tea extract protected the testicular and epididymal tissues by scavenging ROS [19]. Aboua [1] demonstrated that supplementation with carotenoid and tocopherol rich RPO protects testicular epididymal tissues from LPO.

Several researchers have demonstrated that dietary supplementation with antioxidant-rich oils and foods can increase antioxidant enzyme activity. Alinde [5] showed that supplementation with red palm oil increased SOD, CAT, and GSH activities in cardiac tissues. Awoyini [7] concluded that supplementation with green tea and rooibos tea improved the epididymal sperm quality and increased activity of antioxidant enzymes in oxidative stress-induced rats.

This chapter explores how Buriti (*Mauritia flexuosa*) oil may modulate oxidative stress biomarkers and hormonal function in an *in vitro* experimental animal model using male Wistar rats.

## 10.2 MATERIALS AND METHODS

Atlas Animal Foods (Cape Town, SA) supplied standard rat chow (SRC). Buriti oil was purchased from Phytoterapica (Brazil).

Sigma-Aldrich (Johannesburg, SA) supplied: 2-2Œ-azobis (2-methylpropionamidine) dihydrochloride (AAPH), 6-hydroxydopamine (6-HD), diethylenetriaminepentaacetic acid (DETAPAC), 5-5f-Dithio-bis-(2-nitrobenzoic acid) reagent (DTNB), ethylenediaminetetraacetic acid (EDTA), fluorescein sodium salt, glacial metaphosphoric acid (MPA), glutathione reduced (GSH), glutathione reductase (GR), malondialdehyde (MDA) standard, orthophosphoric acid (O-PA), perchloric acid (PCA), potassium phosphate (KH2PO4), reduced nicotinamide adenine dinucleotide phosphate (NAD(P) H), bicinchoninic acid (BCA), sodium azide, sodium hydroxide (NaOH), sulphuric acid, superoxide dismutase (SOD) standard, tertiary-butyl hydroperoxide (t-BHP), thiobarbituric acid (TBA) and trisodium citrate from.

Merck (Johannesburg, SA) supplied: Hydrochloric acid (HCl), isopropanol, methanol, PCA 70%, sodium acetate, and trifluoroacetic acid (TFA).

Sigma Aldrich (Johannesburg, SA) supplied: Ultrapure MilliQ water (Millipore. Greiner), 96-well flat bottom and Costar 96-well UV flat-bottom microplates.

### *10.2.1 TREATMENT OF ANIMALS*

The 60 male Wistar rats were divided into two groups (n=30) and individually housed to ensure equal access to fresh amounts of supplements daily. The experimental group (n=30) received 200 μL of Buriti oil mixed with SRC and free access to water daily. The control group (n=30) received SRC and free access to water daily. Guidelines by Laboratory Animal Care of the National Society of Medical Research and the National Institutes of Health Guideline for Care and Use of Laboratory Animals of the National Academy of Sciences (National Institutes of Health publication no. 80–23, revised 1978) were followed by the Stellenbosch University animal facility, where animals were housed. Post 6-weeks of treatment, rats were anesthetized using 1ml (± 60mg/kg) of sodium pentobarbitone, weighed for body weights, and blood samples were collected using sterile 10ml disposable syringes with 21G sterile hypodermic needles.

SOD, catalase (CAT) and glutathione peroxidase (GPx) activities were determined in epididymal sperm using kits and assessed using a microplate reader (GloMax® Multi Detection System; Promega, UK).

### 10.2.2 COLLECTION OF BLOOD SAMPLES

Blood samples were collected in EDTA containing tubes and in serum separator clot activator tubes (BD Vacutainers, Plymouth, UK), and these samples were placed in ice. Plasma and serum were obtained by centrifugation at 4000 rpm at 4°C for 10min; and within 6 hours of collection these were stored at –80°C until analysis. Testes and epididymis were harvested, weighed, and washed in a phosphate buffer solution before being frozen in liquid nitrogen and stored at –80°C.

### 10.2.3 ESTRADIOL DETERMINATION

The amount of estradiol in the plasma was measured using an ELISA kit (DRG Diagnostics Inc., Germany). All reagents were brought to room temperature before use. The 25μL of each standard, control, and the sample was dispensed into appropriate wells. The 200μL of enzyme conjugate was added to each well, mixed well, and the microtiter plate was incubated for 120 minutes at room temperature. The contents in the wells were discarded and rinsed three times by using the diluted wash solution of 400μL per well. Wash solution was removed by beating the plate on absorbent paper. The 100μL of substrate solution was added to each well and incubated for 15 minutes at room temperature. The 50μL of stop solution was added to each well to stop the reaction. The absorbance of each well was determined at 450nm using a microplate reader (Thermo Electron Corporation, Multiskan spectrum, USA). Plasma estradiol concentrations were expressed in ng/ml.

### 10.2.4 TESTOSTERONE DETERMINATION

Plasma testosterone concentrations were measured using the testosterone kit (Demeditec diagnostics, Germany). The 10μL of each calibrated, and the control sample was dispensed into appropriate wells. The 100μL of incubation buffer was then added to each well followed by 50μL of enzyme conjugate, and the microplate was incubated for 60 minutes at room temperature. The contents in the wells were discarded and were rinsed four times by using the diluted wash solution of 300μL per well. Wash solution was removed by beating the plate on absorbent paper. The 200μL of substrate solution was added to each well and incubated for 30 minutes at room temperature in the dark and 50μL of stop solution was then added to stop the reaction.

The absorbance of each well was determined at 450nm using a microplate reader (Thermo Electron Corporation, Multiskan spectrum, USA). Results were expressed as ng/ml.

### 10.2.5 LIPID PEROXIDATION (LPO)

The 250μL of tissue homogenate was combined with 31.25μL of 4mM of cold BHT/ethanol and 250μL of 0.2M orthophosphoric acid in Eppendorf tubes. Tubes were vortexed for 10 seconds to mix all contents. The 31.25μL of TBA reagent (0.11M in 0.1M NaOH) was then added, and the tube was vortexed again for 10 seconds. Tubes were placed in a water-bath and heated up to 100°C for 2 minutes, and then lids were opened and closed in order to prevent them from popping and left in the water bath for an hour to allow for the reaction to take place (which results in pink color). After exactly one hour, tubes were placed on ice for 2 minutes to allow rapid cooling. Tubes were placed at room temperature for 5 minutes, followed by the addition of 750μL of n-butanol and followed by 100μL of saturated NaCl to aid in the separation of phases. Tubes were vortexed for 10 seconds and micro-fused at 12,000 rpm for 2 minutes at 4°C. The 200μL of the supernatant of the butanol phase was added to each well in triplicates on a 96 well plate and read at $A_{532}$-$A_{572}$ using a microplate reader (Thermo Electron Corporation, Multiskan spectrum, USA)

### 10.2.6 PROTEIN DETERMINATION

The BCA assay was used. Five albumin standards of different concentrations were prepared, and 25μL of each standard or sample was added in duplicate to a microplate well. The 200μL of working reagent was added to each well and mixed thoroughly on a plate shaker for 30 seconds then incubated at 37°C for 30 minutes. The plate was cooled at room temperature, and absorbance was measured at 562 nm on a plate reader (Thermo Electron Corporation, Multiskan spectrum, USA). The protein concentrations were quantified by using the standard curve and expressed as μg/ml.

### 10.2.7 SUPEROXIDE DISMUTASE (SOD) DETERMINATION

The 170μl DETAPAC solution (0.1mM) was added to 6μl lysate; and 24μl of SOD buffer was added to each well. Each sample was run in triplicate.

Fifteen microliters of stock 6-HD was finally added to the previous mixture and read immediately at 490nm for 4min at 1min intervals. The activity of SOD was calculated from a linear calibration curve, in the range of 2 to 20U/mg.

### *10.2.8 CATALASE (CAT)*

The 170μL of phosphate buffer was added to a 96 well plate followed by 75μL of $H_2O_2$ stock solution. The first triplicate was of distilled water for the blank; and the following two triplicates were of 10μL of samples of homogenates. The plate contents were mixed well, and a linear $A_{240}$ decrease/minute was recorded for at least 1 minute in 15 seconds intervals.

### *10.2.9 GLUTATHIONE PEROXIDASE (GPX)*

In a 96-well UV Costar plate, 215μl assay buffer (AB: 50mM potassium phosphate, 1mM EDTA, pH 7.0), 5μl GSH (30.7mg/ml in water), 5μl GR (0.1U/ml in AB), 20μl sample were read before adding 5μl NAD(P)H. Two readings were recorded. The first reading recorded the t-BHP non-dependent NAD(P)H oxidation at 340nm for 3min in 30sec intervals for samples (A1) and blank (A1b). The second reading was performed after adding 50μl of t-BHP. This reading monitored the decrease of t-BHP due to NAD(P)H oxidation at 340nm for 2min in 30sec intervals for the same samples (A2) and blank (A2b). Samples were run in triplicate.

## 10.3 RESULTS

Data are represented as mean ± SD; and (*) indicates a significant difference with $p<0.05$ (Figures 10.1–10.13). Control group received water and SRC diet; Experimental group received water, SRC, and Buriti oil diet.

### *10.3.1 TOTAL BODY WEIGHTS AFTER TREATMENT*

Figure 10.1 indicates that no significant difference was found between the control group (338.2g ± 32.68g) and the Buriti oil supplemented group (336.1g ± 29.24g) in body weights, on the day of sacrifice.

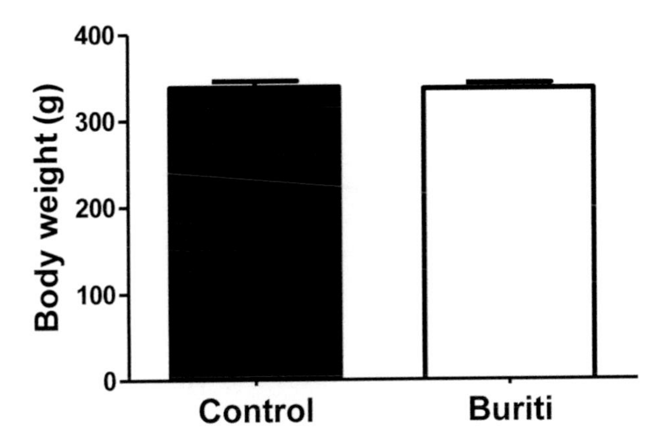


**FIGURE 10.1** Total body weights after treatment.

### *10.3.2 EVALUATION OF TESTICULAR AND EPIDIDYMAL WEIGHTS*

Figure 10.2 indicates that rats fed with Buriti oil had a higher testicular weight (3.577g ± 0.368g) compared to the control group (3.328g ± 0.344g). Likewise, the epididymis of rats fed with Buriti oil was heavier (0.789g ± 0.055g) compared to the control group (0.733g ± 0.095g) as indicated in Figure 10.3. The increase in both testicular and epididymal weights in rats fed with Buriti oil was significant ($p<0.05$).

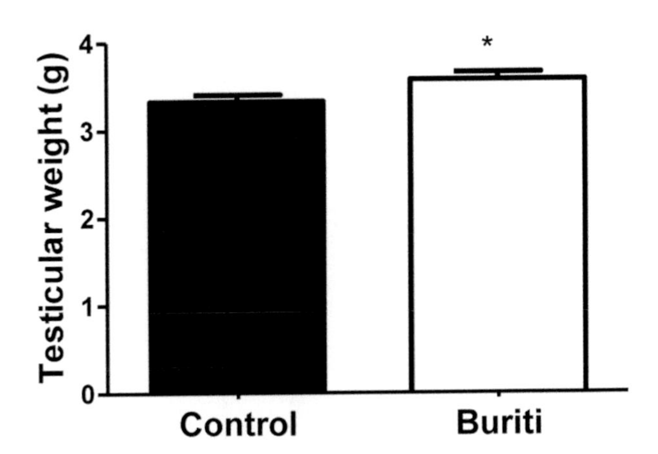


**FIGURE 10.2** Testicular weights.

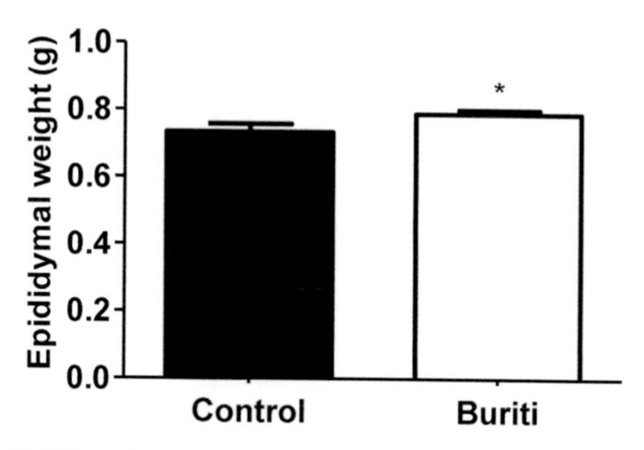


**FIGURE 10.3** Epididymal weights.

### *10.3.3 ASSESSMENT OF TESTICULAR AND EPIDIDYMAL LIPID PEROXIDATION (LPO)*

MDA is a product of LPO, and the increased levels are an indication of oxidative stress damage. Although this study is a baseline study, yet the concentration of MDA in the testis from the Buriti oil fed group in Figure 10.4 (0.103µmol/g ± 0.022 µmol/g) was significantly lower than the control group (0.122 µmol/g ± 0.029 µmol/g). The concentration of MDA in the epididymis of the control group was significantly higher than the Buriti oil fed group. Figure 10.5 shows that the epididymal MDA concentration in the control group was 0.089µmol/g ± 0.057µmol/g, and the Buriti oil fed group was 0.076µmol/g ± 0.056µmol/g. Both testicular and epididymal control MDA levels were higher than the Buriti oil fed group.

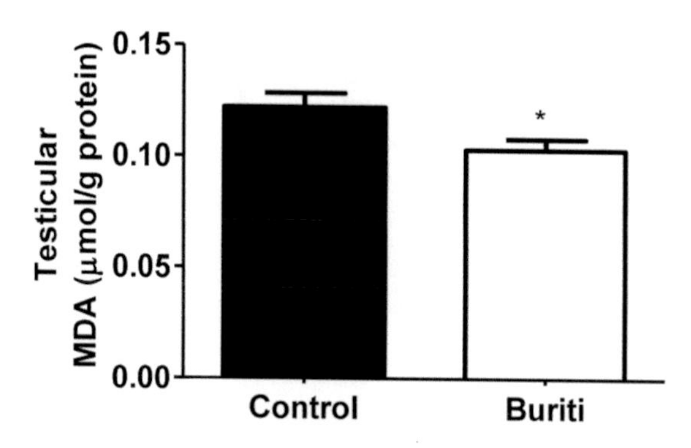


**FIGURE 10.4** Testicular MDA concentration.

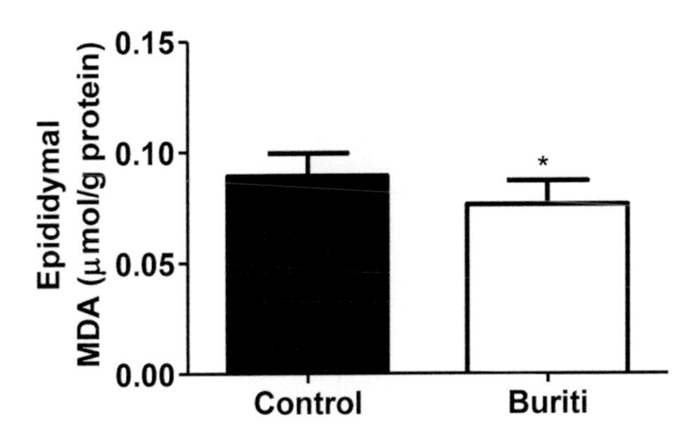


**FIGURE 10.5** Epididymal tissue MDA of control and Buriti diet rats.

### *10.3.4 ASSESSMENT OF TESTICULAR AND EPIDIDYMAL SOD ACTIVITY*

Testicular SOD activities are shown in Figure 10.6. The control group (47.20μmol/mg ± 3.001μmol/mg) had a significantly low SOD activity compared to the Buriti oil group (54.77μmol/mg ± 2.082μmol/mg). The epididymal SOD activity is represented in Figure 10.7. The control group (42.67μmol/mg ± 5.197μmol/mg) has significantly low SOD activity compared to the Buriti oil fed group (46.11μmol/mg ± 3.193μmol/mg). The difference in the concentrations between the control and Buriti oil fed group in both testicular and epididymal tissue was significant ($p<0.05$).

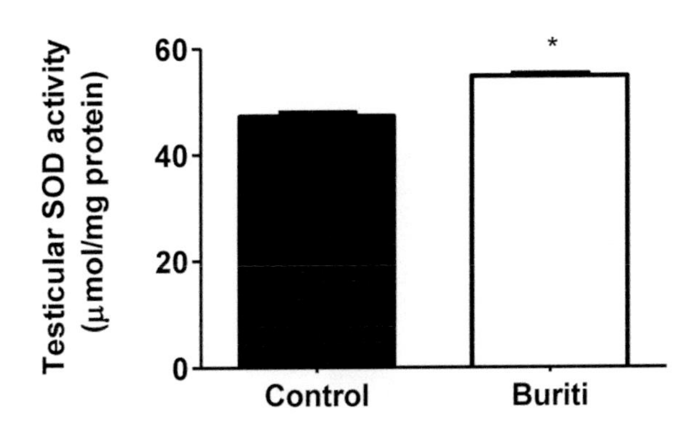


**FIGURE 10.6** Testicular SOD activity.

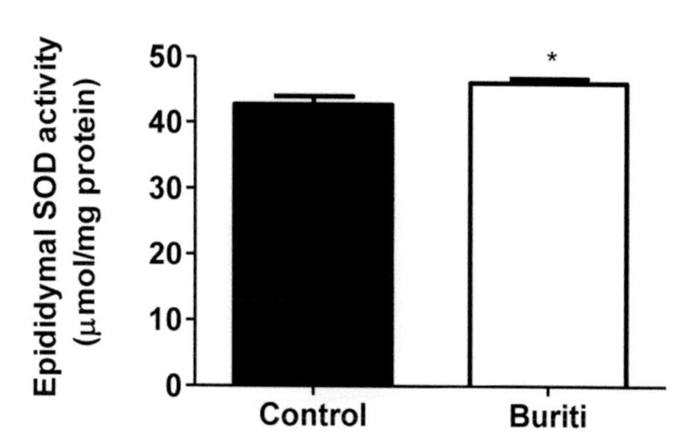


FIGURE 10.7 Epididymal SOD activity.

### 10.3.5 EVALUATION OF TESTICULAR AND EPIDIDYMAL CATALASE (CAT) ACTIVITY

Figures 10.8 and 10.9 represent testicular and epididymal CAT activities, respectively. It can be seen from Figure 10.8 that the control group (34.76µmol/g ± 3.181µmol/g) had lower testicular CAT activity compared to the Buriti oil fed group (40.53µmol/g ± 7.624µmol/g). Figure 10.9 shows that the epididymal CAT activity was higher in the Buriti oil fed group (45.82µmol/g ± 9.184µmol/g) than in the control group (33.30µmol/g ± 5.044µmol/g). In both the epididymal and testicular tissues, there was a significant increase in CAT activity in Buriti oil fed group compared to the control, ($p<0.05$).

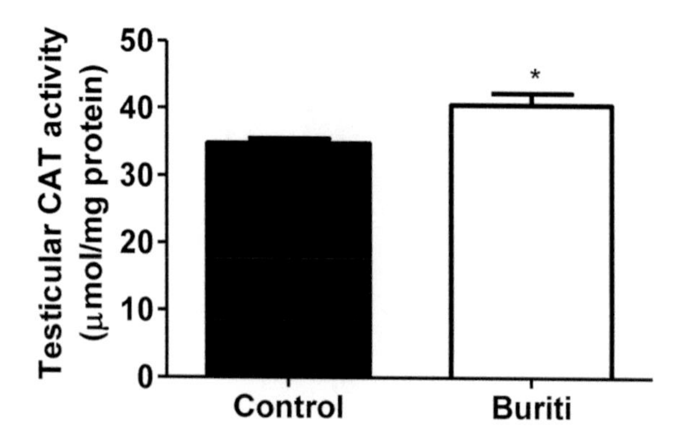


FIGURE 10.8 Testicular catalase activity.

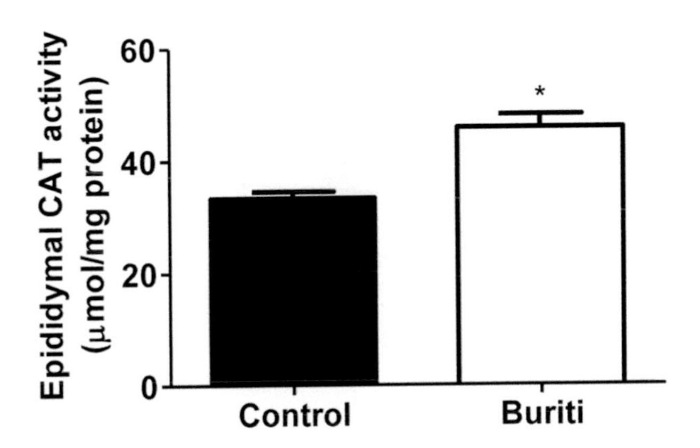


**FIGURE 10.9** Epididymal catalase activity.

### *10.3.6 ASSESSMENT OF TESTICULAR AND EPIDIDYMAL GSH ACTIVITY*

Figures 10.10 and 10.11 show testicular and epididymal GSH, respectively. Figure 10.10 shows that the control group (57.79µmol/mg ± 7.659µmol/mg) had a lower GSH concentration than in the testicular tissue of the Buriti oil fed group (74.67 µmol/mg ± 9.62µmol/mg). The epididymal GSH concentrations are represented in Figure 10.11, which shows that the control group (37.98µmol/mg ± 4.742µmol/mg) had a lower GSH concentration than the Buriti oil fed group (44.35µmol/mg ± 5.245µmol/mg).

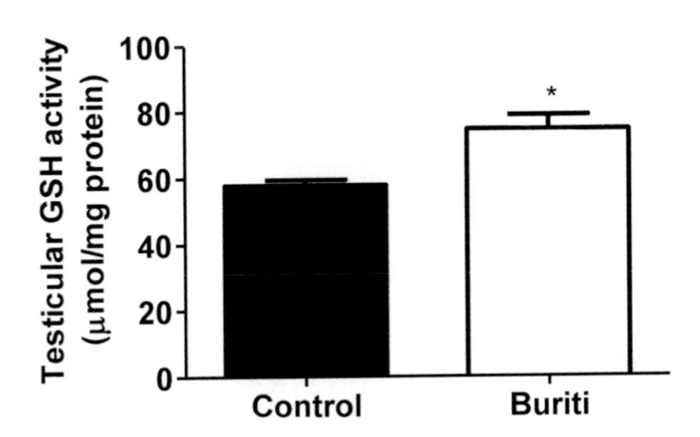


**FIGURE 10.10** Testicular GSH activity.

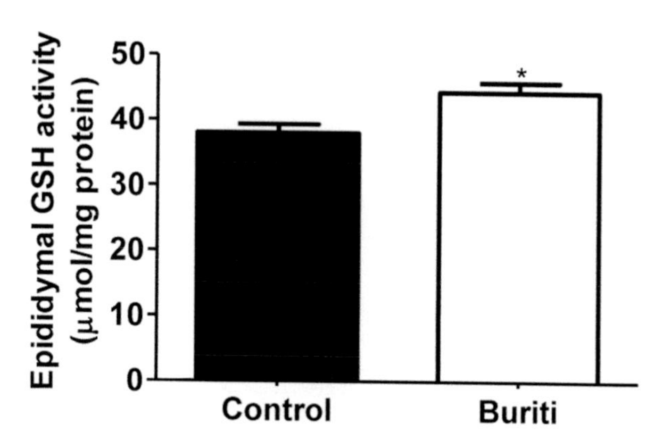


**FIGURE 10.11** Epididymal GSH activity.

### *10.3.7 EVALUATION OF PLASMA TESTOSTERONE CONCENTRATIONS*

The concentrations of plasma testosterone in the control group and Buriti oil supplemented group is shown in Figure 10.12. The plasma testosterone of the control group (1.472ng/ml ± 0.1896 ng/ml) was lower than the Buriti oil group (1.829ng/ml ± 0.2764 ng/ml) and the difference was significant ($p<0.05$).

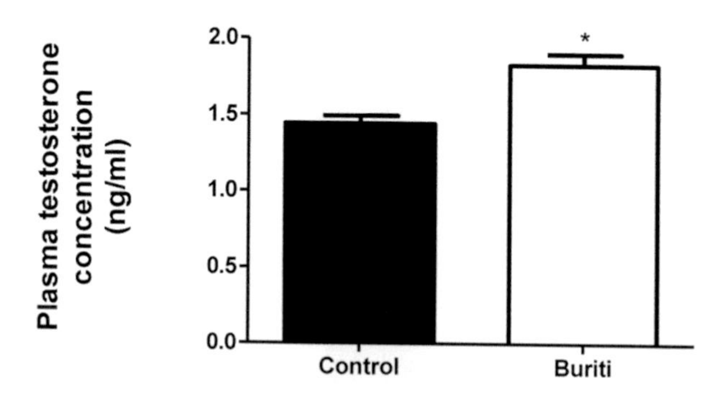


**FIGURE 10.12** Plasma testosterone concentration.

### *10.3.8 EVALUATIONOFPLASMAESTRADIOLCONCENTRATIONS*

The different estradiol concentrations between the control and Buriti oil fed group are illustrated in Figure 10.13. The Buriti oil fed group had a concentration of 1.75ng/ml, while the control was 1.6ng/ml. The results show that the difference in plasma estradiol concentrations between the control and Buriti oil fed group was insignificant (p>0.05).

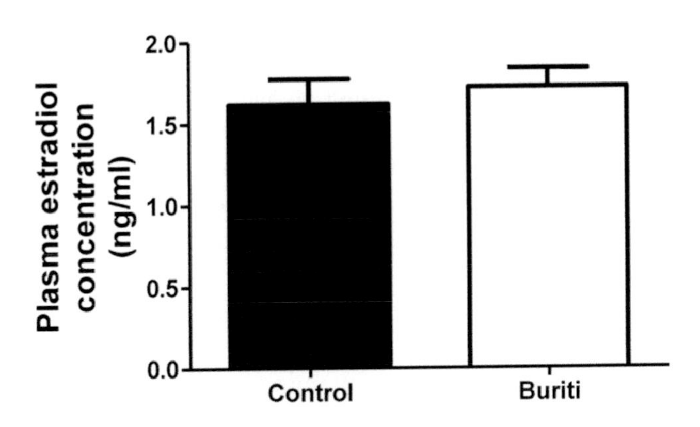


**FIGURE 10.13** Plasma estradiol concentration.

## 10.4 DISCUSSION

Several studies have investigated the effects of oral antioxidant supplementation on male reproductive function. Aboua and his colleagues [1] investigated the effects of red palm oil supplementation on male reproductive function while Awoniyi et al., [7] observed the effects of rooibos tea and green tea on male reproductive function. There are numerous studies conducted on Buriti oil consumption; however, no documented studies on the effects of Buriti oil on male reproductive function have been reported [6, 10]. According to the authors of this chapter, their study might be the first to shed light on the effects of Buriti oil on male Wistar rat reproductive function. Their study was a baseline study; therefore, no external sources of ROS contributed to oxidative stress state. Both groups experienced physiological oxidative stress from metabolic reactions that took place in the body. The body has antioxidant enzymes in place to combat ROS; however, these antioxidants are usually overpowered by ROS that continues to increase due to poor nutrition, exposure to pollutants, and aging. It is therefore important to supplement with dietary antioxidants such as Buriti oil.

### 10.4.1 EFFECTS OF BURITI OIL SUPPLEMENTATION ON BODY, TESTICULAR AND EPIDIDYMAL WEIGHT

There was no significant difference in weights between the Buriti oil supplemented rats and the control group. The control group weighed 338.2g ± 32.68g while the Buriti oil group weighed 336.1g ± 29.24g. And there was no particular trend in the manner in which the rats grew that could indicate that they had two different diets. Somnez [40] investigated the effects of vitamin C supplementation on sperm quality, LPO, testosterone concentration and body weights of male Wistar rats; and they concluded that vitamin C supplementation improved sperm quality but had no effect on body weights. Results of this chapter showed that supplementation with Buriti oil had no significant effect on body weights. This confirmed the study of Sonmez [40] who found no significant increase in body weights of male Wistar rats supplemented with vitamin C.

Jargar [21] investigated the effects of vitamin E supplementation on nickel-induced oxidative stress in rats; and they concluded that nickel induction increased oxidative stress and decreased body weights. Rats supplemented with vitamin E displayed an increase in body weight due to the antioxidative and protective effects of vitamin E. Moreover, Aboua [1] investigated the effects of RPO supplementation on sperm parameters, testicular, and epididymal tissue and body weights of male Wistar rats; and found that supplementation with RPO to rats improved sperm motility, protected against tissue LPO and increased the body weights. The results in this chapter showed no significant increase in body weights of rats supplemented with Buriti oil. Although Buriti oil contains a high vitamin E concentration; yet the results of this chapter did not confirm these above studies. Authors, therefore, postulate that the non-increase in body weight may be due to the very little to the non-existent concentration of tocotrienols in Buriti oil.

The testes and epididymis of experimental rats fed with Buriti oil were significantly larger compared to the control. These findings were similar to the findings of Yuce [50], who demonstrated that supplementation with cinnamon oil caused an increase in reproductive organ weights of rats. Jargar [21] supplemented diabetic rats with vitamin E, and they concluded that the supplemented group gained both body weight and organ weight because vitamin E protected tissues against LPO. The significant increase in epididymal and testicular weight of rats supplemented with Buriti oil indicated that the high vitamin E content of Buriti oil enabled it to protect against oxidative stress damage.

The study by authors of this chapter was a baseline study, and both the control and Buriti oil group had no externally induced ROS. Santos [39] reported that Buriti oil is rich in carotenoids and tocopherols, which protected against oxidative damage. Therefore, the significantly higher epididymal and testicular weight in the experimental Buriti oil fed group compared to the control group indicated that Buriti oil was able to protect organ tissues against oxidative damage, which can cause a decrease in organ size. Authors, therefore, postulate that antioxidant enzymes in the control group might have alleviated ROS that was produced from metabolic reactions in the body while the Buriti oil supplemented group was supplied extra protection with carotenoids and tocopherols and caused the cells to produce more endogenous antioxidant enzymes.

Aboua [1] found that the group fed with RPO had heavier epididymal and testicular weights than the group that was induced with oxidative stress. Authors of this chapter confirmed the data that rat supplemented with Buriti oil showed an increase in weight of testis and epididymis compared to the control group.

### *10.4.2 EFFECTS OF BURITI OIL SUPPLEMENTATION ON LIPID PEROXIDATION (LPO) OF TESTICULAR AND EPIDIDYMAL TISSUE*

The level of MDA reflects the amount of testicular damage caused by ROS. The damage caused by ROS elevated the MDA levels in the tissues. Studies have demonstrated that LPO disrupts lipid membranes and tissues in organs such as the heart, testis, and lung [20, 26, 47]. Antioxidants and antioxidant-rich foods have been used to lower LPO caused by oxidative stress. Alinde [5] showed that dietary supplementation with RPO could decrease LPO in cardiac tissues.

In this chapter, LPO was quantified by measuring the formation of thiobarbituric acid reactive substances (TBARS) and was expressed as nmol of MDA formed/mg of testicular/epididymal tissue. The amount of MDA was significantly higher in the epididymal and testicular tissues of the control group compared to the experimental group.

Buriti oil is rich in lipid-soluble antioxidants (carotenoids and tocopherols), which help to protect against LPO [39]. Carotenoids remove free radicals before they interact with the cell membrane by transferring a hydrogen atom with a single electron to a free radical. Carotenoids have a high free radical scavenging rate, and carotenoid-derived radicals have been shown to be stable and relatively unreactive. Carotenoid-derived radicals may further undergo bimolecular decay to generate non-radical products [9, 12].

Heikal [19] concluded that green tea extract protects against LPO caused by oxidative stress. Rats used in their study were exposed to pesticides, which caused LPO, decreased testicular weight, and impaired sperm motility. They found that rats fed with water and SRC experienced more LPO than the group supplemented with green tea extract. Their results also indicated that soluble polyphenols in green tea extract protected testicular and epididymal tissue by scavenging ROS that can potentially cause LPO. Results in this chapter agree with those by Heikal [19], who found that tocopherols and carotenoids in Buriti oil provided protection against physiological ROS and LPO.

Testicular and epididymal tissues from the control group showed a significant increase in LPO compared to the Buriti oil group. Buriti oil-treated rats presented a significant decrease in LPO in their testicular and epididymal tissues, thus indicating that carotenoids present in Buriti oil were able to protect against oxidative stress by scavenging free radicals hence protecting the tissue from LPO. Findings by authors of this chapter were similar to Aboua [1], who concluded that supplementation with carotenoid, and tocopherol rich RPO protected testicular and epididymal tissues from LPO. Although Aboua induced the oxidative stress; yet there search in this chapter is a baseline study, and no oxidative stress was artificially induced. In the research study of this chapter, the control and experimental Buriti oil-fed groups had only physiological ROS sourced from metabolic reactions in the cells. The increase in MDA production in the control group indicated that the group had less protection against free radicals hence experiencing more LPO than the experimental Buriti oil fed group. This may also show that carotenoids in Buriti oil were able to quench free radicals before they could interact with lipid membranes hence protecting the tissues against LPO.

### *10.4.3 EFFECTS OF BURITI OIL SUPPLEMENTATION ON GSH, SOD, AND CATALASE (CAT) ON TESTICULAR AND EPIDIDYMAL TISSUE*

SOD, CAT, and GSH can maintain ROS scavenging potential in the male reproductive tract and protect against oxidative stress. The level of production and activity of these antioxidant enzymes is affected by physiological and pathological conditions. Decreased levels of these antioxidant enzymes can lead to decreased protection of tissues against oxidative stress damage. Each antioxidant enzyme has a unique mechanism of action against free radicals [3].

Several researchers have demonstrated that dietary supplementation with antioxidant-rich oils and foods can increase antioxidant enzyme activity. Aboua [1] showed that dietary supplementation with red palm oil improved the antioxidant enzyme activities in the male reproductive function of oxidative stress-induced rats. Alinde [5] also showed that supplementation with red palm oil increased the SOD, CAT, and GSH activities in cardiac tissues. Awoyini [7] concluded that supplementation with green tea and rooibos tea improved the epididymal sperm quality and increased activity of antioxidant enzymes in oxidative stress-induced rats. Authors of this chapter have confirmed all findings in this paragraph. However, Luo [30] showed that SOD concentrations were significantly decreased in ROS injured rats.

SOD is the main enzyme involved in the dismutation of $O_2^-$ to $H_2O_2$ and oxygen. In this chapter, there was a significant increase in epididymal and testicular SOD concentration in rats supplemented with dietary Buriti oil compared to the control group. This indicates that the SOD available was not enough to scavenge all ROS present, in physiological conditions. The increased SOD concentration in testicular and epididymal tissue of the Buriti oil-fed rats indicated that antioxidants in Buriti oil increased the antioxidant activity of SOD.

CAT neutralizes $H_2O_2$ to water and superoxide. In this chapter, CAT activity was significantly increased in rats supplemented with Buriti oil. Both testicular and epididymal CAT activity were increased. The control group had significantly low CAT activity due to an overpowering CAT by ROS the tissues. The increase in CAT activity indicated that Buriti oil has antioxidant properties to scavenge free radicals. Findings in this chapter confirmed results by Ayeleso [8], who observed the increase in testicular CAT enzyme activity due to supplementation with RPO and rooibos tea.

The function of GPx is to remove $H_2O_2$ generated by metabolic action. The activity of GPx depends on GSH concentration. GSH scavenges peroxynitrite to HO and converts $H_2O_2$ to water with the help of GPx. Results in this chapter indicate that dietary Buriti oil supplementation increased GSH activity in both testicular and epididymal tissues. The reduced GSH concentration in the control group of testicular and epididymal tissue indicated a high participation of reduced glutathione in $H_2O_2$ peroxidation [28]. Buriti oil is rich in tocopherols, which elevated GSH in epididymal and testicular tissues.

The findings by authors of this chapter agree with those by Abouaand his colleagues [1], who found that RPO supplementation improved GSH activity in epididymal and testicular tissues of rats induced with oxidative rats. Data by authors of this chapter also confirmed the study by Kanter et al.,[24], who investigated the protective effects of quercetin against oxidative stress in diabetic rats and concluded that quercetin increased GSH activity in testis. Alinde [5] showed that supplementation with RPO decreased oxidative stress and increased GSH activity in cardiovascular tissue.

Oxidative stress in this chapter was not induced; therefore, it indicates that dietary Buriti oil supplementation can increase the activity of SOD, CAT, and GSH in a baseline state. The epididymal control groups indicated that sperms did indeed experience oxidative stress when in transit. Authors of this chapter, therefore, concluded that Buriti oil improved the activity of antioxidant enzymes and protected against oxidative stress.

### *10.4.4 EFFECTS OF BURITI OIL SUPPLEMENTATION ON TESTOSTERONE AND ESTRADIOL*

Testosterone is said to be the key hormone for spermatogenesis regulation. Normal concentrations of testosterone influence sexual behavior, muscle mass, energy, cardiovascular health, and bone integrity [32]. Buriti oil is rich in vitamin E, and many studies have shown the importance of vitamin E supplementation in increasing testosterone levels [15, 48]. Estradiol is a natural estrogen found in high concentrations in females and very low concentrations in males [49]. Testosterone and estrogen are highly involved in the regulation of fertility. The enzyme aromatase converts testosterone to estradiol [45]. Estrogens were female hormones, and their presence in males remained questionable, but Dorrington [11] later demonstrated the ability of testis to produce estrogens.

Several studies investigated the effects of vitamin E supplementation on testosterone, estrogen, or both, and they all concluded that it increases the concentration of the hormones. Jargar [21] supplemented male rats with vitamin E and found that it protected testicular tissue and significantly increased testosterone levels. They concluded that tocopherols in vitamin E protected and increased Leydig cell numbers hence increasing testosterone release. Findings of this chapter support the results by Jargar [21], who indicated that supplementation with vitamin E rich Buriti oil significantly increased testosterone concentration and protected testicular tissue.

Garcia [15] conducted an animal study using rats to demonstrate the effects of silver nanoparticles (NPs) on Leydig cell function and testosterone levels. They hypothesized that silver NPs would cause oxidative stress damage on Leydig cells and decrease testosterone levels. However, they found that silver NPs increased Leydig cell size and testosterone levels.

The results in this chapter are contrary to those by Garcia [15], who indicated that physiological oxidative stress damage in the control group caused a decrease of testosterone concentrations and that Buriti oil is rich in antioxidants that may improve testosterone concentration in plasma.

Hartman [18] proved that an increase in serum α-tocopherol resulted in an increase in serum androgens and estrogens in older men. However, the Alpha-Tocopherol Beta-Carotene (ATBC) cancer prevention study contradicts Hartman [18] as it is concluded that serum α-tocopherol is inversely associated with testosterone levels among men who received vitamin E supplements. The current study, however, partially agrees with Hartman [18] study as authors of this chapter found that the Buriti oil supplemented group had a significant increase in testosterone compared to the control group, but the estradiol concentration was not significantly increased.

Turk [48] found that plasma testosterone levels in male rats were increased in the group on pomegranate juice instead of water. In the study of the chapter, the Buriti oil supplemented group showed an increase in testosterone levels, and this shows that tocopherols in the oil offered protection to testicular tissue and Leydig cells, which secrete testosterone [39].

Animal studies show that estrogen deficiency or estrogen resistance impairs sperm motility and can cause infertility [22]. The effect of antioxidant supplementation on estradiol in the study by authors of this chapter was not as hypothesized in the beginning. Buriti oil supplementation did not have a significant effect on the estradiol hormone concentrations. The difference in estradiol concentrations between the control group and Buriti oil group was not significant ($p>0.05$). Therefore, the results in this chapter contradict those by Selvakumar [41], who showed that alpha-tocopherol supplementation in male rats increased estradiol concentrations.

## 10.5 BURITI OIL: POSSIBLE MECHANISM OF ACTION

Buriti oil has a high content of tocopherols, carotenoids, and mono-unsaturated fatty acids (MUFAs), which enabled it to provide protection against oxidative stress in testicular and epididymal tissues. There are different mechanisms by which these contents act in order to protect tissues.

Authors of this chapter postulated that Buriti oil can protect against LPO and increase antioxidant enzyme activity due to its high oxidative stability. Patel [38] reported that Buriti oil consists of both tocopherols and carotenoids, which may increase its oxidative stability. Sen [42] stated that tocopherols are saturated while tocotrienols are unsaturated. The ability of tocopherols to interact with free radicals is increased by their structure, which consists of a Chromatol ring and a 15-Carbon tail. Nishio [35] compared the rate of tocopherol and tocotrienol uptake by cells and concluded that tocopherols have a slower uptake rate and provide more protection against LPO than tocotrienols despite the slow rate of uptake by cells.

The high antioxidant capacity of Buriti oil is due to the presence of carotenoids and tocopherols. These antioxidants are absorbed into the body at different rates and different metabolic routes.

Tocopherols and tocotrienols have different absorption rates [2]. Fairus et al., [13] administered equal amounts of tocotrienol and tocopherol extracted from palm oil to healthy subjects. They found that tocotrienols were mainly detected in the high-density lipoprotein (HDL) cholesterol at 4 to 8 hours before clearance; compared to tocopherols that were distributed equally in all the lipoprotein fractions. Tocopherols in the study by Fairus [13] were detected in plasma even after 24 hours, suggesting that tocotrienols are absorbed faster than tocopherols and that tocotrienols may go through alternative metabolism pathway that requires further investigation. Another possible explanation of the low tocotrienol concentration in plasma is the low affinity of α-tocopherol transport protein (α-TTP) for tocotrienols [16]. Authors of this chapter, therefore, postulated that supplementation with Buriti oil offers longer protection against free radicals because it is rich in tocopherols, which can be detected in plasma for a longer period of time compared to tocotrienols.

Packer [36] suggested that tocotrienols are more effective than tocopherols in reducing LPO because they are absorbed faster and have higher intramembrane activities. However, Singh et al., [43] demonstrated that tocotrienols disappeared in the plasma within 24 hours of consumption while tocopherols were increased, and they suggested that tocotrienols were converted to tocopherols by hydrogenation. Alpha-tocotrienol was suggested to be secreted by small HDL particles whilst alpha-tocopherol is exclusively secreted in chylomicrons. The small HDL particles selectively distributed α-tocotrienol to organs and tissues high in adipose content: epididymal fat, perirenal fat, and skin. In contrast, α-tocopherol was reported to be more evenly distributed because LDL receptors are available in all tissues. The even distribution of alpha tocopherols makes Buriti oil powerful free radical scavenging oil in a variety of tissues, and this was confirmed by the protection of lipid membranes in testicular and epididymal tissues.

The β-carotene is a dietary source of vitamin A for humans, and its bioavailability depends on food processing techniques, size, and dose. In humans, the absorption of β-carotene from plant sources ranges from 5 to

65%. Intestinal conversion of β-carotene to vitamin A decreases when the oral dose is increased, and this makes β-carotene a safe source of vitamin A. Buriti oil is, therefore, a safe source of vitamin A as it contains a high concentration of β-carotene [25].

The *in vivo* intestinal absorption of carotenoids takes place through a series of steps. Carotenoids are up-taken by intestinal mucosal cells and are incorporated into chylomicrons where they are transported to various cells in the body. β-carotene is fat-soluble, and it scavenges free radicals in lipid membranes better than non-lipid soluble antioxidants such as vitamin C [17]. The high concentration of β-carotene in Buriti oil increased its ability to protect against LPO, and this was characterized by the low MDA concentration in the experimental group.

There is a high concentration of MUFAs in Buriti oil [39]. MUFAs play an important role in the maintenance of serum lipid profiles and cell membrane [6, 37]. Consumption of MUFAs reduces systemic and cellular oxidative stress.

Buriti oil is rich in carotenoids that are fat-soluble phytochemicals with provitamin A and possess antioxidant role. Carotenoids have a long-chain conjugated polyene structure, which makes them susceptible to oxidation, therefore, increasing their interactions with free radicals. Carotenoids in Buriti oil were able to react with free radicals before they interacted with testicular and epididymal lipid membranes, thereby protecting against oxidative stress [25].

Bone density and muscle mass both influence the bodyweight [33]. A study conducted by Maniam [31] concluded that palm oils are rich in tocotrienols that are known to decrease osteoporosis, therefore, increasing bone density. Increased bone density contributes to bodyweight increase. Tocotrienols have been proven to offer more protection against free radical-induced damage on rat bone. Santos [39] reported that Buriti oil contains a very low concentration of tocotrienols compared to other palm oils. Rats supplemented with Buriti oil in this chapter did not have a significant weight gain; therefore it is postulated that bones of Buriti oil-fed rats were not protected against osteoporosis due to lack of tocotrienols; and this caused lack of body weight increase. The study by Aboua et al., [1] showed that RPO supplementation to rats increased their body weight. The outcome in this chapter was different from theirs, because RPO is rich in both tocopherols and tocotrienols while Buriti oil is rich in tocopherols but low on tocotrienols.

This chapter confirms that testosterone concentrations were increased after supplementation with tocopherol-rich Buriti oil. Supplementation of rats with tocopherol for one month before exposure to oxidative stress significantly prevents the stress-induced decrease of testosterone [29, 46]. As rats age, they lose their muscle mass. Sinha [44] supplemented aged rats with testosterone and found that testosterone supplementation increased muscle mass. Testosterone concentration in the study by authors of this chapter was increased, and this means that muscle mass was also increased; however, it did not show significant increases in body weights of rats fed with Buriti oil. This lack of body weight increase could be due to low concentrations of estradiol, which could have resulted in low bone density. Other researchers have demonstrated the role of estradiol in protecting against bone degradation and resorption [14, 23, 27]. From the findings of this chapter and previous literature, it can safely be postulated that adequate levels of both testosterone and estradiol are necessary for an increase in body weight.

Santos [39] reported that Buriti oil has a low concentration of tocotrienols and a high concentration of tocopherols; it was also confirmed by authors of this chapter. Muhammad et al., [34] found that tocotrienol supplementation was able to increase estrogen levels in rats that suffered from severe osteoporosis; and their study, therefore, suggested that tocotrienol deficiency can decrease estrogen levels. From the information here, it can, therefore, be postulated that low tocotrienol concentrations in Buriti oil caused the non-significant increase of estradiol concentration. Moreover, Muhammad et al., [34] concluded that estrogen reduces bone resorption; therefore, its deficiency can cause increased osteoporosis. The study in this chapter suggests that body weights of rats fed with Buriti oil did not increase as hypothesized because there was not enough protection against bone degeneration.

The Alpha-tocopherol has a direct stimulatory effect on enzymes of gonadal steroid biosynthesis and can exert modulatory action on gonadotropins synthesis and secretion [4]. The study in this chapter indicated that rats on Buriti oil supplementation had increased testosterone concentrations in the plasma, and this could be due to the high concentrations of alpha tocopherols in Buriti oil. Tocopherols protected cells against oxidative stress by scavenging free radicals. Leydig cells in the Buriti oil fed group were protected against oxidative stress hence were able to secrete more testosterone compared to the control group.

## 10.6 SUMMARY

The study in this chapter explored how Buriti oil can modulate oxidative stress biomarkers and hormonal function in an *in vitro* experimental animal model using male Wistar rats. The findings in this chapter indicate that: (1) Dietary supplementation of Buriti oil could have a protective effect against physiological oxidative stress and increase testosterone secretion, LPO damages by restoring MDA levels of oxidative stress and increasing antioxidant enzyme activities (SOD, CAT, and GSH); (2) Antioxidants in Buriti oil can influence Leydig cell function hence may cause an increase of testosterone in the plasma, and estradiol concentration in plasma was not increased by Buriti oil consumption.

More studies with advanced technologies are recommended to investigate the effects of Buriti oil supplementation on all male reproduction parameters.

## ACKNOWLEDGMENTS

The financial assistance of the Cape Peninsula University of Technology University Research Fund (URF) towards this research is acknowledged. The Oxidative Stress Research Centre (OSRC) facilities were used for the determination of antioxidant content and activity of Buriti oil and blood sample analysis under the supervision of Mr. F. Rautenbach (OSRC Laboratory Manager). Authors thank the Stellenbosch University for housing our rats during the study.

## KEYWORDS

- **β-carotene**
- **antioxidant enzyme activities**
- **Buriti oil**
- **epididymal status**
- **estradiol concentrations**
- **lipid peroxidation damage**
- **male Wistar rats**
- **plasma testosterone**
- **scavenging free radicals**
- **sperm motility**
- **testosterone secretion**

## REFERENCES


1. Aboua, Y., Brooks, N., Mahfouz, R., Agarwal, A., & Du Plessis, S., (2012). Red palm oil diet can reduce the effects of oxidative stress on rat spermatozoa. *Andrologia, 44*, 32–40.
2. Abuasal, B. S., Qosa, H., Sylvester, P. W., & Kaddoumi, A., (2012). Comparison of the intestinal absorption and bioavailability of γ-tocotrienol and α-tocopherol: *In vitro, in situ* and *in vivo* studies. *Biopharmaceutics & Drug Disposition, 33*(5), 246–256.
3. Agarwal, A., Gupta, S., & Sikka, S., (2006). The role of free radicals and antioxidants in reproduction. *Current Opinion in Obstetrics and Gynecology, 18*(3), 325–332.
4. Al-Damegh, M. A., (*2014*). Stress-induced changes in testosterone secretion in male rats: Role of oxidative stress and modulation by antioxidants. *Open Source Online Journal of Animal Sciences, 4*(2), 70–77.
5. Alinde, O. B. L., Esterhuyse, A. J., & Oguntibeju, O. O., (2012). Role of reactive oxygen species in the pathogenesis of cardiovascular disease. *Scientific Research and Essays, 7*(49), 4151–4159.
6. Aquino, J. D. S., Pessoa, D. C. N. D. P., Oliveira, C. E. V. D., Cavalheiro, J. M. O., & Stamford, T. L. M., (2012). Making cookies with buriti oil (*Mauritia flexuosa* L.): Alternative source of dietary vitamin A in school meals. *Revista de Nutrição, 25*(6), 765–774.
7. Awoniyi, D. O., Aboua, Y. G., Marnewick, J. L., Du Plesis, S. S., & Brooks, N. L., (2013). Protective effects of rooibos (*Aspalathuslinearis*), green tea (*Camellia sinensis*) and commercial supplements on testicular tissue of oxidative stress induced rats. *African Journal of Biotechnology, 10*(75), 17317–17322.
8. Ayeleso, A. O., Oguntibeju, O. O., Aboua, Y. G., & Brooks, N. L., (2014). Effects of red palm oil and rooibos on sperm motility parameters in streptozotocin induced diabetic rats. *African Journal of Traditional, Complementary and Alternative Medicines, 11*(5), 8–15.
9. Böhm, F., Edge, R., & Truscott, G., (2012). Interactions of dietary carotenoids with activated (singlet) oxygen and free radicals: Potential effects for human health. *Molecular Nutrition & Food Research, 56*(2), 205–216.
10. Darnet, S. H., Silva, L. H. M. D., Rodrigues, A. M. D. C., & Lins, R. T., (2011). Nutritional composition, fatty acid and tocopherol contents of buriti (*Mauritia flexuosa*) and patawa (*Oenocarpusbataua*) fruit pulp from the amazon region. *Food Science and Technology (Campinas), 31*(2), 488–491.
11. Dorrington, J. H., Fritz, I. B., & Armstrong, D. T., (1978). Control of testicular estrogen synthesis. *Biology of Reproduction, 18*(1), 55–64.
12. Everett, S. A., Dennis, M. F., Patel, K. B., Maddix, S., Kundu, S. C., & Willson, R. L., (1996). Scavenging of nitrogen dioxide, thiyl, and sulfonyl free radicals by the nutritional antioxidant-carotene. *Journal of Biological Chemistry, 271*(8), 3988–3994.
13. Fairus, S., Nor, R. M., Cheng, H. M., & Sundram, K., (2006). Postprandial metabolic fate of tocotrienol-rich vitamin E differs significantly from that of alpha-tocopherol. *American Journal of Clinical Nutrition, 84*(4), 835–842.
14. Falahati-Nini, A., Riggs, B. L., Atkinson, E. J., O'Fallon, W. M., Eastell, R., & Khosla, S., (2000). Relative contributions of testosterone and estrogen in regulating bone resorption and formation in normal elderly men. *Journal of Clinical Investigation, 106*(12), 1553.

15. Garcia, T. X., Costa, G. M., Franca, L. R., & Hofmann, M. C., (2014). Sub-acute intravenous administration of silver nanoparticles in male mice alters Leydig cell function and testosterone levels. *Reproductive Toxicology*, *45*, 59–70.
16. Gee, P. T., (2011). Vitamin E - essential knowledge for supplementation. *Lipid Technology*, *23*(4), 79–82.
17. Harrison, E. H., (2012). Mechanisms involved in the intestinal absorption of dietary vitamin A and provitamin A carotenoids. *Biochimica et Biophysica Acta (BBA) Molecular and Cell Biology of Lipids*, *1821*(1), 70–77.
18. Hartman, T. J., Dorgan, J. F., Virtamo, J., Tangrea, J. A., Taylor, P. R., & Albanes, D., (1999). Association between serum α-tocopherol and serum androgens and estrogens in older men. *Nutrition and Cancer*, *35*(1), 10–15.
19. Heikal, T. M., Mossa, A. H., Ibrahim, A. W., & Abdel-Hamid, H. F., (2014). Oxidative damage and reproductive toxicity associated with cyromazine and chlorpyrifos in male rats: The protective effects of green tea extract. *Research Journal of Environmental Toxicology*, *8*(2), 53–67.
20. Ibrahim, R. Y. M., & Ghoneim, M. A. M., (2014). Study of some biochemical and molecular changes induced by radiation hormesis in testicular tissues of male rats. *International Journal*, *2*(7), 397–407.
21. Jargar, J. G., Yendigeri, S., Dhundasi, S. A., & Das, K. K., (2014). Protective effect of Vitamin E (a-tocopherol) on nickel-induced alteration of testicular pathophysiology in alloxan-treated diabetic rats. *International Journal of Clinical and Experimental Physiology*, *1*(4), 290.
22. Joseph, A., Shur, B. D., & Hess, R. A., (2011). Estrogen, efferent ductules, and the epididymis. *Biology of Reproduction*, *84*(2), 207–217.
23. Juntal, M. C., Krust, A., Chambon, P., & Mark, M., (2008). Sterility and absence of histopathological defects in non-reproductive organs of a mouse ERβ-null mutant. *Proceedings of the National Academy of Sciences*, *105*(7), 2433–2438.
24. Kanter, M., Aktas, C., & Erboga, M., (2012). Protective effects of quercetin against apoptosis and oxidative stress in streptozotocin-induced diabetic rat testis. *Food and Chemical Toxicology*, *50*(3), 719–725.
25. Karuppanapandian, T., Moon, J. C., Kim, C., Manoharan, K., & Kim, W., (2011). Reactive oxygen species in plants: Their generation, signal transduction, and scavenging mechanisms. *Australian Journal of Crop Science*, *5*(6), 230–235.
26. Kharwar, R. K., & Haldar, C., (2012). Daily variation in antioxidant enzymes and lipid peroxidation in lungs of a tropical bird *Perdiculaasiatica*: Role of melatonin and nuclear receptor RORα. *Comparative Biochemistry and Physiology Part A: Molecular & Integrative Physiology*, *162*(4), 296–302.
27. Khosla, S., Melton, L. J., & Riggs, B. L., (2011). The unitary model for estrogen deficiency and the pathogenesis of osteoporosis. *Journal of Bone and Mineral Research*, *26*(3), 441–451.
28. Krishnamoorthy, G., Selvakumar, K., & Elumalai, P., (2011). Protective role of lycopene on polychlorinated biphenyls (Aroclor 1254)-induced adult rat Sertoli cell dysfunction by increased oxidative stress and endocrine disruption. *Biomedical Preventative Nutrition, 1*, 116–125.
29. Lodhi, G. M., Latif, R., Hussain, M. M., Naveed, A. K., & Aslam, M., (2014). Effect of ascorbic acid and alpha tocopherol supplementation on acute restraint stress induced changes in testosterone, corticosterone and nor epinephrine levels in male Sprague Dawley rats. *Journal of Ayub Medical College, 26*(1), 7–10.

30. Luo, Q., Li, Z., Huang, X., Yan, J., Zhang, S., & Cai, Y. Z., (2006). Lyciumbarbarum polysaccharides: Protective effects against heat-induced damage of rat testes and H 2 O 2-induced DNA damage in mouse testicular cells and beneficial effect on sexual behavior and reproductive function of hemicastrated rats. *Life Sciences*, *79*(7), 613–621.
31. Maniam, S., Mohamed, N., Shuid, A. N., & Soelaiman, I. N., (2008). Palm tocotrienol exerted better antioxidant activities in bone than α-tocopherol. *Basic & Clinical Pharmacology & Toxicology*, *103*(1), 55–60.
32. Menke, A., Guallar, E., Rohrmann, S., Nelson, W. G., Rifai, N., Kanarek, N., & Platz, E. A., (2010). Sex steroid hormone concentrations and risk of death in US men. *American Journal of Epidemiology*. E-article: doi: 10.1093/aje/kwp415415.
33. Mithal, A., Bonjour, J. P., Boonen, S., Burckhardt, P., Degens, H., & Fuleihan, G. E. H., (2013). IOF-CSA Nutrition Working Group, Impact of nutrition on muscle mass, strength, and performance in older adults. *Osteoporosis International*, *24*(5), 1555–1566.
34. Muhammad, N., Luke, D. A., Shuid, A. N., Mohamed, N., & Soelaiman, I. N., (2013). Tocotrienol supplementation in postmenopausal osteoporosis: Evidence from a laboratory study. *Clinics*, *68*(10), 1338–1343.
35. Nishio, K., Horie, M., Akazawa, Y., Shichiri, M., Iwahashi, H., Hagihara, Y., & Niki, E., (2013). Attenuation of lipopolysaccharide (LPS)-induced cytotoxicity by tocopherols and tocotrienols. *Redox Biology*, *1*(1), 97–103.
36. Packer, L., Weber, S. U., & Rimbach, G., (2001). Molecular aspects of tocotrienol antioxidant action and cell signaling. *Journal of Nutrition*, *131*, 369S-373S.
37. Pantsi, W. G., Bester, D. J., Esterhuyse, A. J., & Aboua, G., (2014). Dietary antioxidant properties of vegetable oils and nuts: The race against cardiovascular disease progression, Chapter 9. In: Oluwafemi, O., (ed.), *Antioxidant-Antidiabetic Agents and Human Health* (pp. 209–238). ISBN 978–953–51–1215–0, IntechOpen.com.
38. Patel, V., Rink, C., Khanna, S., & Sen, C. K., (2011). Tocotrienols: The lesser known form of natural vitamin E. *Indian Journal of Experimental Biology*, *49*(10), 732–738.
39. Santos, M. F. G., Alves, R. E., & Ruíz-Méndez, M. V., (2013). Minor components in oils obtained from Amazonian palm fruits. *Grasas Aceites*, *64*(5), 531–536.
40. Sönmez, M., Türk, G., & Yüce, A., (2005). The effect of ascorbic acid supplementation on sperm quality, lipid peroxidation and testosterone levels of male Wistar rats. *Theriogenology*, *63*(7), 2063–2072.
41. Selvakumar, K., Banu, L. S., Krishnamoorthy, G., Venkataraman, P., Elumalai, P., & Arunakaran, J., (2011). Differential expression of androgen and estrogen receptors in PCB (Aroclor 1254)-exposed rat ventral prostate: Impact of alpha-tocopherol. *Experimental and Toxicologic Pathology*, *63*(1), 105–112.
42. Sen, C. K., Khanna, S., Roy, S., & Packer, L., (2000). Molecular basis of vitamin E action tocotrienol potently inhibits glutamate-induced pp60c-Src Kinase activation and death of ht4 neuronal cells. *Journal of Biological Chemistry*, *275*(17), 13049–13055.
43. Singh, V. K., Beattie, L. A., & Seed, T. M., (2013). Vitamin E: Tocopherols and tocotrienols as potential radiation countermeasures. *Journal of Radiation Research*, *54*(6), 973–988.
44. Sinha, I., Sinha-Hikim, A. P., Wagers, A. J., & Sinha-Hikim, I., (2014). Testosterone is essential for skeletal muscle growth in aged mice in a heterochronic parabiosis model. *Cell and Tissue Research*, *357*(3), 815–821.
45. Smith, L. B., & Saunders, P. T., (2011). The skeleton: New controller of male fertility? *Cells*, *144*(5), 642–643.

46. Smith, L. B., & Walker, W. H., (2014). The regulation of spermatogenesis by androgens. *Seminars in Cellular & Developmental Biology, 30*, 2–13.
47. Thamahane-Katengua, E. T. M., (2013). *Effect of Rooibos and Red Palm Oil Supplementation, Alone or in Combination, on Cardiac Function After Exposure to Hypertension and Inflammation in an Ischaemia/Reperfusion Injury Model* (p. 243). Unpublished DTech Thesis, Cape Peninsula University of Technology, Cape Town.
48. Türk, G., Sönmez, M., Aydin, M., Yüce, A., Gür, S., Yüksel, M., & Aksoy, H., (2008). Effects of pomegranate juice consumption on sperm quality, spermatogenic cell density, antioxidant activity and testosterone level in male rats. *Clinical Nutrition, 27*(2), 289–296.
49. Wildman, R. P., Wang, D., Fernandez, I., Mancuso, P., Santoro, N., Scherer, P. E., & Sowers, M. R., (2013). Associations of testosterone and sex hormone binding globulin with adipose tissue hormones in midlife women. *Obesity, 21*(3), 629–636.
50. Yüce, A., Türk, G., Çeribaşi, S., Sönmez, M., Çiftçi, M., & Güvenç, M., (2013). Effects of cinnamon (*Cinnamomumzeylanicum*) bark oil on testicular antioxidant values, apoptotic germ cell and sperm quality. *Andrologia, 45*(4), 248–255.

CHAPTER 11

# HERBAL MANAGEMENT FOR POLYCYSTIC OVARIAN SYNDROME

HUMA BADER-UL AIN, FARHAN SAEED,
MUHAMMAD UMAIR ARSHAD, and HAFIZ ANSAR RASUL SULERIA

## ABSTRACT

Polycystic ovarian syndrome (PCOS) is one of the most common reproductive endocrine aberrations, and it is characterized by a broad spectrum of clinical disorders including type 2 diabetes, hyperandrogenism, menstrual abnormalities, polycystic ovary, and obesity. PCOS issue is rising due to an array of factors, such as insulin resistance, obesity, hormonal disturbances, gene predisposition, stress, fatigue, and oral contraceptives, but the exact etiology of this syndrome is still unknown. This chapter focuses on the review of etiology, epidemiology of PCOS along with related disorders and diagnostic tools. Particularly the combined effect of different types of dietary interventions and lifestyle management is the limelight of this chapter.

## 11.1 INTRODUCTION

Polycystic ovarian syndrome (PCOS) is one of the most common reproductive endocrine disorders with a broad spectrum of clinical disorders (i.e., disturbance in the menstrual cycle, obesity, excess of androgen, infertility, ovary cyst, insulin resistance and impaired glucose tolerance in women of reproductive age). Among these aberrations, the most commonly seen disorder is a menstrual disturbance that includes oligomenorrhea, amenorrhea, and prolonged erratic menstrual bleeding. About 50% of women with PCOS have insulin resistance and obesity problems [13, 32, 38, 73, 77, 100, 123, 125, 140, 147, 153]. PCOS can be defined as a clinical aberration with heterogeneously excessive androgen along with varying degrees of reproductive and metabolic abnormalities. There are varying clinical expressions of PCOS including oligo-ovulation (infrequent or irregular ovulation), an ovulation (failure of the ovary to release ovary ova over a period generally exceeding three months), and hyperandrogenism with polycystic ovaries [12, 74]. According to the 2003 Rotterdam criteria, polycystic ovaries are defined as either 12 or more follicles measuring 2–9 mm in diameter, and/or increased the ovarian volume of>10 ml [33, 84].

The worldwide prevalence of PCOS is 5–10% [52, 147], while few recent investigations showed that PCOS prevalence is about 9–21% [62, 72], whereas the prevalence of polycystic ovary is about 40.9% in Pakistani infertile women [15]. A recorded percentage of women having PCOS is approximately 15–20% and 1 in 10, 1 in 15, and 1 in 20 women of child-bearing age are affected [116, 136]. Recent studies showed that PCOS is affecting about 1 in 5 women of reproductive age [37, 73]. About 7% of females are affected by PCOS in the United States [153]. According to the National Institutes of Health, Rotterdam, and Androgen Excess-PCOS criteria, overall prevalence of PCOS is 6%, 10%, and 10%, respectively [24], whereas an estimate of PCOS in the community suggests a prevalence of 8.7%, 17.8%, and 12.0%, respectively among Australian women [90]. However, among Iranian women, prevalence is 7%, 15.2% and 7.92%, respectively [97], while among Turkish women; prevalence is 6.1%, 19.9% and 15.3%, respectively [156]. In Indian women, the prevalence of PCOS is about 9.13% [7, 24, 115]. Table 11.1 summarizes the PCOS, including signs and symptoms, related disorders, causes, and management.

## 11.2 SIGNS AND SYMPTOMS OF PCOS

Patients with PCOS often show multiple gynecologic, metabolic, and dermatologic manifestations with some principal signs of acne, excessive growth of body or facial hairs (hirsutism), androgenetic alopecia, whereas significant symptoms include menstrual abnormalities or disturbance, pain in pelvic region, more luteinizing hormone (LH), fewer follicle-stimulating hormone (FSH) and difficulty becoming pregnant (infertility).

Obesity is also a key trait contributor in metabolic abnormalities [20, 28, 94, 153]. The symptoms of PCOS generally embark on just about menarche [50, 67]. Homburg [65] listed the hirsutism, acne, seborrhea, alopecia, obesity, and acanthosis nigricans as the cutaneous presentations of increased androgenic hormones. Phenotypic expression of PCOS is highly variable and is affected by several internal and external factors:

- **Internal Factors:** These include more insulin production and adrenal steroid genesis;
- **External Factors** These are exercise, quality & quantity of food, body weight, and lifestyle [88].
- **Other Factors:** Affecting the phenotype are life stage, ethnicity, and genotype [110, 146]. From adolescence to postmenopausal age, phenotypic expression of PCOS is changed and highly influenced by obesity and metabolic alterations [4, 64].

**TABLE 11.1** Summary of Polycystic Ovarian Syndrome: Signs & Symptoms, Complications, Etiology, and Management

| Parameters | Descriptions | References |
|---|---|---|
| Signs | Acne | [20, 28, 120] |
| | Excessive growth of body or facial hairs (hirsutism) | |
| | Androgenetic alopecia | |
| Symptoms | Menstrual abnormalities or disturbance | [13, 37, 89, 121, 154] |
| | Pain in pelvic region | |
| | More luteinizing hormone (LH) | |
| | Fewer follicle-stimulating hormone (FSH) | |
| | Difficulty becoming pregnant (infertility) | |
| Related disorders | Obesity | [13, 37, 89, 121, 154] |
| | Clinical manifestation of hyperandrogenism | |
| | Infertility | |
| | Insulin resistance | |
| | Cancer | |
| | Cardiovascular diseases | |
| | Menstrual abnormalities | |
| | Hypertension | |
| | Anxiety | |
| | Depression | |
| Causes | Genetic predisposition | [29, 33, 38, 88, 74] |
| | Hormonal imbalance | |
| | Environmental factors | |
| Management | Dietary interventions | [40, 77, 104, 153, 155] |
| | Antioxidant supplementation | |
| | Pharmaceutical treatment | |
| | Lifestyle management | |

## 11.3 RELATED DISORDERS OF PCOS

PCOS is described as a hormonal irregulation with widely unstable clinical symptoms. PCOS includes group of clinical abnormalities such as obesity, clinical manifestation of hyperandrogenism, infertility, insulin resistance, cancer, cardiovascular diseases, menstrual abnormalities, hypertension, anxiety, depression [13, 25, 29, 30, 37, 73, 77, 89, 110, 121, 125, 144, 149, 154].

### *11.3.1 HIRSUTISM AND ACNE*

Hirsutism is a widespread medical appearance of hyperandrogenism in 70% of women with PCOS [8, 47]. Hirsutism is the presence of excess body hair of male type pattern on the face and body of women [101]. Adrenal or ovarian disorder is the main contributor of this abnormal growth of hairs [158]. The prevalence of hirsutism in PCOS is high (5–10% worldwide) and varies with age, body weight, menopausal status, and ethnic origin [85]. It causes considerable psychological suffering depending on the social and ethnic norms. Hirsutism is significantly associated with fundamental disorders and co-morbidities [9, 137]. Through ultrasound, about 50% of women with normal menstrual cycle having hirsutism can be identified [139]. Zandi et al., [157] concluded that PCOS is a common disorder associated with acne in Iranian women, and not necessarily related to some clinical signs and symptoms such as obesity or hirsutism. In women with acne, the most important predictors of PCOS were menstrual disorder and LH to FSH ratio.

Seirafi et al., [130] showed that the prevalence of PCOS in women with acne was about 40%. PCOS could be observed in 5.5% of cases of acne. Essah et al., [43] concluded that a common manifestation of hyperandrogenemia is acne vulgaris. Various factors account for the development of acne. Among these factors, the androgenic stimulation of sebaceous glands is the most important factor. Clinically, excessive hair growth and acne are consequences of excessive androgen production by adrenal glands, whereas, an ovulation results in subfertility and excessive or absence of periods [44].

### *11.3.2 OBESITY*

Obesity and excess weight are the most significant clinical aberrations having a significant effect on PCOS expression and related metabolic abnormalities. Obesity, along with insulin resistance, is associated with the worsening of *diabetes mellitus* type 2, cardiovascular diseases and may others reproductive and metabolic features of PCOS [13, 17, 86, 89, 110]. It is revealed from family studies that in a predisposed population, weight gain may act as the major promotor of PCOS genotype. Moreover, with reference to symptoms, generally, obesity is related to worsen the symptoms of PCOS, whereas weight loss reduces the occurrence of symptoms and related hormonal abnormalities [13, 81]. The factors involved in the exacerbation of obesity are lifestyle, diet, and economic status, individual physiological and anatomical variations. Persons with poor socioeconomic status have a strong influence of estrogen on fat accumulation resulted in the development of greater adiposity [41, 58]. The prevalence of obesity/overweight varies in different countries, and a study in the US has shown that 42% of the PCOS population were obese (BMI>30kg/m$^2$) and 24% were overweight (BMI 25–29.9 kg/m$^2$) [10, 55, 119].

### *11.3.3 INSULIN RESISTANCE*

The chance of metabolic syndrome is higher in PCOS women than in normal healthy women, and women with PCOS are four times more vulnerable to type II diabetes than the common population [153]. Insulin resistance is the most common complication, symptom, and etiology of PCOS; and it is raised from the obesity and PCOS. During insulin resistance, large production of insulin stimulates adrenal glands to produce more androgen. Then this androgen plays a critical role in the development of the normal male phenotype. Beta cells released more insulin, and in turn, hyperinsulinemia and this hyperinsulinemia along with insulin resistance have a synergistic effect on all features of PCOS, such as hyperandrogenism, menstrual abnormalities, infertility, acne, hirsutism, and hormonal disturbances. If beta cells secrete less insulin, then such insulin insufficiency will lead to glucose intolerance and type II diabetes. Furthermore, in the case of pregnant PCOS women, insulin resistance is considered as a major cause for gestational diabetes [16, 78, 122, 134, 144, 152]. Literature shows that the prevalence of glucose intolerance, *diabetes mellitus*, and insulin resistance in women with PCOS is 10.0%, 35.0%, and 85%, respectively [142]. Studies demonstrated that there is a low level of insulin action and circulating adiponectin in women suffering from PCOS [6, 129].

### 11.3.4 INFERTILITY

Boomsma et al., [22] introduced PCOS as the most common hormonal disorders causing one of the most significantly alarming associated morbidities referred as an ovulatory infertility in the women of reproductive age. During childbearing age, PCOS causes the menstrual abnormalities, which further lead to infertility [93, 134]. It affects about 90–95% of anovulatory women with PCOS. Moreover, in PCOS, infertility ratio is about 6–15% [47]. However, about 40% of women with PCOS are affected by this infertility [93]. The two important features of PCOS, such as insulin resistance and visceral fat, are the main contributors of infertility. Obesity negatively affects the conception rates of childbearing women [89]. Primordial follicles are in normal concentration in case of polycysts possessing women, whereas these women have a larger number of primary and secondary follicles. Although owing to imbalances in factors for normal follicular development, follicular growth becomes detained as follicles reach a diameter of 4–8mm. As a dominant follicle does not build up, ovulation is not ensured [26, 146]. Moreover, the chances of spontaneous abortion in women with PCOS vary from 42–73% [53, 70]. In women with PCOS, chances of infertility are higher, particularly in obese women. In such women, the chances and occurrence of many complications are more when they are pregnant. Major pregnancy-related complications are hypertensive aberrations, gestational diabetes, premature delivery, and delivery through cesarean section. The progeny of PCOS women have amplified menace of inherited deformities and hospitalization in childhood. Clinicians should be conscious with respect to the enhanced chance of abnormalities and should know the prevention and management of these aberrations [73, 143].

### 11.3.5 HYPERCHOLESTEROLEMIA

In PCOS, there are higher levels of systolic blood pressure, diastolic blood pressure, low-density lipoprotein cholesterol, and fasting plasma glucose, which lead to the cardiovascular diseases. In such cases, risk factors of cardiovascular disease such as a family history of coronary artery disease in first degree relatives, systemic hypertension, diabetes mellitus, abdominal obesity, and the metabolic syndrome are more ubiquitous [47, 150].

Romero et al., [127] found that insulin resistance, hyperandrogenism, and hyperinsulinemia contribute to the coronary artery disease. Macut et al.,

[87] investigated the relation of clinical markers and risk factors of cardiovascular disease in PCOS patients. It was analyzed that PCOS patients over 30 years had a higher waist-to-hip ratio, systolic, and diastolic blood pressure, total cholesterol, high-density lipoprotein cholesterol, low-density lipoprotein cholesterol, triglycerides, and more risk of hypertension and atherosclerosis.

### 11.3.6 CANCER

PCOS is one of the most common endocrinal disorders associated with the consequent development of gynecological cancers such as uterine, breast, and ovarian cancer [19, 76, 133]. According to the prospective studies, it is found that PCOS is associated with increased risk of endometrial, kidney, colon, and brain cancer [31, 56, 120]. In PCOS women, there is an augmented pervasiveness of endometrial hyperplasia and carcinoma [14, 63]. This enhancement is attributed to the unrelenting prompt of endometrial tissue by estrogen (mainly estrone) without the progesterone-induced inhibition of proliferation and differentiation to secretory endometrium that occurs after ovulation. In PCOS patients, there are various factors such as obesity and type 2 diabetes, which are associated with endometrial carcinoma [36, 61].

### 11.3.7 NEUROPSYCHOLOGICAL ABNORMALITIES

A study on PCOS reveals that there are apparent similarities with other aberrations such as obesity, Cushing's syndrome, ovarian, and adrenal neoplasms, and congenital adrenal hyperplasia due to which 70% of cases remain undiagnosed [23]. Farrell and Antoni [44] studied that mood dysfunction and psychiatric problems are more noteworthy in women with PCOS than in normal healthy women. Women are taking anti-androgens (71%) experience a higher depression rate than women without anti-androgen intake (67%). Barry et al., [19] observed that BMI has a direct link with anxiety and depression. Lower BMI reduces the chances of anxiety and depression. Women with PCOS are also at increased risk of number of mental health disorders, including depression, bipolar disorder, anxiety, and eating disorders [136]. Diabetes, cardiovascular disease, endometrial cancer, and psychiatric disorders are risk factors for women with PCOS, while in non-PCOS women; there is a lower risk of these entities [45].

## 11.4 ETIOLOGY

The exact etiology of the PCOS is unclear, but the modern approaches have listed several factors, which are responsible for the whole spectrum of abnormalities. These etiological factors include gene predisposition (gene susceptibility), strong stimulation of adrenal glands to produce more androgen, increased insulin production due to cell resistance to it, oral contraceptives, hormonal disturbance as well as stress and fatigue [29, 38, 74, 87, 118, 123, 124].

### *11.4.1 GENETIC PREDISPOSITION*

Genetic predisposition (gene susceptibility) is considered as the main cause of polycystic ovary syndrome development [41, 32]. Regarding the origin of PCOS, environmental factors such as prenatal exposure to androgens and weight gain have been discussed as major causative factors. Along with environmental factors, there are various genetic factors which act as risk factors for the manifestation of PCOS or for increasing the chances of PCOS. Among these genetic factors, the most important one is the presence and exposure of the fetus to PCOS in an early age from its mother, i.e., hereditary PCOS. Hickey et al., (29) examined that increased level of androgen is responsible for the occurrence of PCOS in offspring/children, based on experimental data from animal studies as well as clinical material of pathological conditions in human populations (i.e., congenital adrenal hyperplasia).

PCOS is considered as a heritable disorder based on a cluster of cases in the family. An elevated incidence of PCOS is indicative of genetic manipulations. Li and Baek [92] found that if gene susceptibility is early diagnosed in case of PCOS, then it may prevent the risk of obesity, type II diabetes, and cardiovascular diseases.

### *11.4.2 HORMONAL IMBALANCE*

According to Dunne and Slater [39], motivators of PCOS are many hormones such as testosterone-producing androgen, cortisol, estrogens (female hormone), FSH, insulin (carrier of blood glucose to other cells), LH, progesterone (female hormone), prolactin, and thyroid hormones disturbances. In retort to prompt by LH, the ovarian theca-cells amalgamate androgens. It is revealed from *in-vitro* and *in-vivo* studies that chances of conversion of androgenic signs into testosterone are higher in case of PCOS women with abnormal theca cells in their ovaries as compared to normal women with normal theca cells [114].

Among hormones, LHs stimulates the production of more theca cells (in abnormal number) whereas, modification of aroma activity of granulose cells is done through FSH that is the indication for estimation of estrogen number produced from androgenic signs. As the number of LH gets bigger than FSH, the synthesis of androgen in ovaries is significantly increased. Insulin plays both a direct and indirect role in the pathogenesis of hyperandrogeniemia in the polycystic ovary syndrome. Insulin acts synergistically with LH to augment the androgen production of theca cells. In PCOS women, the numbers of testosterone hormones that are male hormone are more than in normal women without PCOS [19].

Azziz et al., [9] described that the eminent androgen levels impinge on approximately 60–80% of PCOS women and can upshot in the clinical signs: hirsutism, acne, and to some degree alopecia [108].

### *11.4.3 ENVIRONMENTAL FACTORS*

An assortment of environmental factors potentially involves the cause, incidence, and intonation of polycystic ovary syndromes such as environmental toxins, diet, and nutrition, socioeconomic status, and geography. The environmental toxin is mainly imperative consideration causing PCOS and disrupting the reproductive health. Most important toxins are plasticizers like bisphenol or phthalates (belonging to the group of endocrine-disrupting chemicals and advanced glycation end-products). Exposure time to endocrine-disrupting chemicals is the most crucial parameter for adverse effects on health. The most vulnerable groups are fetuses, infants, and young children, especially in the early period of development. If such exposure is hereditary, then it impersonates endogenous hormones followed by altered fetal programming and consequently PCOS. The long-term exposure leads to many hormonal abnormalities, metabolic complications such as obesity, insulin resistance, hyperinsulinemia, type II diabetes, and cardiovascular disease [98, 128].

## 11.5 POTENTIAL STRATEGIES TO PREVENT POLYCYSTIC OVARIAN SYNDROME (PCOS)

### 11.5.1 DIETARY INTERVENTIONS

To provide an ample amount of macro, micronutrients, and energy for the most favorable health condition and reduce the menace of the diet associated disease development is the endeavor of dietary treatment. The dietary modification should be based on a balanced diet by considering the glycemic index, carbohydrate, lipid, protein, and energy level. It may have a beneficial effect on hormonal and metabolic profile [104]. In case of PCOS, the main target of dietary modification is to reduce the body mass especially the visceral fat followed by the decrease in insulin resistance and improvement of hormonal and metabolic profile and almost all the symptoms of PCOS. Short term dietary modifications reduce the body mass and regulate the menstrual cycle. Whereas, long-term intervention results in maintaining reduced weight and decrease the risk of type 2 *diabetes mellitus* and coronary heart diseases and cancer development [92] and restoration of regular menses [117].

Wong et al., [152] investigated that dietary modification is an important tool in reducing and controlling weight, but it has no beneficial effects on hyperandrogenism. Post-prandial serum glucose, insulin concentration, serum triglyceride, and HDL-cholesterol levels are unfavorably affected, and free androgen index is increased by high simple carbohydrate diets. Low glycemic index diets improve insulin sensitivity, regulate the menstrual cycle, increase the HDL-cholesterol concentration [91], reduce the body mass as well as its maintenance, decrease the risk of type 2 *diabetes mellitus*, cardiovascular disease development, endometrial cancer development indirectly by lowering BMI, fibrinogen, hyperinsulinemia, hyperandrogenemia, and total and high-density lipoprotein (HDL) cholesterol in PCOS women and improve the quality of life [21, 106], frequent restoration of regular menses and have favorable effect on the lipid profile [92].

Marsh et al., [92] revealed that carbohydrates with a high glycemic index increase insulin resistance. Change in the glycemic index of diet is beneficial in improving the satiety and reducing the PCOS symptoms such as type II diabetes and cardiovascular diseases [103].

Quality and quantity of fat are the most important parameters for the healthy life and against PCOS patients [18]. As far as the low-fat diet is concerned, it is an important parameter in reducing the weight, metabolic, and reproductive symptoms, and improving the insulin sensitivity, long-term maintenance of weight loss [103]. Monounsaturated fat-enriched diet is important for the reduction in weight [106]. High protein diet reduces the weight and the PCOS symptoms [103]. It targets the depression and self-esteem [106]. Morenga et al., [109] compared the effects of a high protein diet and high fiber diet against the PCOS. It was found that both diets effectively reduced the weight, total body fat, waist circumference, total, and HDL cholesterol, triglycerides, blood pressure, and fasting blood glucose. However, a high protein diet was more effective than a high fiber diet in reducing fat and blood pressure.

Diet low in saturated fats, glycemic index, and high in fiber is normally good for regulation of blood glucose and insulin level and reduce the metabolic risk [71]. Altieri et al., [5] observed that ovulation in PCOS women can be disturbed by high protein and high glycemic index diet. Low carbohydrate is more effective in reducing weight and risk of cardiovascular diseases than a low-fat diet [21]. Hypocaloric diet (high protein and low glycemic index) along with sibutramine after six months of administration is effective in reducing the weight, androgen, testosterone, FAI, and TG levels and improvement of insulin sensitivity [96].

Sorensen et al., [138] investigated that best weight loss is obtained by the replacement of carbohydrates with protein. Farshchi et al., [46] found that balanced diet treatment with exercise improves the endocrine features, reproductive function, and cardiometabolic risk factors. McGrievy et al., [95] examined that whole grains, iron, fiber, and healthy eating habits are beneficial for infertility, healthy body weight, and good quality of life.

Moran et al., [107] investigated that for short-term treatment of PCOS, meal replacement is important. Hu et al., [66] compared the effects of low carbohydrate diet and low-fat diet on the PCOS complications and found that low carbohydrate diet significantly caused lower reduction in total cholesterol and low-density lipoprotein cholesterol but greater increase in high-density lipoprotein cholesterol and greater decrease in triglycerides compared to low-fat diet [132, 151]. Nutritional counseling is also an important parameter for PCOS [23]. Keeping all these studies in consideration, it may be concluded that PCOS symptoms and related metabolic disorders can be managed by the administration of balanced diet containing high carbohydrates with the low glycemic index, high protein, and low fat.

Nowadays, plant-based remedies have captured greater interest, and the trend of their consumption against different complications and aberrations have been increased. According to a survey by WHO, about 80% of diseased people are using plant-based medicines (functional foods) [2]. These functional and medicinal foods are considered important owing to the presence of an array of bioactive moieties, which are associated with many beneficial effects [1]. As being low in cost, these compounds can provide shelter against certain chronic diseases in developing countries [113]. Hormonal imbalance and PCOS can be managed by certain types of natural remedies like liquorice reduces serum testosterone, flaxseed which has the anti-androgenic effect, cinnamon that reduces insulin resistance, aloe-vera that restores ovarian steroid status and fenugreek- cholesterol-lowering herb [112].

### 11.5.2 ANTIOXIDANT SUPPLEMENTATION

In the recent era, the utilization of antioxidants in the treatment of PCOS patients has captured greater interest. Major complications of PCOS, including hyperandrogenism, diabetes mellitus type II, and obesity can lead to the development of oxidative stress in women with PCOS [111]. Such patients have significantly low levels of antioxidants and vitamins in serum that plays a key role in increasing the risk of oxidative status [3], cardiovascular disease, insulin resistance, hypertension, central obesity, and dyslipidemia [48, 102, 131]. The use of antioxidant in the management of women with PCOS has significantly improved insulin sensitivity and other health threating conditions [3, 131].

### 11.5.3 PHARMACEUTICAL TREATMENTS

Pharmacologic therapy is a vital thread in the management of patients with metabolic syndrome when lifestyle modifications are not enough to accomplish the therapeutic goals [4]. Therapeutic intervention may include contemporary therapies, combined oral contraceptives, antiandrogen agents, and insulin-sensitizing agents [28, 68, 69, 75]. PCOS is the combination of an array of complications. The main aim of pharmacological therapy is the use of specific medication for target aberration. The first-line medications for ovulatory infertility are clomiphene and letrozole. Metformin is considered as second-line medication for ovulatory infertility. These medications might result in better pregnancy outcomes and higher live-birth. Exogenous gonadotropins, *in vitro* fertilization (IVF) and ovarian drilling, are usually recommended in case of infertility [20, 34, 35, 40, 51, 79, 80, 142, 153].

In the case of irregular menstruation, hormonal contraceptives and combined oral contraceptives are first-line therapy to regulate the menstrual cycle and to provide endometrial protection and contraception. The second line medication is metformin in women with insulin resistance and dysglycemia [140, 83]. Furthermore, metformin is considered as the first-line medication for the women with hyperglycemia, insulin resistance, and cardiovascular risk. Glitazones is second-line therapy in this case [153]. Varied types of ovarian surgery have been affianced (wedge resection, electrocautery, laser vaporization, multiple ovarian biopsies, and others) and all procedures upshot in a distorted endocrine profile following surgery [11, 59, 60].

As far as obesity is concerned, the best and first-line medication or therapy is the diet and lifestyle management. After this, the second important therapy is metformin. Metformin helps to reduce the body weight as well as improve some other related disorders like insulin resistance, ovulation in obese women. Anti-obesity drugs also play an important role in body mass reduction. In the case of severely obese women, bariatric surgery is recommended. There is an inverse relationship between weight loss and various treatments such as laparoscopic ovarian diathermy, clomiphene citrate, and gonadotropins [42, 110, 125].

For hirsutism, both therapies with medication and cosmetics are implied. The medications encompass oral contraceptives, anti-androgen drugs, flutamide, spironolactone, dexamethasone; or prednisone, metformin, cyproterone acetate, and eflornithine are used to reduce the androgen level and block the effect of androgen by suppressing ovarian androgen production and increasing sex hormone-binding globulin. Electrolysis and lasers, especially alexandrite and diode lasers, are used for the enduring removal of unwanted hair [54, 140, 159]. Jeopardy and reimbursement of treatment must be warily well thought out and conversed with the patient. Expectations for efficacy should be aptly set. A minimum of 6 months is requisite to perceive benefit from pharmacotherapy; and lifelong treatment is often obligatory for persistent gain [137].

### *11.5.4 LIFESTYLE MANAGEMENT*

With special reference to the PCOS, lifestyle is referred as the eating habits, psychological behaviors, physical activity, and socio-economical status of a person. Lifestyle management is first life treatment and most successful strategy for the PCOS symptoms and related complications.

The principal role of lifestyle management is bodyweight reduction, which can improve the insulin resistance, hyperandrogenism, menstrual function, fertility, body composition, ovulatory function, and pregnancy rate. Lifestyle intervention further reduces the risk of diabetes and the metabolic syndrome. Moreover, it improves the levels of FSH, SHBG, FAI, FG, total testosterone, and androstenedione level in women with PCOS. It can be concluded that lifestyle modifications (awareness of the disease, exercise, good eating habits) can ameliorate reproductive and metabolic health, reduce adverse obstetric outcomes, stress, future metabolic concerns related to a genetic predisposition and worsened by an unhealthy lifestyle [20, 62, 77, 122, 140].

Among all treatments (dietary interventions, medications, lifestyle modifications), exercise is the most important strategy for quality of life. Intensive exercise is a preeminent tool to combat PCOS by targeting its symptoms. Exercise improves the insulin sensitivity, pregnancy rate, endothelial function, and ovulation, hormonal profile (sex-hormone binding globulin, LH, and follicle-stimulating hormone) and reduces the risk of infertility, cardiovascular disease, obesity, insulin resistance. It has beneficial effects on waist circumference and weight. Therefore, it is concluded that intensive exercise can combat the metabolic and reproductive abnormalities of PCOS [86, 141, 126].

The reduction of body weight is important for the management of PCOS. Moran et al., [102] recommended lifestyle modifications such as changes of dietary habits and increasing physical activity for the treatment of overweight or obese PCOS women, it decreases body mass especially visceral fat and is considered to maintain this effect [117]. These modifications improve the hormonal profile and result in the restoration of regular menses by decreasing insulin resistance [104]. Intake of high protein, low energy diet, a moderately low carbohydrate and fat with or with exercise is important in reducing body weight that further experienced a significant reduction in blood pressure, serum fasting glucose, insulin, and testosterone levels and significantly increase the circulation of sex hormone-binding globulin levels and restoration of ovulatory cycles [57, 148].

## 11.6 SUMMARY

This chapter articulates a general viewpoint of endocrinologists on PCOS, endeavoring to assist clinicians in the management of this intricate and versatile disorder. Further research is also considered necessary to delineate the etiological features of PCOS, which should embrace not only inherent and, almost certainly, epigenetic factors, but also early proceedings accountable for the development of different phenotypes. In turn, this exertion could errand deterrent approaches, anyway practicable before or during puberty. Diagnosis of PCOS might also be potentially augmented using new biomarkers of androgen excess and ovarian dysfunction.

## KEYWORDS


- **functional foods**
- **genetic predisposition**
- **herbs**
- **hormonal disturbances**
- **hyperandrogenism**
- **infertility**
- **menstrual abnormalities**
- **nutraceuticals**
- **polycystic ovarian syndrome**


## REFERENCES


1. Ahmed, A., Arshad, M. U., Saeed, F., Ahmed, R. S., & Chatha, S. A. S., (2016). Nutritional probing and HPLC profiling of roasted date pit powder. *Pak. J. Nutr.*, *15*(3), 229–237.
2. Alagesaboopathi, C., (2011). Ethnobotanical studies on useful plants of Kanjamalaihills of Salem district of Tamil Nadu, Southern India. *Ar. App. Sci. Res.*, *3*(5), 532–539.
3. Al-kataan, M. A., Ibrahim, M. A., Al-jammas, M. H. H., Shareef, Y. S., & Sulaiman, M. A., (2010). Serum antioxidant vitamins changes in women with polycystic ovarian syndrome. *J. Bahrain Med. Sci.*, *22*, 68–71.
4. Allahbadia, G. N., & Merchant, R., (2008). Polycystic ovary syndrome in the Indian subcontinent. *Semin. Reprod. Med.*, *26*(1), 22–34.
5. Altieri, P., Cavazza, C., Pasqui, F., Morselli, A. M., Gambineri, A., & Pasquali, R., (2013). Dietary habits and their relationship with hormones and metabolism in overweight and obese women with polycystic ovary syndrome. *Clin. Endocrinol (Oxf.)*, *78*, 52–59.
6. Aroda, V., Ciaraldi, T. P., Chang, S. A., Dahan, M. H., Chang, R. J., & Henry, R. R., (2008). Circulating and cellular adiponectin in polycystic ovary syndrome: Relationship to glucose tolerance and insulin action. *Fertil. Steril.*, *89,* 1200–1208.
7. Asuncion, M., Calvo, R. M., Millan, J. L. S., Sancho, J., Avila, S., & Morreale, H. F. E., (2000). A prospective study of the prevalence of the polycystic ovary syndrome in unselected Caucasian women from Spain. *J. Clin. Endocrinol. Metab.*, *85*(7), 2434–2438.

8. Atmaca, M., Seven, I., Ucler, R., Alay, M., Barut, V., Dirik, Y., & Sezgin, Y., (2014). An interesting cause of hyperandrogenemic hirsutism. *Case Rep. Endocrinol.*, 987272.
9. Azziz, R., Carmina, E., Dewailly, D., & Kandarakis, E. D., (2009). The Androgen excess and PCOS society criteria for the polycystic ovary syndrome: The complete task force report. *Fertil. Steril.*, *91*(2), 456–488.
10. Azziz, R., Woods, K. S., Reyna, R., Key, T. J., Knochenhauer, E. S., & Yildiz, B. O., (2004). The prevalence and features of the polycystic ovary syndrome in an unselected population. *J. Clin. Endocrinol. Metab.*, *89*(6), 2745–2749.
11. Badawy, A., & Elnashar, A., (2011). Treatment options for polycystic ovary syndrome. *Int. J. Women Health*, *3*, 25–35.
12. Balaji, A. B., Kulkarni, P. K., Farees, N., & Deepthi, M., (2014). Life style exercise modification effect in infertility women with PCOS diseases. *Ind. J. App. Res.*, *4*(9).
13. Baldani, D. P., Skrgati, L., & Ougouag, R., (2015). Polycystic Ovary syndrome: Important under recognized cardiometabolic risk factor in reproductive-age women. *Int. J. Endocrin.*, 17–21.
14. Balen, A., (2001). Polycystic ovary syndrome and cancer. *Hum. Reprod.*, *7*, 522–525.
15. Baqai, Z., Khanam, M., & Parveen, S., (2010). Prevalence of PCOS in infertile patients. *Med. Chanel*, *16*(3), 255–260.
16. Barber, T. M., & Franks, S., (2012). The link between polycystic ovary syndrome and both type 1 and Type 2 diabetes mellitus: What do we know today? *Womens Health (Lond).*, *8*(2), 147–154.
17. Barber, T. M., McCarthy, M. I., Wass, J. A., & Franks, S., (2006). Obesity and polycystic ovary syndrome. *Clin. Endocrinol. (Oxf).*, *65*(2), 137–145.
18. Barr, S., Hart, K., Reeves, S., Sharp, K., & Jeans, Y. M., (2011). Habitual dietary intake, eating pattern and physical activity of women with polycystic ovary syndrome. *Eur. J. Clin. Nutr.*, *65*(10), 1126–1132.
19. Barry, J. A., Azizia, M. M., & Hardiman, P. J., (2014). Risk of endometrial, ovarian and breast cancer in women with polycystic ovary syndrome: A systematic review and meta-analysis. *Hum Reprod Update.*, *20*(5), 748–758.
20. Bates, G. W. J., & Propst, A. M., (2012). Polycystic ovarian syndrome management options. *Obstet. Gynecol. Clin. North Am.*, *39*(4), 495–506.
21. Bazzano, L. A., Hu, T., Reynolds, K., & Yao, L., (2014). Effects of low-carbohydrate and low-fat diets: a randomized trial. *Ann. Intern. Med.*, *161*(5), 309–318.
22. Boomsma, C. M., Fauser, B. C., & Macklon, N. S., (2008). Pregnancy complications in women with polycystic ovary syndrome. *Semin. Reprod. Med.*, *26*, 72–84.
23. Boyle, J., & Teede, H. J., (2012). Polycystic ovary syndrome-an update. *Aust. Fam. Physician.*, *41*(10), 752–756.
24. Bozdag, G., Mumusoglu, S., Zengin, D., Karabulut, E., & Yildiz, B. O., (2016). The prevalence and phenotypic features of polycystic ovary syndrome: A systematic review and meta-analysis. *Hum Reprod.* *31*(12): 2841–2855.
25. Brahm, J., Brahm, M., & Segovia, R., (2011). Acute and fulminant hepatitis induced by flutamide: Case series report and review of the literature. *Ann. Hepatol.*, *10*, 93–98.
26. Brassard, M., AinMel, Y., & Baillargeon, J. P., (2008). Basic infertility including polycystic ovary syndrome. *Med. Clin. North Am.*, *92*, 1163–1192.
27. Bruner, B., Chad, K., & Chizen, D., (2006). Effects of exercise and nutritional counseling in women with polycystic ovary syndrome. *Appl. Physiol. Nutr. Metab.*, *31*(4), 384–391.

28. Buzney, E., Sheu, J., Buzney, C., & Reynolds, R. V., (2014). Polycystic ovary syndrome: A review for dermatologists: Part II, Treatment. *J. Am. Acad. Dermatol., 71*(5), 859, 1–859.
29. Caldwell, A. S., Middleton, L. J., & Jimenez, M., (2014). Characterization of reproductive, metabolic, and endocrine features of polycystic ovary syndrome in female hyperandrogenic mouse models. *Endocrin., 155*(8), 3146–3159.
30. Cascella, M., Magistrato, A., Tavernelli, I., Carloni, P., & Rothlisberger, U., (*2006*). Role of protein frame and solvent for the redox properties of azurin from *Pseudomonas aeruginosa. PNAS., 103(52), 19641–19646.*
31. Chittenden, B. G., Fullerton, G., Maheshwari, A., & Bhattacharya, S., (2009). Polycystic ovary syndrome and the risk of gynecological cancer: a systematic review. *Reprod Biomed Online., 19*(3), 398–405.
32. Cui, L., Li, G., Zhong, W., Bian, Y., & Su, S., (2015). Polycystic ovary syndrome susceptibility single nucleotide polymorphisms in women with a single PCOS clinical feature. *Hum. Reprod., 30*(3), 732–736.
33. Dewailly, D., Lujan, M. E., Carmina, E., Cedars, M. I., Laven, J., Norman, R. J., & Morreale, E. H. F., (2014). Definition and significance of polycystic ovarian morphology: A task force report from the androgen excess and polycystic ovary syndrome society. *Hum. Reprod. Update., 20*(3), 334–352.
34. Diamanti-Kandarakis, E., Spina, G., Kouli, C., & Migdalis, I., (2001). Increased endothelin-1 levels in women with polycystic ovary syndrome and the beneficial effect of metformin therapy. *J. Clin. Endocrinol. Metab., 86,* 3595–3598.
35. Dikensoy, E., Balat, O., Pence, S., Akcali, C., & Cicek, H., (2009). The risk of hepatotoxicity during long-term and low-dose flutamide treatment in hirsutism. *Arch. Gynecol. Obstet., 279,* 321–327.
36. Dumesic, D. A., & Lobo, R. A., (2013). Cancer risk and PCOS. *Steroids., 78*(8), 782–785.
37. Dumitrescu, R., Mehedintu, C., Briceag, I., Purcarea, V. L., & Hudita, D., (2015). The polycystic ovary syndrome: An update on metabolic and hormonal mechanisms. *J. Med. Life., 8*(2), 142–145.
38. Dunaif, A., (2016). Perspectives in polycystic ovary syndrome: From hair to eternity. *J. Clin. Endocrinol. Metab., 101*(3), 759–768.
39. Dunne, N., & Slater, W., (2006). The natural diet solution for PCOS and Infertility: How to manage polycystic ovary syndrome naturally. *Natural Solutions for PCOS* (p. 552). Health Solutions Press, New York.
40. Ecklund, L. C., & Usadi, R. S., (2015). Endocrine and reproductive effects of polycystic ovarian syndrome. *Obstet. Gynecol. Clin. North Am., 42*(1), 55–65.
41. Eleftheriadou, M., Stefanidis, K., Lykeridou, K., Iliadis, I., & Michala, L., (2015). Dietary habits in adolescent girls with polycystic ovarian syndrome. *Gynecol. Endocrinol., 31*(4), 269–271.
42. Esfahanian, F., Zamani, M. M., Heshmat, R., & Moini, F., (2013). Effect of metformin compared with hypocaloric diet on serum C-reactive protein level and insulin resistance in obese and overweight women with polycystic ovary syndrome. *J. Obstet. Gynaecol. Res., 39*(4), 806–813.
43. Essah, P. A., Wickham, E. P., Nunley, J. R., & Nestler, J. E., (2006). Dermatology of androgen-related disorders. *Clin. Dermatol., 24,* 289–298.
44. Farrell, K., & Antoni, M. H., (2010). Insulin resistance, obesity, inflammation, and depression in polycystic ovary syndrome: Biobehavioral mechanisms and interventions. *Fertil. Steril., 94*(5), 1565–1574.

45. Farrell-Turner, K. A., (2011). Polycystic ovary syndrome: Update on treatment options and treatment considerations for the future. *Clinical Medicine Insights: Women's Health, 4*, 67–81.
46. Farshchi, H., Rane, A., Love, A., & Kennedy, R. L., (2007). Diet and nutrition in polycystic ovary syndrome (PCOS): Pointers for nutritional management. *J. Obstet. Gynaecol., 27*(8), 762–773.
47. Fauser, B., Tarlatzis, B., & Rebar, R., (2012). Consensus on women's health aspects of polycystic ovary syndrome (PCOS): The Amsterdam ESHRE/ASRM. Sponsored 3rd PCOS consensus workshop group. *Fertile Steril., 97*(1), 28–38.
48. Fenkci, V., Fenkci, S., Yilmazer, M., & Serteser, M., (2003). Decreased total antioxidant status and increased oxidative stress in women with polycystic ovary syndrome may contribute to the risk of cardiovascular disease. *Fertil Steril., 80*, 123–127.
49. Fogel, R. B., Malhotra, A., Pillar, G., Pittman, S. D., Dunaif, A., & White, D. P., (2001). Increased prevalence of obstructive sleep apnea syndrome in obese women with polycystic ovary syndrome. *J. Clin. Endocrinol. Metab., 86*, 1175–1180.
50. Franks, S., (2002). Adult polycystic ovary syndrome begins in childhood. *Best Pract Res Clin Endocrinol Metab., 16*, 263–272.
51. Gallo, M. F., Lopez, L. M., Grimes, D. A., Schulz, K. F., & Helmerhorst, F. M., (2011). Combination contraceptives: Effects on weight. *Cochrane Database Syst. Rev., 9*, CD003987.
52. Garg, D., & Tal, R., (2016). The role of AMH in the pathophysiology of polycystic ovarian syndrome. *Reprod. Biomed Online, 33*(1), 15–28.
53. Glueck, C., Phillips, H., Cameron, D., Sieve-Smith, L., & Wang, P., (2001). Continuing metformin throughout pregnancy in women with polycystic ovary syndrome appears to safely reduce first-trimester spontaneous abortion: a pilot study. *Fertil. Steril., 75*(1), 46–52.
54. Goodman, N. F., Cobin, R. H., & Futterweit, W., (2015). American association of clinical endocrinologists, american college of endocrinology, and androgen excess and pcos society disease state clinical review: Guide to the best practices in the evaluation and treatment of polycystic ovary syndrome--Part 1. *Endocr. Pract., 21*(11), 1291–1300.
55. Gopal, M., Duntley, S., Uhles, M., & Attarian, H., (2002). The role of obesity in the increased prevalence of obstructive sleep apnea syndrome in patients with polycystic ovarian syndrome. *Sleep Med., 3*, 401–404.
56. Gottschau, M., Kiaer, S. K., Jensen, A., Munk, C., & Mellemkjaer, L., (2015). Risk of cancer among women with polycystic ovary syndrome: A Danish cohort study. *Gynecol Oncol., 136*(1), 99–103.
57. Gower, B. A., Chandler-Laney, P. C., & Ovalle, F., (2013). Favourable metabolic effects of a eucaloric lower-carbohydrate diet in women with PCOS. *Clin. Endocrinol. (Oxf)., 79*, 550–557.
58. Grantham, J. P., & Henneberg, M., (2014). The estrogen hypothesis of obesity. *PLoS One., 9*(6), e99776.
59. Hannaford, P. C., Iversen, L., & Macfarlane, T. V., (2010). Mortality among contraceptive pill users: Cohort evidence from Royal College of General Practitioners' Oral Contraception Study. *BMJ., 340*, c927.
60. Hannaford, P. C., Selvaraj, S., Elliott, A. M., Angus, V., Iversen, L., & Lee, A. J., *(2007)*. Cancer risk among users of oral contraceptives: Cohort data from the Royal College of General Practitioner's oral contraception study. *BMJ., 335*, 651.

61. Haoula, Z., Salman, M., & Atiomo, W., (2012). Evaluating the association between endometrial cancer and polycystic ovary syndrome. *Hum. Reprod., 27*(5), 1327–1331.
62. Haqq, L., McFarlane, J., Dieberg, G., & Smart, N., (2014). Effect of lifestyle intervention on the reproductive endocrine profile in women with polycystic ovarian syndrome: A systematic review and meta-analysis. *Endocr Connect., 3*(1), 36–46.
63. Hardiman, P., Pillay, O. C., & Atiomo, W., (2003). Polycystic ovary syndrome and endometrial carcinoma. *Lancet., 362*, 1082–1089.
64. Hart, R., Hickey, M., & Franks, S., (2004). Definitions, prevalence and symptoms of polycystic ovaries and polycystic ovary syndrome. *Best Pract. Res. Clin. Obstet Gynaecol., 18*(5), 671–683.
65. Homburg, R., (2008). Polycystic ovary syndrome. *Best Pract. Res. Clin. Obstet. Gynaecol., 22*, 261–274.
66. Hu, T., Mills, K. T., Yao, L., & Demanelis, K., (2012). Effects of low-carbohydrate diets versus low-fat diets on metabolic risk factors: A meta-analysis of randomized controlled clinical trials. *Am. J. Epidemiol., 176*(7), S44–S54.
67. Ibanez, I., Valls, C., Potau, N., Marcos, M. V., & De Zegher, F., (2001). Polycystic ovary syndrome after precocious pubarche: Ontogeny of the low birthweight effect. *Clin. Endocrinol. (Oxf)., 55*, 667–672.
68. Ibanez, L., & Zegher, F., (2003). Low-dose combination of flutamide, metformin and an oral contraceptive for non-obese, young women with polycystic ovary syndrome. *Hum. Reprod., 18*, 57–60.
69. Isikoglu, M., Ozgur, K., & Oehninger, S., (2007). Extension of GnRH agonist through the luteal phase to improve the outcome of intracytoplasmic sperm injection. *J. Reprod. Med., 52*, 639–644.
70. Jakubowicz, D. J., Iuorno, M. J., Jakubowicz, S., Roberts, K., & Nestler, J. E., (2002). Effects of metformin on early pregnancy loss in the polycystic ovary syndrome. *J. Clin. Endocrinol. Metab., 87*(2), 524–529.
71. Jeans, Y. M., Barr, S., Smith, K., & Hart, K. H., (2009). Dietary management of women with polycystic ovary syndrome in the United Kingdom: The role of dietitians. *J. Hum. Nutr. Diet., 22*(6), 551–558.
72. Joham, A. E., Nanayakkara, N., & Ranasinha, S., (2016). Obesity, polycystic ovary syndrome and breastfeeding: An observational study. *Acta Obstet. Gynecol. Scand., 95*(4), 458–466.
73. Joham, A. E., Palomba, S., & Hart, R., (2016). Polycystic ovary syndrome, obesity, and pregnancy. *Semin. Reprod. Med., 34*(2), 93–101.
74. Jonard, C. S., & Dewailly, D., (2013). Pathophysiology of polycystic ovary syndrome: The role of hyperandrogenism. *Front. Horm. Res., 40*, 22–27.
75. Kemmeren, J. M., Algra, A., & Grobbee, D. E., (2001). Third generation oral contraceptives and risk of venous thrombosis: Meta-analysis. *BMJ., 323*, 131–134.
76. Kim, J., Mersereau, J. E., Khankari, N., & Bradshaw, P. T., (2016). Polycystic ovarian syndrome (PCOS), related symptoms/sequelae, and breast cancer risk in a population-based case-control study. *Cancer Causes Control., 27*(3), 403–414.
77. Ko, H., Teede, H., & Moran, L., (2016). Analysis of the barriers and enablers to implementing lifestyle management practices for women with PCOS in Singapore. *BMC Res Notes., 9*, 311–318.
78. Lakkakula, B. V., (2013). Genetic variants associated with insulin signaling and glucose homeostasis in the pathogenesis of insulin resistance in polycystic ovary syndrome: A systematic review. *J. Assist. Reprod. Genet., 30*(7), 883–895.

79. Legro, R. S., Brzyski, R. G., Diamond, M. P., & Coutifaris, C., (2014). Letrozole versus clomiphene for infertility in the polycystic ovary syndrome. *N. Engl. J. Med., 371*(2), 119–129.
80. Legro, R. S., Dodson, W. C., Kunselman, A. R., & Stetter, C. M., (2016). Benefit of delayed fertility therapy with preconception weight loss over immediate therapy in obese women with PCOS. *J. Clin. Endocrinol. Metab., 101*(7), 2658–2666.
81. Legro, R. S., Kunselman, A. R., & Demers, L., (2002). Elevated dehydroepiandrosterone sulfate levels as the reproductive phenotype in the brothers of women with polycystic ovary syndrome. *J. Clin. Endocrinol. Metab., 87*, 2134–2138.
82. Li, L., & Baek, K. H., (2015). Molecular genetics of polycystic ovary syndrome: An update. *Curr. Mol. Med., 15*(4), 331–342.
83. Lindh, I., Ellstrom, A. A., & Milsom, I., (2011). The long-term influence of combined oral contraceptives on body weight. *Hum. Reprod., 26*, 1917–1924.
84. Lujan, M. E., Jarrett, B. Y., & Brooks, E. D., (2013). Updated ultrasound criteria for polycystic ovary syndrome: Reliable thresholds for elevated follicle population and ovarian volume. *Hum. Reprod., 28*(5), 1361–1368.
85. Lumachi, F., & Basso, S. M., (2010). Medical treatment of hirsutism in women. *Curr. Med. Chem., 17*(23), 2530–2538.
86. Lundgren, K. M., Romundstad, L. B., & During, V. V., (2016). Exercise prior to assisted fertilization in overweight and obese women (FertilEX): Study protocol for a randomized controlled trial. *Trials., 17*(1), 268–276.
87. Macut, D., Bacevic, M., Antic, I. B., Macut, J. B., & Civcic, M., (2015). Predictors of subclinical cardiovascular disease in women with polycystic ovary syndrome: Interrelationship of dyslipidemia and arterial blood pressure. *Int. J. Endocrinol.*, 812610.
88. Macut, D., Pfeifer, M., & Yildiz, B. O., (2013). Polycysticovary syndrome: Novel insights into causes and therapy. *Front Horm. Res. Basel Karger., 40*, 1–21.
89. Mani, H., Kkunti, K., Daly, H., Barnett, J., & Davies, M., (2015). Education and self-management for women with polycystic ovary syndrome, a narrative review of literature. *Ibn. J. Med. Biomed Sci.*, *7*(1), 213–218.
90. March, W. A., Moore, V. M., & Willson, K. J., (2010). The prevalence of polycystic ovary syndrome in a community sample assessed under contrasting diagnostic criteria. *Hum. Reprod., 25*(2), 544–551.
91. Marsh, K., & Miller, J. B., (2005). The optimal diet for women with polycystic ovary syndrome? *Br. J. Nutr., 94*(2), 154–165.
92. Marsh, K. A., Steinbeck, K. S., Atkinson, F. S., Petocz, P., & Miller, J. C. B., (2010). Effect of a low glycemic index compared with a conventional healthy diet on polycystic ovary syndrome. *Am. J. Clin. Nutr., 92*, 83–92.
93. Mascarenhas, M. N., Flaxman, S. R., & Boerma, T., (2012). National, regional, and global trends in infertility prevalence since 1990: A systematic analysis of 277 health surveys. *PLoS Med., 9*, e1001356.
94. McCartney, C. R., & Marshall, J. C., (2016). Polycystic ovary syndrome. *Eng. J. Med., 375*, 54–64.
95. McGrievy, G. T., Davidson, C. R., & Billings, D. L., (2015). Dietary intake, eating behaviors, and quality of life in women with polycystic ovary syndrome who are trying to conceive. *Hum. Fertil. (Camb)., 18*(1), 16–21.
96. Mehrabani, H. H., Salehpour, S., Amiri, Z., Farahani, S. J., Meyer, B. J., & Tahbaz, F., (2012). Beneficial effects of a high-protein, low-glycemic-load hypocaloric diet in

overweight and obese women with polycystic ovary syndrome: A randomized controlled intervention study. *J. Am. Coll. Nutr., 31*(2), 117–125.

97. Mehrabian, F., Khani, B., Kelishadi, R., & Ghanbari, E., (2011). The prevalence of polycystic ovary syndrome in Iranian women based on different diagnostic criteria. *Endokrynol. Pol., 62*(3), 238–242.
98. Merkin, S. S., Phy, J. L., Sites, C. K., & Yang, D., (2016). Environmental determinants of polycystic ovary syndrome. *Fertil. Steril., 106*(1), 16–24.
99. Mirza, S. S., Shafique, K., Shaikh, A. R., Khan, N. A., & Qureshi, M. A., (2014). Association between circulating adiponectin levels and polycystic ovarian syndrome. *J. Ovarian Res., 7*, 18–24.
100. Mobeen H, Afzal N, Kashif M. Polycystic Ovary Syndrome May Be an Autoimmune Disorder. *Scientifica (Cairo).* 2016;2016; E-article ID: 4071735; doi:10.1155/2016/4071735.
101. Mofid, A., Alinaghi, S. A. S., Zandieh, S., & Yazdani, T., (2008). Hirsutism. *Int. J. Clin. Pract., 62*, 433–443.
102. Mohan, S. K., & Priya, V. V., (2009). Lipid peroxidation, glutathione, ascorbic acid, vitamin E, antioxidant enzyme and serum homocysteine status in patients with polycystic ovary syndrome. *Biol. Med., 1*, 44–49.
103. Moran, L., & Norman, R. J., (2004). Understanding and managing disturbances in insulin metabolism and body weight in women with polycystic ovary syndrome. *Best Pract. Res. Clin. Obstet. Gynaecol., 18*(5), 719–736.
104. Moran, L. J., Brinkworth, G. D., & Norman, R. J., (2008). Dietary therapy in polycystic ovary syndrome. *Semin. Reprod. Med., 26*(1), 85–92.
105. Moran, L. J., Hutchison, S. K., Norman, R. J., & Teede, H. J., (2011). Lifestyle changes in women with polycystic ovary syndrome. *Cochrane Database Syst. Rev.*, 7, CD007506.
106. Moran, L. J., Ko, H., & Misso, M., (2013). Dietary composition in the treatment of polycystic ovary syndrome: A systematic review to inform evidence-based guidelines. *J. Acad. Nutr. Diet., 113*(4), 520–545.
107. Moran, L. J., Noakes, M., Clifton, P. M., Wittert, G. A., Williams, G., & Norman, R. J., (2006). Short-term meal replacements followed by dietary macronutrient restriction enhance weight loss in polycystic ovary syndrome. *Am. J. Clin. Nutr., 84*(1), 77–87.
108. Moran, L. J., Pasquali, R., Teede, H. J., Hoeger, K. M., & Norman, R. J., (2009). Treatment of obesity in polycystic ovary syndrome: A position statement of the androgen excess and polycystic ovary syndrome society. *Fertil. Steril., 92*, 1966–1982.
109. Morenga, L. A. T., Levers, M. T., Williams, S. M., Brown, R. C., & Mann, J., (2011). Comparison of high protein and high fiber weight-loss diets in women with risk factors for the metabolic syndrome: A randomized trial. *Nutr. J., 10*, 40–46.
110. Motta, A. B., (2012). The role of obesity in the development of polycystic ovary syndrome. *Curr. Pharm. Des., 18*(17), 2482–2491.
111. Murri, M., Luque-Ramırez, M., Insenser, M., Ojeda-Ojeda, M., & Escobar-Morreale, H. F., (2013). Circulating markers of oxidative stress and polycystic ovary syndrome (PCOS): A systematic review and meta-analysis. *Hum. Reprod. Update., 19*, 268–288.
112. Nagarathna, P. K. M., Rajan, P. R., & Koneri, R., (2013). A detailed study on poly cystic ovarian syndrome and its treatment with natural products. *Int. J. Toxi. Pharma. Res., 5*(4), 109–120.

113. Namsa, N. D., Mandal, M., Tangjang, S., & Mandal, S. C., (2011). Ethnobotany of the Monpa ethnic group at Arunachal Pradesh, India. *J. Ethnobio. Ethnomed., 7*, 31–37.
114. Nelson, V. L., Qin, K. N., & Rosenfield, R. L., (2001). The biochemical basis for increased testosterone production in theca cells propagated from patients with polycystic ovary syndrome. *J. Clin. Endocrinol. Metab., 86*, 5925–5933.
115. Nidhi, R., Padmalatha, V., Nagarathna, R., & Amritanshu, R., (2011). Prevalence of polycystic ovarian syndrome in Indian adolescents. *J. Pediatr. Adolesc. Gynecol., 24*(4), 223–227.
116. Norman, R. J., Dewailly, D., Legro, R. S., & Hickey, T. E., (2007). *Polycystic Ovary Syndrome, 370* (9588), 685–697.
117. Nybacka, A., Carlstrom, K., Stahle, A., Nyren, S., Hellstrom, P. M., & Hirschberg, A. L., (2011). Randomized comparison of the influence of dietary management and/or physical exercise on ovarian function and metabolic parameters in overweight women with polycystic ovary syndrome. *Fertil. Steril., 96*, 1508–1513.
118. Parahuleva, N., Pehlivanov, B., Orbecova, M., Deneva, T., & Uchikova, E., (2013). Serum levels of anti-muller hormone in women with polycystic ovary syndrome and healthy women of reproductive age. *Akush. Ginekol (Sofiia)., 52*(1), 16–23.
119. Pasquali, R., Gambineri, A., & Pagotto, U., (2006). The impact of obesity on reproduction in women with polycystic ovary syndrome. *BJOG: Int. J. Obs. Gynae., 113*(10), 116–126.
120. Peigne, M., & Dewailly, D., (2014). Long term complications of polycystic ovary syndrome (PCOS). *Ann Endocrinol (Paris). 75*(4), 194–199,
121. Petrikova, J., Lazurova, I., Dravecka, I., Vrbikova, J., Kozakova, D., Figurova, J., Vaczy, Z., & Rosocha, J., (2015). The prevalence of non-organ specific and thyroid autoimmunity in patients with polycystic ovary syndrome. *Biomed Pap. Med. Fac. Univ. Palacky. Olomouc Czech Repub., 159*(2), 302–306.
122. Platt, A. M., (2015). Insulin resistance, metabolic syndrome, and polycystic ovary syndrome in obese youth. *NASN., 30*(4), 213–218.
123. Rajaeieh, G., Marasi, M., Shahshahan, Z., Hassanbeigi, F., & Safavi, S. M., (2014). The relationship between intake of dairy products and polycystic ovary syndrome in women who referred to Isfahan University of Medical Science Clinics in 2013. *Int. J. Prev. Med., 5*(6), 687–694.
124. Ramanand, S. J., Ghongane, B. B., & Ramanand, J. B., (2013). Clinical characteristics of polycystic ovary syndrome in Indian women. *Ind. J. Endocrinol. Metab., 17*(1), 138–145.
125. Ravn, P., Haugen, A. G., & Glintborg, D., (2013). Overweight in polycystic ovary syndrome. An update on evidence based advice on diet, exercise and metformin use for weight loss. *Minerva Endocrinol., 38*(1), 59–76.
126. Roessler, K. K., Birkebaek, C., Ravn, P., Andersen, M. S., & Glintborg, D., (2013). Effects of exercise and group counseling on body composition and VO2max in overweight women with polycystic ovary syndrome. *Acta Obstet. Gynecol Scand., 92*(3), 272–277.
127. Romero, G. G., & Morreale, H. F. E., (2006). Hyperandrogenism, insulin resistance and hyperinsulinemia as cardiovascular risk factors in diabetes mellitus. *Curr. Diabetes Rev., 2*(1), 39–49.
128. Rutkowska, A. Z., & Kandarakis, E. D., (2016). Polycystic ovary syndrome and environmental toxins. *Fertil. Steril., 106* (4), 948–958.

129. Salpeter, S. R., Greyber, E., Pasternak, G. A., & Salpeter, E. E., (2010). Risk of fatal and nonfatal lactic acidosis with metformin use in type 2 diabetes mellitus. *Cochrane Database Syst. Rev.*, CD002967.
130. Seirfi, H., Farnaghi, F., & Vasheghani-farahani, A., (2007). Assessment of androgens in women with adult-onset acne. *Int. J. Dermatol., 46*, 1188–1191.
131. Sekhon, L. H., Sajal, G., Yesul, K., & Ashok, A., (2010). Female infertility and antioxidants. *Curr. Women's Health Rev., 6*, 84–95.
132. Sharman, M. J., Gomez, A. L., Kraemer, W. J., & Volek, J. S., (2004). Very low-carbohydrate and low-fat diets affect fasting lipids and postprandial lipemia differently in overweight men. *J. Nutr., 134*(4), 880–885.
133. Shen, C. C., Yang, A. C., Hung, J. H., Hu, L. Y., & Tsai, S. J., (2015). A nationwide population-based retrospective cohort study of the risk of uterine, ovarian and breast cancer in women with polycystic ovary syndrome. *Oncologist., 20*(1), 45–49.
134. Shlomo, B. I., & Younis, J. S., (2014). Basic research in PCOS: Are we reaching new frontiers? *Reprod Biomed Online., 28*(6), 669–683.
135. Sim, S. Y., Chin, S. L., Tan, J. L., Brown, S. J., Cussons, A. J., & Stuckey, B. G., (2016). Polycystic ovary syndrome in type 2 diabetes: Does it predict a more severe phenotype? *Fertil. Steril., 106*(5), 1258–1263.
136. Sirmans, S. M., & Pate, K. A., (2013). Epidemiology, diagnosis, and management of polycystic ovary syndrome. *Clin. Epid., 6*, 1–13.
137. Somani, N., & Turvy, D., (2014). Hirsutism: an evidence-based treatment update. *Am. J. Clin. Dermatol., 15*(3), 247–266.
138. Sorensen, L. B., Soe, M., Halkier, K. H., Stigsby, B., & Astrup, A., (2012). Effects of increased dietary protein-to-carbohydrate ratios in women with polycystic ovary syndrome. *Am. J. Clin. Nutr., 95*(1), 39–48.
139. Soulter, I., Sanchez, l., Perez, M., Bartolucci, A., & Azziz, R., (2004). The prevalence of androgen excess among patients with minimal unwanted hair growth. *Am. J. Obstet. Gynecol., 191*, 1914–1920.
140. Spritzer, P. M., Motta, A. B., Petermann, S. T., & Kandarakis, D. E., (2015). Novel strategies in the management of polycystic ovary syndrome. *Minerva Endocrinol., 40*(3), 195–212.
141. Sprung, V. S., Cuthbertson, D. J., Pugh, C. J., Aziz, N., Kemp, G. J., Daousi, C., Green, D. J., Cable, N. T., & Jones, H., (2013). Exercise training in polycystic ovarian syndrome enhances flow-mediated dilation in the absence of changes in fatness. *Med. Sci. Sports Exerc., 45*(12), 2234–2242.
142. Stepto, N. K., Cassar, S., Joham, A. E., Hutchison, S. K., Harrison, C. L., Goldstein, R. F., & Teede, H. J., (2013). Women with polycystic ovary syndrome have intrinsic insulin resistance on euglycaemic-hyperinsulaemic clamp. *Hum. Reprod., 28*(3), 777–784.
143. Sterling, L., Liu, J., Okun, N., Sakhuja, A., Sierra, S. A., & Greenblatt, E., (2016). Pregnancy outcomes in women with polycystic ovary syndrome undergoing *in vitro* fertilization. *Fertil. Steril., 105*(3), 791.e2–797.e2.
144. Tang, T., Lord, J. M., Norman, R. J., Yasmin, E., & Balen, A. H., (2012). Insulin-sensitizing drugs (metformin, rosiglitazone, pioglitazone, D-chiro-inositol) for women with polycystic ovary syndrome, oligo amenorrhoea and subfertility. *Cochrane Database Syst Rev.*, 5, CD003053.
145. Taponen, S., Ahonkallio, S., & Martikainen, H., (2004). Prevalence of polycystic ovaries in women with self-reported symptoms of oligomenorrhoea and/or hirsutism: Northern Finland Birth Cohort 1966 Study. *Hum. Reprod.*, 1–6.

146. Teede, H., Deeks, A., & Moran, L., (2010). Polycystic ovary syndrome: A complex condition with psychological, reproductive and metabolic manifestations that impacts on health across the lifespan. *BMC Med.*, *8*, 41.
147. Thathapudi, S., Kodati, V., Erukkambattu, J., Katragadda, A., Addepally, U., & Hasan, Q., (2014). Anthropometric and biochemical characteristics of polycystic ovarian syndrome in south indian women using AES-2006 Criteria. *Int. J. Endocrinol. Metab.*, *12*(1), e12470.
148. Thomson, R. L., Buckley, J. D., Noakes, M., Clifton, P. M., Norman, R. J., & Brinkworth, G. D., (2008). The effect of a hypocaloric diet with and without exercise training on body composition, cardiometabolic risk profile, and reproductive function in overweight and obese women with polycystic ovary syndrome. *J. Clin. Endocrinol. Metab.*, *93*, 3373–3380.
149. Vgontzas, A. N., Legro, R., Bixler, E. O., Grayev, A., Kales, A., & Chrousos, G. P., (2001). Polycystic ovary syndrome is associated with obstructive sleep apnea and daytime sleepiness: Role of insulin resistance. *J. Clin. Endocrinol. Metab.*, 517–520.
150. Vipin, V. P., Dabadghao, P., Shukla, M., Kapoor, A., Raghuvanshi, A. S., & Ramesh, V., (2016). Cardiovascular disease risk in first-degree relatives of women with polycystic ovary syndrome. *Fert Ster.*, *105*(5), 1338.e3–1344.e3.
151. Volek, J. S., Sharman, M. J., Gomez, A.L, Scheett, T. P., & Kraemer, W. J., (2003). An isoenergetic very low carbohydrate diet improves serum HDL cholesterol and triacylglycerol concentrations, the total cholesterol to HDL cholesterol ratio and postprandial pipemic responses compared with a low-fat diet in normal weight, normolipidemic women. *J. Nutr.*, *133*(9), 2756–2761.
152. Wang, F., Pan, J., Liu, Y., Meng, Q., Lv, P., Qu, F., et al., (2015). Alternative splicing of the androgen receptor in polycystic ovary syndrome. *Proc. Natl. Acad. Sci. USA.*, *112*(15), 4743–4748.
153. Williams, T., Mortada, R., & Porter, S., (2016). Diagnosis and treatment of polycystic ovary syndrome. *Am. Fam. Phys.*, *94*(2), 106–113.
154. Witchel, S. F., Oberfield, S., Rosenfield, R. L., & Coder, E., (2015). The diagnosis of polycystic ovary syndrome during adolescence. *Horm. Res. Paediatr.*, *83*, 376–389.
155. Wong, J. M., Gallagher, M., Gooding, H., Feldman, H. A., Gordon, C. M., Ludwig, D. S., & Ebbeling, C. B., (2016). A randomized pilot study of dietary treatments for polycystic ovary syndrome in adolescents. *Pediatr. Obes.*, *11*(3), 210–220.
156. Yildiz, B. O., Bozdag, G., Yapici, Z., Esinler, I., & Yarali, H., (2012). Prevalence, phenotype and cardiometabolic risk of PCOS under different diagnostic criteria. *Hum. Reprod.*, *27*(10), 3067–3073.
157. Zandi, S., Farajzadeh, S., & Safari, H., (2010). Prevalence of polycystic ovary syndrome in women with acne: Hormone profiles and clinical findings. *J. Pak. Assoc. Dermat.*, *20*, 194–198.
158. Zuuren, E. J. V., & Pijl, H., (2007). Hirsutism. *Ned Tijdschr. Geneeskd.*, *151*(42), 2313–2318.
159. Zuuren, E. J. V., Fedorowicz, Z., Carter, B., & Pandis, N., (2015). Interventions for hirsutism (excluding laser and photoepilation therapy alone). *Cochrane Database Syst Rev.*, 4, CD010334.

CHAPTER 12

# ROLE OF HERBAL MEDICINES DURING PREGNANCY AND LABOR

FANUEL LAMPIAO, IBRAHIM CHIKOWE, MAYESO GWEDELA, and KONDWANI KATUNDU

## ABSTRACT

Herbal remedies in pregnancy pose a danger to both the mother and the fetus, considering that they possess pharmacologically active ingredients but are not subject to the rigorous regulations of synthetic drugs. Women are often unaware of the potential dangers that herbal medicines have, ranging from teratogenicity to interactions with other drugs. It is indeed of concern that women often use herbal medicines concurrently with other prescribed conventional medicine, posing a risk for drug-drug interaction. There has been an increased rate of the wide use of herbal medicines throughout the world, but caution needs to be taken since some of the herbs may cause direct end-organ damage, especially those containing cantharidin. Indeed, it is important that women realized that the safety of herbal medicines should always be questioned and taken into consideration. Therefore, it is recommended that pregnant woman should use the uncertified herbal product to a minimum and preferably not at all. The studies should be conducted to carry out quality assurance, efficacy, and toxicity testing on herbal medicines, which will help to strengthen the knowledge base and build evidence around locally used herbal medicine for pregnancy.

## 12.1 INTRODUCTION

Many women use therapeutic herbs during pregnancy and childbirth, throughout the world. Unfortunately, most of these herbs during pregnancy and childbirth have not been clinically tested to certify their safety and efficacy, even though anecdotal evidence may support their use. In Africa in the absence of adequate biomedical services, traditional healers in rural African communities tend to be first professionals consulted by 80–85% of the women [51] with health problems because of easy access geographically and culturally accepted treatments. Moreover, they are cost-effective compared to conventional hospitals [39]; and these have credibility, acceptance, and respect among the community persons [39].

In African societies, there is a group of traditional healers specialized in treating pregnant women. They are called traditional birth attendants (midwives) and are usually women with high respect for their obstetric and ritual expertise. Their area of specialty is to look after pregnant women and assist in the delivery of a child. They have the duty to teach the pregnant mothers to avoid behavior that will put their pregnancy at risk, ritual bathing of the mother, ritual disposal of the placentas, and provision of medicines that help mothers to heal after delivery. They also have the knowledge of postpartum, cord care, and provide advice on contraception and fertility.

High maternal mortality rates remain a great challenge for most developing countries. Some of the causes of maternal mortality include obstructed labor, prepartum, and postpartum hemorrhage, and uterus rupture [32]. The use of herbal medicines in pregnancy and labor has been reported to be a contributing factor to the high maternal mortality among women, especially in poor resource countries.

This review chapter discusses the reasons:

1. Why pregnant women use herbal medicines during pregnancy and labor;
2. Effects of herbs on embryonic and fetal development;
3. Effects of herbs on the outcomes of pregnancy.

## 12.2 WHY DO PREGNANT WOMEN USE MEDICINAL HERBS?

The use of herbal medicine during pregnancy is highly prevalent nowadays, both in developed and developing countries. The high usage may be due to the promotion of natural living and the use of natural products as herbal medicines [37, 40, 47, 48]. Studies from case studies in Europe, America, and Australia have shown that women tend to seek information about herbal medication from untrained and unprofessional sources [2, 18, 24, 46]. In

African countries, studies have shown that the older generations transfer the knowledge on traditional herbs during pregnancy to the younger generation [35]. For those women in Africa who deliver at a traditional birth attendant, the use of herbal medicine during the delivery is often high [30].

Cost and accessibility are other factors that may influence the use of herbal medicines in other settings. Unlike in developed countries where herbal medicines may be costly, in developing settings such as most African countries, accessibility, and reduced cost of buying herbal medicine seem to be an important motivator for the use of herbs compared to conventional medicine [4]. The perception that herbal medicines are safe is another misconception but rather an important factor that influences pregnant women to use herbal medicine.

Safety has been one of the common reasons given for the use of herbal medicine in several studies [1, 3, 13, 40]. It is a concern that not only do most women consider herbal medication use safe and more effective than orthodox medication; but even those — who consider them as harmful — still take herbal products [2, 13, 46].

To compound the problem of taking herbal medicine among pregnant women is the fact that discloser of herbal medicine use by the health care provider is very low in many settings and is not often reported [7, 14, 20, 42]. Lack of disclosure is even a challenge in developed countries [17]. These may be the reasons why women do not disclose the use of herbal products to clinicians [26]. It appears that lack of confidence in the conventional health-care providers and their negative attitude towards herbal medicine inhibits these women from disclosure [20]. Another challenge is that these herbal products are considered usually as dietary supplements and not medication per se; hence, regulation and monitoring are difficult [5].

### *12.2.1 PREVENTION OF MISCARRIAGE*

About 25% of all pregnancies do not continue to term [53]. Loss of pregnancies may be due to several reasons such as rejection of the fetus by the mother's immune system, a loose cervix, and a nonviable fetus. Pregnant women facing a threatened miscarriage have few options to avert the situation. Maliwichi-Nyirenda et al., [32] sampled Mulanje district in Malawi to investigate medicinal plants during pregnancy and documented twenty-one plant species to treat abdominal pains and miscarriage.

### 12.2.2 FACILITATING LABOR

Literature review has shown that about 185 medicinal plants in Africa are used by pregnant women to induce labor. Other uses included expulsion of the placenta and management of postpartum hemorrhage [50]. A study by Borokini et al., [6] in Oyo State - Nigeria revealed that herbal preparations were used for ante-natal care and safe delivery. Among the Brou, Saek, and Kry ethnic groups in Lao People`s Democratic Republic, some herbs are used to facilitate childbirth and to assist recovery after a miscarriage [10].

### 12.2.3 FETAL HEALTH

Some research studies in Africa have indicated that some women wanted to have healthy babies [15, 31] with adequate weight. Therefore, these women take herbs for good fetal development. Similarly, in Malaysia, women are said to use herbs to promote fetal physical health and intelligence [45].

### 12.2.4 EXPECTANT MOTHER'S HEALTH

Pregnant women have used herbs to treat pregnancy-related or pregnancy-unrelated problems [43]. Herbs have been used to treat colds, respiratory illnesses, and skin problems during pregnancy [38]. Pregnancy-related dermatological conditions (such as stretch marks and cellulitis) have also been reported to be treated with herbs [9]. In addition, women are reported to use herbal therapies against urinary tract infections [3]. Cranberry (*Vaccinium macrocarpon*) has antibiotic properties against bacteria such as *E. coli* and *Pseudomonas aeruginosa*; and is thus used against urinary tract infections during pregnancy [29]. Some herbs are used for the expulsion of the placenta and management of postpartum hemorrhage [50].

## 12.3 EFFECTS OF HERBS ON DEVELOPMENT OF EMBRYO AND FETAL

Although herbs are widely accepted as safe and with few side effects because they are natural [22], yet data is very limited on their safety in pregnancy use [29]. In fact, most of the herbs circulating worldwide have unknown standardized active ingredients. Toxicity testing has, therefore, been a way of revealing and extrapolating some risks associated with them [22]. Research has shown that herbal medicines are associated with risk factors such as toxic potential, consumer-related factors (age, disease factors, and pregnancy), date of expiration, contamination, adulteration, and lack of regulation [27]. Primary risks associated with herbal use in pregnancy are toxicity to the mother with the potential to affect the fetus indirectly, fetotoxicity, teratogenesis, increased miscarriage risks, and poor neonatal health [22].

Though some herbals are safe and important for the sustainability of pregnancy, yet others are also harmful and pose the considerable potential to harm the fetus (fetotoxicity) and embryo (embryotoxicity) during pregnancy [44, 54]. Therefore, traditional medicines must conform to modern safety requirements if they are to be accepted. The safety of herbal medicines (western and traditional) during pregnancy, however, is questionable due to insufficient data [54]. Most known knowledge about herbal medicine today has been derived from historical use, empirical and observational evidences and limited pharmacologic and animal studies. Producing evidence for harm caused by herbs is a challenging task; and establishing the causal-effect relationship is even more challenging [22].

As per tradition with conventional medicines, herbal, and traditional medicine toxicity tests start with preclinical testing, such as animal tests and/or non-animal tests (of crude extracts and purified compounds). Examples of preclinical tests include:

- Cell-based cytotoxicity testing.
- Herbal toxicokinetics.
- Toxicogenomic screening tools (DNA microassays, proteomics, metabolomics).
- High throughput next-generation sequencing.

Development of toxicity testing has been dominated using rats, rodents, zebrafish, roundworm models, and stem cell culture. The sufficient preliminary tests lead to clinical testing (clinical/safety trials) [22].

Apart from testing for direct toxicity of herbs on embryo and fetal effects, herbs, and traditional medicines should also be evaluated for toxicities against vital processes that might play an important role in embryo and fetus development, such as, genetic toxicity, endocrine action, and apoptosis effects, and ability to cross the blood-brain-barrier (BBB) and blood-placenta barrier. Furthermore, developmental neurotoxicity is also evaluated especially for herbs that target the nervous system because the developing nervous system of the fetus is susceptible to toxic insults and even lower level exposure can induce functional changes [52]. Various studies have been conducted to determine the toxic effects of herbs on the embryo and fetus, and they have not been given a final authority on the safety and risks of these medicines. Selected case studies on the embryotoxicity and fetotoxicity of herbs are discussed in this section.

Lin-Yan Li et al., [28] assessed the embryotoxicity of four commonly used Chinese herbs in China during pregnancy: *Rhizoma Atractylodes macrocephala, Radix isatidis, Coptis chinensis,* and *Flos Genkwa.* The herbs were studied using embryonic stem cell test (EST) and prediction models based on concentrations of compounds inhibiting the proliferation of 50% embryonic stem cells (ESCs; $IC_{50}$ES), concentrations inhibiting 50% of 3T3 cells ($IC_{50}$3T3) and concentrations inhibiting differentiation of 50% ESCs ($IC_{50}$ES). *Rhizoma Atractylodes macrocephala* and *Radix Isatidis* did not show embryotoxicity, while *Coptischinensis* and *Flos Genkwa* showed weak embryotoxicity and strong embryotoxicity, respectively [28].

Golalipour *et al.,* [16] conducted an embryotoxicity study of *Mentha piperita L.* on Balb/c mice during the organogenesis period. The study found no significant differences regarding bone ossification (no delay), congenital malformations (not gross) between treatment and control groups; though the mean weight of fetuses was decreased in the treatment group compared to control group.

Moallen and Niapour [34] conducted an embryotoxicity study of *Perovskia abrotanoides* in mice during organogenesis. This herbal medicine is used as "antistress" tranquilizer, energizer drink, fighting common cold and bronchitis, especially in pregnancy. Ethanolic (0.125g/kg) and aqueous extracts (0.25g/kg) were administered *in vivo.* Results showed toxic effects like minimal maternal toxicity and significant resorption, stillbirth, and fetal malformations like polydactyly, spina bifida, aglossia, tarsal extensor, gastroschisis, and numerous skeletal abnormalities mainly with ethanolic extracts. Specific skeletal abnormalities based on the type of extract were also observed.

Animal studies on the effects of pyrrolizidine alkaloid-containing herbs on the fetus have shown the possibility of the compound to undergo trans-placental passage and transfer to the breast milk. A case of fatal neonatal liver injury has also been reported in the offspring of a mother, who had used pyrrolizidine-containing herbal tea for cough during pregnancy [41].

*Aloe vera* herb, a widely used and accepted herb worldwide, has been described as potentially unsafe for internal use in pregnancy due to its purgative properties. A toxicity study of aloe vera-derived ingredients in mice and rats through acute oral, parenteral, and intravenous studies have shown negative results for the development of toxicity. Furthermore, ethanolic extracts of *Aloe Vera* concentrations 100mg/kg ingested orally in rats for a chronic period have shown reproductive toxicity, sperm damage, and inflammation effects when compared to the control group [23].

The study by Xiao et al., [55] revealed the toxicity of herbal medicine Cauwo on the embryo of a rat. Embryos isolated during the organogenesis period (gestational day 9.5) were exposed to different concentrations of the herb. The growth and differentiation rates were observed after 48 hours. Results showed positive teratogenic effects, such as cardiac defects, irregular somites, and brain malformations (narrow brain vesicles).

The fetotoxicity study of retrorsine on mice embryos, in which different concentrations of the herb were administered into the whole embryos revealed that the herbal ingredients had the potential for fetal toxicity. Among significant toxicity, observed effects were: delayed embryo development, malformations in all organic morpho-differentiation indexes, especially in Otic-olfactory organ, branchial arch, maxillary, mandible, and bud [19].

In a study of embryotoxicity and teratogenicity of artemisinin, SD rats were administered with different doses of artemisinin orally during the gestation period (7–17 days). Results showed positive results for developmental retardation (low body weight, short body height, and short tail), increased embryo-lethality, post-implantation loss, and skeletal abnormalities. These results showed embryotoxicity and teratogenicity effects in rats and potential risk for humans [11].

Experimental studies, evaluating the female reproductive and/or developmental toxicity of some herbal medicines recommended for use in Brazil, showed both positive and negative results [52]. The common negative effects observed in the studies were:

- Maternal toxicity;
- Embryonic loss before implantation;
- Resorption or malformations;
- Estrogenic effect in utero-trophic bioassays;
- Early, and mid-term abortion;
- Morphological abnormalities in the head;
- Trunk and limbs, preimplantation loss or resorptions;
- Fetal death;
- Dose-related maternal weight loss;
- Dose-related alteration of normal blastocyst formation;
- Embryotoxicity and teratogenicity;
- Ossification;

- Fetotoxicity;
- External malformations;
- Pulps physical development;
- Disruption of estrogen cycle and fetal growth retardation [52].

## 12.4 EFFECTS OF HERBS ON PREGNANCY OUTCOMES

Herbal remedies, particularly in pregnancy, pose a danger to both the mother and the fetus, considering that they possess pharmacologically active ingredients but are not subject to the rigorous regulations [13]. Women are often unaware of the potential dangers that herbal medicines have, ranging from teratogenicity to interactions with other drugs [37, 46]. It is indeed of concern that women often use herbal medicines concurrently with other prescribed conventional medicine, posing a risk for drug-drug interaction [35]. Furthermore, it is a concern that there is a considerable proportion of women using herbal preparations during the first trimester, considering that this is a critical period of organ development [7]. Several studies and case reports have offered some evidence linking the use of herbal medicine in pregnancy to poor pregnancy outcomes, maternal morbidity, and neonatal mortality.

The Taiwanese cross-sectional study found that a possible link between taking herbal medicine namely *Huang Lian* and *An-Tai-Yin* in the first trimester was associated with giving birth to babies with congenital malformations of the nervous system and musculoskeletal and connective tissue, respectively [8]. Though such a study could not display a cause and effect relationship due to its observational nature, yet it had the strength of a large sample size and offered a hint of necessary caution when using such herbal medicines especially in the first trimester, during which organogenesis is prominent.

It is not only the first trimester, which is very critical; there is still a risk of taking herbal medicine in the second and third trimester to poor pregnancy outcome [36]. The case-control study showed increased odds of giving birth to a preterm child given the use of flax, but the use of chamomile, peppermint, and green tea did not seem to show the same effect [36]. A similar observation was made from a multicenter retrospective study in Italy where after adjusting for other confounders, the use of almond oil as an herbal supplement for at least 3 months was associated with increased odds of preterm birth [12].

## 12.5 SUMMARY

Worldwide pregnant women use medicinal herbs for various reasons. While herbal medicines play a significant role during pregnancy, especially among the poor rural women, yet little research has been done on these herbs to demonstrate their efficacy and safety. It is high time that studies should be conducted to carry out quality assurance, efficacy, safety, and toxicity testing on herbal medicines, which will help to strengthen the knowledge base and build evidence around locally used herbal medicine during pregnancy.

## KEYWORDS


- **contraception**
- **efficacy**
- **fertility**
- **herbal medicine**
- **embryonic stem cell test**
- **toxicity**
- **traditional healers**


## REFERENCES


1. Achema, G., Emmanuel, A., & Oguche, M., (2012). Evaluation of the use of herbal drugs by pregnant women in Nigeria. *African Journal of Midwifery & Women's Health, 6*, 78–83.
2. Adams, J., Chi-Wai, L., Sibbritt, D., Broom, A., Wardle, J., Homer, C., & Beck, S., (2009). Women's use of complementary and alternative medicine during Pregnancy: A critical review of the literature. *Birth Issues in Perinatal. Care, 36*, 237–245.
3. Al-Ramahi, R., Jaradat, N., & Adawi, D., (2013). Use of herbal medicines during pregnancy in a group of Palestinian women. *Journal of Ethnopharmacology, 150*, 79–84.
4. Bayisa, B., Tatiparthi, R., & Mulisa, E., (2014). Use of herbal medicine among pregnant women on antenatal care at Nekemte Hospital, Western Ethiopia. *Jundishapur Journal of Natural Pharmaceutical Products, 9*, e17368–e17368.
5. Born, D., & Barron, M. L., (2005). Herb use in pregnancy: What nurses should know? *MCN: The American Journal of Maternal Child Nursing, 30*, 201–208.
6. Borokini, T. I., Ighere, D. A., & Clement, M., (2013). Ethnobiological survey of traditional medicine practice for women's health in Oyo State. *Journal of Medicinal Plants Studies, 1*, 17–29.

7. Broussard, C. S., Louik, C., Honein, M. A., & Mitchell, A. A., (2010). Herbal use before and during pregnancy. *American Journal of Obstetrics and Gynecology, 202*, 443, e1–443.e6.
8. Chuang, C., Chang, P., Hsieh, W., Tsai, Y., Lin, S., & Chen, P., (2009). Chinese herbal medicine use in Taiwan during pregnancy and the postpartum period: A population-based cohort study. *International Journal of Nursing Studies, 46*, 787–795.
9. Cuzzolin, L., Francini-Pesenti, F., & Verlato, G., (2010). Use of herbal products among 392 Italian pregnant women: Focus on pregnancy outcome. *Pharmacoepidemiology and Drug Safety*, *9*, 1151–1158.
10. De Boer, H., & Lamxay, V., (2009). Plants used during pregnancy, childbirth and postpartum healthcare in Lao PDR: A comparative study of the Brou, Saek and Kry ethnic groups. *Journal of Ethnobiology and Ethnomedicine, 5*, 25–27.
11. El-Dakdoky, M. H., (2009). Evaluation of the developmental toxicity of artemether during different phases of rat pregnancy. *Food Chem. Toxicol.*, *47*, 1437–1441.
12. Facchinetti, F., Pedrielli, G., Benoni, G., Joppi, M., Verlato, G., Dante, G., Balduzzi, S., & Cuzzolin, L., (2012). Herbal supplements in pregnancy: Unexpected results from a multicenter study. *Human Reproduction, 27*, 3161–3167.
13. Fakeye, T., Adisa, R., & Musa, I. E., (2009). Attitude and use of herbal medicines among pregnant women in Nigeria. *BMC Complementary and Alternative Medicine, 9*, 53–57.
14. Gardiner, P., Kemper, K. J., Legedza, A., & Phillips, R. S., (2007). Factors associated with herb and dietary supplement use by young adults in the United States. *BMC Complementary and Alternative Medicine, 7*, 39–41.
15. Gbadamosi, I. T., & Otobo, E. R., (2014). Assessment of the nutritional qualities of ten botanicals used in pregnancy and child delivery in Ibadan, Nigeria. *International Journal of Phytomedicine, 6*, 16–22.
16. Golalipour, M. J., Ghafari, S., Maleki, A., Kiani, M., Asadu, E., & Farsi, M., (2011). Study of embryotoxicity of *Mentha piperita* L. during organogenesis in Balb/c mice. *Int. J. Morphol.*, *29*, 862–867.
17. Hall, H., & Jolly, K., (2014). Women's use of complementary and alternative medicines during pregnancy: A cross-sectional study. *Midwifery, 30*, 499–505.
18. Hall, H. G., Griffiths, D. L., & McKenna, L. G., (2011). The use of complementary and alternative medicine by pregnant women: Literature review. *Midwifery, 27*, 817–824.
19. Han, J., Liang, A., & Yi, Y., (2011). Developmental toxicity of retrorsive on mouse embryos *in vitro*. *Zhongguo Zhong Yao Za Zhi*, *36*, 1901–1904.
20. Holst, L., Wright, D., Haavik, S., & Nordeng, H., (2009). The use and the user of herbal remedies during pregnancy. *Journal of Alternative & Complementary Medicine, 15*, 787–792.
21. Holst, L., Nordeng, H., & Haavik, S., (2008). Use of herbal drugs during early pregnancy in relation to maternal characteristics and pregnancy outcome. *Pharmaco Epidemiology and Drug Safety, 17*, 151–159.
22. Ifeoma, O., & Oluwakanyinsola, S., (2013). Screening of herbal medicines for potential toxicities. *New Insights into Toxicity and Drug Testing*, *CC-BY*, e-article: doi: 10.5772/54493.
23. Kalder, M., Knoblauch, K., Hrgovic, I., & Münstedt, K., (2011). Use of complementary and alternative medicine during pregnancy and delivery. *Archives of Gynecology & Obstetrics, 283*, 475–482.

24. Kennedy, D. A., Lupattelli, A., Koren, G., & Nordeng, H., (2013). Herbal medicine use in pregnancy: Results of a multinational study. *BMC Complementary and Alternative Medicine, 13*, 355–355.
25. Ko, R. J. U. S., (2004). Perspective on the adverse reactions from traditional Chinese medicines. *Journal of the Chinese Medical Association, 67*, 109–116.
26. Kochhar, K., Saywell, R. J., Zollinger, T. W., Mandzuk, C. A., Haas, D. M., Howell, L. K., Martir, J., & Reger, M. K., (2010). Herbal remedy use among Hispanic women during pregnancy and while breastfeeding: Are physicians informed? *Hispanic Health Care International, 8*, 93–106.
27. Lanini, J., Duarte-Almeida, J. M., Nappo, S. A., & Carlini, E. A., (2012). Are medicinal herbs safe?: The opinion of plant vendors from Diadema. *Revista Brasileira de Farmacognosia, 22*, 102–105.
28. Li, L. Y., Cao, F. F., Su, Z. J., Zhang, Q. H., Dai, X. Y., Xiao, X., Huang, Y. D., Zheng, Q., & Xu, H., (2015). Assessment of the embryo toxicity of four Chinese herbal extracts using the embryonic stem cell test. *Mol Med Rep.*, E-pub., doi: 10.3892/mmr.2015.3598.
29. Low, T., (2009). The use of botanicals during pregnancy and lactation. *Alternative Therapies, 15*, 54–58.
30. Mabina, M. H., Moodley, J., & Pitsoe, S. B., (1997). The use of traditional herbal medication during pregnancy. *Tropical Doctor, 27*, 84–86.
31. Malan, D. F., & Neuba, D., (2011). Traditional practices and medicinal plants use during pregnancy by Anyi-Ndenye women (Eastern Côte d'Ivoire). *African Journal of Reproductive Health March, 15*, 85–94.
32. Maliwichi-Nyirenda, C. P., & Maliwichi, L. L., (2010). Traditional methods used in family planning and conception in Malawi: A case study of Mulanje district. *Indilinga–African Journal of Indigenous Knowledge Systems, 9*, 230–237.
33. Maliwichi-Nyirenda, C. P., & Maliwichi, L. L., (2010). Medicinal plants used for contraception and pregnancy-related cases in Malawi: A case study of Mulanje district. *Journal of Medicinal Plants Research, 4*, 3024–3030.
34. Moallen, S. A., & Niapour, M., (2008). Study of embryotoxicity of Perovskiaabrotanoides, an adulterant in folk-medicine, during organogenesis in Mice. *J. Ethnopharmacol., 117*, 108–114.
35. Mothupi, M. C., (2014). Use of herbal medicine during pregnancy among women with access to public healthcare in Nairobi, Kenya: A cross-sectional survey. *BMC Complementary and Alternative Medicine, 14*, 432–436.
36. Moussally, K., & Bérard, A., (2010). Exposure to herbal products during pregnancy and the risk of preterm birth. *European Journal of Obstetrics & Gynecology and Reproductive Biology, 150*, 107–108.
37. Nordeng, H., & Havnen, G. C., (2005). Impact of socio-demographic factors, knowledge and attitude on the use of herbal drugs in pregnancy. *Acta Obstetricia Et Gynecologica Scandinavica, 84*, 26–33.
38. Olowokere, A. E., & Olajide, O., (2013). Women's perception of safety and utilization of herbal remedies during pregnancy in a local government area in Nigeria. *Clinical Nursing Studies, 1*, 1–14.
39. Peltzer, K., & Mngqundaniso, N., (2008). Patients consulting traditional health practitioners in the context of HIV/AIDS in urban areas in Kwazulu-Natal, South Africa. *African Journal of Traditional, Complementary and Alternative Medicines, 5*, 370–379.

40. Rahman, A. A., Sulaiman, S. A., Ahmad, Z., Daud, W. N. W., & Hamid, A. M., (2008). Prevalence and pattern of use of herbal medicines during pregnancy in Tumpat district, Kelantan. *The Malaysian Journal of Medical Sciences, 15*, 40–48.
41. Rasenack, R., Muller, C., Kleinschmidt, M., Rasenack, J., & Wiedenfeld, H., (2003). Veno-occlusive disease in a fetus caused by pyrrolizidine alkaloids of food origin. *Fetal Diagn. Ther., 18*, 223–225.
42. Refuerzo, J. S., Blackwell, S. C., Sokol, R. J., Lajeunesse, L., Firchau, K., & Kruger, M., (2005). Use of over-the-counter medications and herbal remedies in pregnancy. *American Journal of Perinatology, 22*, 321–324.
43. Sawalha, A. F., (2007). Consumption of prescription and non-prescription medications by pregnant women: Cross sectional study in Palestine. *The Islamic University Journal, 15*, 41–57.
44. Shinde, P., Patil, P., & Bairagi, V., (2012). Herbs in pregnancy and lactation: Review appraisal. *International Journal of Pharmaceutical Sciences and Research, 12*, 3001–3006.
45. Sooi, L. K., & Keng, S. L., (2013). Herbal medicines: Malaysian women's knowledge and practice. *Evidence-Based Complementary and Alternative Medicine, 13*, 43–49.
46. Steel, A., Adams, J., Sibbritt, D., Broom, A., Frawley, J., & Gallois, C., (2014). Relationship between complementary and alternative medicine use and incidence of adverse birth outcomes: An examination of a nationally representative sample of 1835 Australian women. *Midwifery, 30*, 1157–1165.
47. Tiran, D., (2005). NICE guideline on antenatal care: Routine care for the healthy pregnant woman–recommendations on the use of complementary therapies do not promote clinical excellence: National Institute for Clinical Excellence (NICE). *Complementary Therapies in Clinical Practice, 11*, 127–129.
48. Tiran, D., (2003). The use of herbs by pregnant and childbearing women: Risk-benefit assessment. *Complementary Therapies in Nursing & Midwifery, 9*, 176–181.
49. Tovar, R. T., & Petzel, R. M., (2009). Herbal toxicity. *Disease-a-Month, 55*, 592–641.
50. Tripathi, V., Stanton, C., & Anderson, F., (2013). Traditional preparations used as uterotonics in Sub-Saharan Africa and their pharmacological effects. *International Journal of Gynecology and Obstetrics, 120*, 16–22.
51. UNAIDS, (2006). *Collaborating with Traditional Healers for HIV Prevention and Care in Sub-Saharan Africa: Suggestions for Program Managers and Field Workers* (p. 120). UNAIDS, Geneva.
52. Verissimo, L. F., Bacchi, A. D., Zaminelli, T., De Paula, G. H., & Moreira, E. G., (2011). Herbs of interest to the Brazilian Federal Government: Female reproductive and developmental toxicity studies. *Revista Brasileira de Farmacognosia, 21*, e-article: http://dx.doi.org/10.1590/s0102–695x2011005000146 (Accessed on 30 July 2019).
53. Westfall, R. E., (2001). Herbal medicine in pregnancy and childbirth. *Advances in Therapy, 18*, 47–55.
54. Wiebrecht, A., Gaus, W., Becker, S., Hummelsberger, J., & Kuhlmann, K., (2014). Safety aspects of Chinese herbal medicine in pregnancy-re-evaluation of experimental data of two animal studies and the clinical experience. *Complementary Therapies in Medicine, 22*, 954–964.
55. Xiao, K., Yan, G. Y., Wang, L., Liu, Y. Q., Peng, C., Duan, J. C., Fu, Z. T., & Li, H. X., (2008). Toxicity of Caowu on *in vivo* cultured rat embryos. *Sichuan Da Xue Xue Bao Yi Xue Ban (Journal of Sichuan University, Medical Science Edition), 39*, 441–444.

# PART III

# Novel Research Techniques in Medicinal Plants

# CHAPTER 13

# OMICS TECHNOLOGY: NOVEL APPROACH FOR SCREENING OF PLANT-BASED TRADITIONAL MEDICINES

ROJITA MISHRA, SATPAL SINGH BISHT, and MAHENDRA RANA

## ABSTRACT

Plant-based bioactive components have been the precursors for contemporary research on natural products research and are the main source for unraveling the lead molecules. Synergistic approaches of bioinformatics, system biology, and omics data (transcriptomics, proteomics, metabolomics, and physiognomics) can be utilized to discover and probe the mode of action of natural products. Omics technologies permit transcriptional profiling and generation of proteomics and metabolomics data. Integration of novel systems biology tools and approaches with traditional medicine philosophies can transform new drug development strategies to the biology-driven drug discovery leading to precise and personal medication. Bioinformatics provides data related to diseases, toxicity-related issues, therapeutic aspects, etc. This book chapter reviews the various approaches for augmentation of 'omics' data and its role in research on natural products.

## 13.1 INTRODUCTION

Application and development of computational tools and techniques for biological sciences are called bioinformatics that has the objectives to manage and analyses biological data. Advancement of various omics technologies (such as genomics, proteomics, and metabolomics) has generated a huge data in molecular biology [1]. Systems biology helps to have knowledge about the biological systems and processes and integration of omics data sources for the analysis of network regulation, system working principles from a broader sense with a synergistic approach [2]. Chinese herbal medicine and various information from Charka Samhita, Ayurveda, and Unani systems harbor ample knowledge of important databases in the world. Voluminous 'omics' data is necessary; on the contrary, little genomic information is known about these medicinal plants:

- To cultivate medicinal plants in harsh environmental conditions.
- To develop a more easy method of authentication and classification of medicinal plants.
- To develop the efficacy of traditional medicinal materials.
- To properly identify of functional genes and enzymes that mainly involved in bioactive compound production.
- To save traditional endangered medicinal plant species.

Next-generation sequencing and its application in medicinal plant genomics, bioinformatics, and system biology are in high demand for future research and product development in the therapeutic sector [3].

Plant metabolites are a rich source of various biochemicals, and each plant has its own metabolite fingerprint. This diversity itself has an analytical challenge of both profiling and quantification of such complex metabolites. With the development of scientific knowledge and techniques, we have just started to learn the role of these metabolites, which are mainly involved in adaption to specific ecological niches and environmental conditions. Various omics technologies more specifically genomics along with metabolomics, transcriptomics, and proteomics provide a basic platform for identifying functional genomics and its synergistic consequences towards chemical fingerprinting of medicinal plants for its application in therapeutics and natural product research.

In this book chapter, methods of omics technologies are described briefly for the management of medicinal plant resources and its sustainable utilization.

## 13.2 TRANSCRIPTOMICS

The transcriptome is the study of RNA transcripts that are produced by the genome in a circumstance or inside a specialized cell. Transcriptomics is the analysis of these transcriptomes with the help of high throughput methods like microarray technique.

### 13.2.1 SAGE OR CAGE ANALYSIS

In 1995, Serial Analysis of Gene Expression (SAGE) was developed, which is based on the sequencing. The basic working principle is an analysis of concatenated random transcripts by Sanger's method. Briefly, in SAGE analysis, cDNA is produced from mRNA then is digested into 11bp tag fragments with the help of restriction endonucleases. The cDNA tags are concatenated head to tail into about 500 bp long strands and sequenced using Sanger sequencing. The sequences are deconvoluted into original 11bp tags. Two approaches are there: (1) If a reference genome is available then tags can be aligned to identify the corresponding genes; (2) In other approaches where a reference genome is not available, and tags are used as diagnostic markers if the tag is expressed differentially in the disease state. The Cap Analysis of Gene Expression (CAGE) is a slight variation of SAGE, where sequence tags are from the 5'-end of the mRNA transcript. If the reference genome is available, then transcriptional starting site of the gene is identified in other words, promoter analysis, and cloning of full-length cDNAs become possible [4].

### 13.2.2 MICROARRAYS

The mRNA isolated from the organism and double-stranded cDNA is formed by using reverse transcriptase. In microarrays, this cDNA is fragmented and labeled, which then is attached to sequentially design complementary oligonucleotides, and these can be detected by their fluorescence intensity. Based on their method of preparation, microarrays are of two types:

1. Low-density spotted arrays; or
2. High-density short probe arrays [4].

### 13.2.3 RNA SEQUENCING

In RNA sequencing, the isolated mRNA is converted to double-stranded cDNA. The ds-cDNA is sequenced and aligned to a reference genome sequence. Then easily the regions, which are being transcribed, can be reconstructed. This information can be used to know about the expressed gene, their comparative level of expression, and about their splice variants [4].

### *13.2.4 NATURAL PRODUCT RESEARCH AND TRANSCRIPTOMICS*

#### *13.2.4.1 TRANSCRIPTIONAL DATA ANALYSIS OF DRUG REPOSITIONING*

Drug repositioning can be possible by transcriptome analysis of disease or similar drugs. Transcriptomics information is utilized for the identification of coherent drug targets. Compounds, which are used for disease treatment before can be better screened and drug repositioning, becomes an easy task using transcriptome analysis or gene expression profile of disease and effect on gene expression by the existing natural isolated compounds [5].

#### *13.2.4.2 CONNECTIVITY MAP FROM TRANSCRIPTOMICS TO DRUG DISCOVERY*

Gene expression profiling helps in quantifying the effects of bioactive compounds at the level of transcription and gives information about the living system. Transcriptomic changes following compound administration can be measured in multiple cell lines. Transcriptomic information is used for analyzing the effects of small molecules in physiology and is possible after the arrival of the connectivity map. Several uses like pathway elucidation, toxicity models, and toxicogenomic classifications are developed based on drug-induced gene expression profiling [6].

#### *13.2.4.3 USE OF SINGLE CELL ANALYTICAL TOOLS FOR TRANSCRIPTOMICS AND DRUG DEVELOPMENT*

Single-cell analytical tools for transcriptomics include mass cytometry, single-cell barcode chips, micro engraving, and single-cell western blotting. These techniques help to face major challenges in the discovery of a drug and its development because these techniques provide better accuracy in designing an accurate disease model, help in the better interpretation of biomarker levels and patients response to specific therapies. Hence this is the best way to develop natural product derived drugs by transcriptomics [7].

## 13.3 METABOLOMICS

It mainly deals with the metabolic profiling of medicinal plants, which are essential for drug discovery and phytomedicines [8–15]. Metabolomics is a vital part of a system biology approach, which connects genotype with the diversified specific targeted phenotype of cells, tissue, and organs [16]. It is a qualitative as well as quantity wise measurement of all the metabolites of an individual plant. Sometimes, it is used as a synonym of metabolite profiling [17]. Metabonomics is a derived branch of metabolomics, which is mainly involved in the quantification of the metabolite towards environment, medication, or disease of a tissue or extracellular fluids [18].

Metabolomics mainly involves targeted and global metabolite analysis. Metabolite analysis or complete metabolome can be analyzed using techniques like GC-MS, LC-MS. Metabolite profiling [19] also involves various TLC, HPLC, NMR spectroscopy, Raman Spectroscopy, etc.

Various methods in metabolomics are shown in Figure 13.1. Metabolomics is an analysis of metabolome by various methods like mass spectrometry, FTIR spectroscopy, NMR spectroscopy, and Raman spectroscopy. Broadly this method can be divided into the analysis of metabolomics through chromatography approach and chromatography free approach. Chromatography approach involves HPLC and its variations and MS and its variations. Chromatography free approach includes all the spectroscopic methods like NMR, IR, etc. The Open Source Access Online Journal [https://www.omicsonline.org/special-issue/natural-product-updates-and-application.php] focuses on updates on plant-based biocompounds.

## 13.4 PROTEOMICS

Proteomics mainly describes about the function, interaction, modification, and targeting as well as regulation of proteome expressed by the cell [20]. The simplest way of analyzing protein expression, function localization is by 2D-gel electrophoresis; subsequently, the cellular fractions of the gel are studied using fluorescent dyes and compared with wild type homologs and analogs for comparative study regarding expression level and post-transcriptional modifications [21–23]. Proteomics includes isolation of sub proteomes based on the cellular localization of isoelectric point [23]; mass spectrometric techniques like isotope-coded affinity tags (ICAT) [24] and multidimensional protein identification technology (MudPIT) [25]. Protein function and interactions are being addressed on a genome-wide scale through the development of protein arrays and protein interaction maps [26].

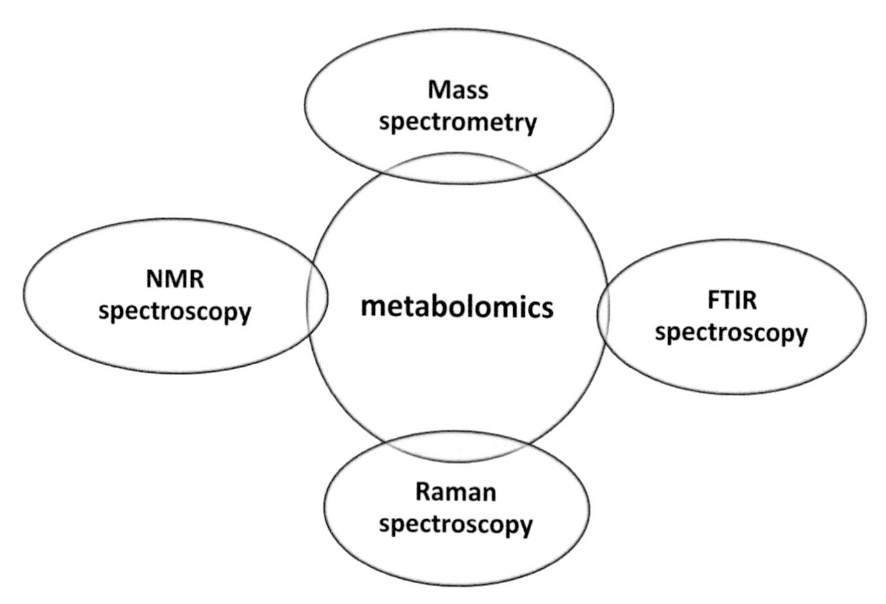


**FIGURE 13.1** Various types of techniques used in metabolomics with the advancement of science.

### *13.4.1 CHEMICAL PROTEOMICS*

Chemical proteomics is an advance technique of drug affinity chromatography, which further along with MS and bioinformatics analysis produces exclamatory results [27]. There are two types of chemical proteomics:

1. Activity-based probe profiling (ABPP); and
2. Compound centric chemical proteomics (CCCP) [28].

## 13.5 GENOMICS

Genomics plays an important role in the herbal drugs and their discovery. More specifically is the development of metagenomics and advance bioinformatics technology for gene mining, which can cluster the diversity of secondary metabolite pathway genes and correlation of these data for their biosynthetic pathways. Mainly if we know the genes and the networking of biosynthetic pathways, which is possible through genomics followed by gene mining with bioinformatics tools, then natural drug production can be channelized through metabolic engineering. Genes can be targeted for expression from inactive clusters, or the pathway can be modified for the production of more useful secondary metabolites.

Epigenomics is a branch of genomics that is essential for understanding the mechanism of the complicated gene expression pattern of medicinal plants. Thus, it plays a vital role in natural product drug discovery [29].

## 13.6 OMICS DATA ANALYSES FOR NATURAL DRUG DISCOVERY

All omics data always need common statistical and machine learning tools. Classification algorithms in medicinal plants for microarray analysis are more useful for authentication, quality maintenance, and characterization of right species. Classification algorithms have a decision tree J48, Random forest (RF), Naivy Bayes (NB), Simple logistic (SL), SVMs, etc. The SVMs usually perform various classifications [30]. This is highly essential to build user-friendly software with a graphical user interface (GUI) to analyze high throughput data for general biologists.

To overcome the problems, various types of user-friendly software have been developed like PMC-SVM (multi-category support vector machines) by Zhang et al., [31]. Similarly, another user-friendly software support vector machine recursive feature elimination (SVMRFE) [32]

## 13.7 AVAILABLE DATABASES AND FUTURE REQUIRED DATABASES

Medicinal plant database mainly provides information regarding taxonomy, cultivation, and therapeutic importance, the geographical distribution of medicinal plants, biologically active compounds, and images of individual species. An example is CMKb, which is a web-based relational prototype database that is used for the integration of Australian aboriginal medicinal plant knowledge. There are databases, which describe medicinal plants for a disease [33–35]. In China, there are many national wide and regional medicinal plant databases, which provide information regarding active/ inactive compounds, chemicals, and targets of medicinal plants. There is a need of medicinal plant genome databases harboring information regarding DNA, mRNA level, and gene function of a group of medicinal plants. Functional omics databases of medicinal plants can be constructed based on functional genomics, proteomics, and metabolomics data of medicinal plants [30].

## 13.8 SUMMARY

Omics studies starting from genomics, transcriptomics, proteomics, epigenomics, metabolomics along with various advanced tools and techniques of bioinformatics can be applied for medicinal plant research towards the development of natural drugs. Now research on natural product drug discovery has started shifting towards multi-omics application and synergistic approach of high-throughput techniques to decrease the cost and decrease the duration of years to develop a full established herbal medicine a new branch of science called reverse pharmacology, similarly pharmacogenomics, nutrigenomics, and metabonomics.

## KEYWORDS

- **chemical proteomics**
- **genomics**
- **metabolomics**
- **omics study**
- **proteomics**
- **SAGE analysis**
- **transcriptomics**

## REFERENCES

1. Andrew, M., & Piggott, P., (2004). Quality, not quantity: The role of natural products and chemical proteomics in modern drug discovery. *Combinatorial Chemistry & High Throughput Screening*, *7*, 607–630.
2. Arulrayan, N., Rangasamy, S., James, E., & Pitchai, D., (2007). A database for medicinal plants used in the treatment of diabetes and its secondary complications. *Bioinformation*, *2*, 22–23.

3. Berggren, K., Steinberg, T. H., & Lauber, W. M., (1999). A luminescent ruthenium complex for ultrasensitive detection of proteins immobilized on membrane supports. *Annals of Biochemistry*, *276*(2), 129–143.
4. Bie, V., Günter, K., & Liesbet, V., (2015). Using transcriptomics to guide lead optimization in drug discovery projects: Lessons learned from the QSTAR project. *Drug Discovery Today*, *20*(5), 505–513.
5. Chen, C., Gonzalez, F. J., & Idle, J. R., (2007).LC–MS-based metabolomics in drug metabolism. *Drug Metabolism Review*, *39*, 581–597.
6. Claudino, W. M., Quattrone, A., Biganzoli, L., Pestrin, M., Bertini, I., & Di Leo, A., (2007). Metabolomics: Available results, current research projects in breast cancer, and future applications. *Journal of Clinical Oncology*, *25*, 2840–2846.
7. Cordwell, S. J., Nouwens, A. S., Verrills, N. M., Basseal, D. J., & Walsh, B. J., (1094). Subproteomics based upon protein cellular location and relative solubilities in conjunction with composite two-dimensional electrophoresis gels. *Electrophoresis*, *21*(6), 1094–1103.
8. Deng, Y., Ai, J., & Xiao, P., (2010). Application of bioinformatics and systems biology in medicinal plant studies. *Chinese Herbal Medicines*, *2*(3), 170–179.
9. Dopazo, J., (2013). Genomics and transcriptomics in drug discovery. *Drug Discovery Today***,** http://dx.doi.org/10.1016/j.drudis.2013.06.003 (Accessed on 30 July 2019).
10. Dunn, W. B., & Ellis, D. I., (*2005*). Metabolomics: Current analytical platforms and methodologies. *Trend Anal. Chem.*, *24*, 285–294.
11. Ellis, D. I., Dunn, W. B., Griffin, J. L., Allwood, J. W., & Goodacre, R., (2007). Metabolic fingerprinting as a diagnostic tool. *Pharmacogenomics*, *8*, 1243–1266.
12. Fiehn, O., Kopka, J., Dormann, P., Altmann, T., Trethewey, R. N., & Willmitzer, L., (2000). Metabolite profiling for plant functional genomics. *Nature Biotechnology*, *18*, 1157–1161.
13. Griffin, J. L., & Kauppinen, R. A., (2007). Tumor metabolomics in animal models of human cancer. *Journal of Proteomics Research*, *6*, 498–505.
14. Griffin, J. L., (2006). Understanding mouse models of disease through metabolomics. *Current Opinion in Chemical Biology*, *10*, 309–315.
15. Gygi, S. P., Rist, B., Gerber, S. A., Turecek, F., Gelb, M. H., & Aebersold, R., (1999). Quantitative analysis of complex protein mixtures using isotope-coded affinity tags. *Nature Biotechno*logy, *17*(10), 994–999.
16. James, R. H., Antoni, R., & Paul, S. M., (2016). Single cell analytic tools for drug discovery and development. *Nature Review of Drug Discovery, 15*(3), 204–216.
17. Jarayaman, K. S., (2000). India pushes ahead with plant database. *Nature*, *405*, 267–277.
18. Kann, M. G., (2010). Advances in translational bioinformatics: Computational approaches for the hunting of disease genes. *Briefing in Bioinformatics*, *11*(1), 96–110.
19. Kasirajan, M., & Gopalakrishnan, R., (2007). A database for medicinal plants used in treatment of asthma. *Bioinformation*, *2*, 105–106.
20. Kell, D. B., (2006). Systems biology, metabolic modeling and metabolomics in drug discovery and development. *Drug Discovery Today*, *11*, 1085–1092.
21. Lie-Fen, S., & Ning-Sun, Y., (2008). Metabolomics for phytomedicine research and drug development. *Current Opinion in Chemical Biology*, *12*, 66–71.
22. Lindon, J. C., Holmes, E., & Nicholson, J. K., (2007). Metabonomics in pharmaceutical R&D. *FEBS J.*, *274*, 1140–1151.

23. Liu, Q. Z., & Sung, A. H., (2009a). Feature selection and classification of MAQC-II breast cancer and multiple myeloma microarray gene expression data. *PLoS One*, *4*(12), e8250.
24. Lowe, R., Shirley, N., Bleackley, M., Dolan, S., & Shafee, T., (2017). Transcriptomics technologies. *PLoS Computational Biology*, *13*(5), e1005457.
25. Mackintosh, J. A., Choi, H. Y., & Bae, S. H., (2273). Fluorescent natural product for ultra-sensitive detection of proteins in one-dimensional and two-dimensional gel electrophoresis. *Proteomics*, *3*(12), 2273–2288.
26. Pirooznia, M., Yang, J. Y., Yang, M. Q., & Deng, Y. P., (2008). Comparative study of different machine learning methods on microarray gene expression data. *BMC Genomics*, *9*, S1–S13.
27. Ray, L. B., Chong, L. D., & Gough, N. R., (2002). Computational biology. *Sci. STKE*, *148*, 10–18.
28. Sumner, L. W., Mendes, P., & Dixon, R., (2003). Plant metabolomics: Large-scale phytochemistry in the functional genomics era. *Phytochemistry*, *62*, 817–836.
29. Ulrich-Merzenich, G., Zeitler, H., Jobst, D., & Panek, D., (2007). Application of the "-omic-" technologies in phytomedicine. *Phytomedicine*, *14*, 70–82.
30. Uwe, R., (2009). Target profiling of small molecules by chemical Proteomics. *Nature Chemical Biology*, *5*(9), 617–625.
31. Walhout, A. J. M., & Vidal, M., (2001). Protein interaction maps for model organisms. *Nature Review of Molecular Cell Biol*ogy, *2*(1), 55–62.
32. Washburn, M. P., Wolters, D., & Yates, J. R., (2001). Large-scale analysis of the yeast proteome by multidimensional protein identification technology. *Nature Biotechnol*ogy, *19*(3), 242–247.
33. Williams, K. L., & Hochstrasser, D. F., (1997). *Proteome Research: New Frontiers in Functional Genomics* (pp. 1–12). Springer-Verlag: Berlin.
34. Xu, Z. N., (2003). Relationship between development of bioinformatics and study of Chinese herbal medicine. *Chinese Traditional Herbal Drugs*, *34*(6), 481–486.
35. Zhang, C. Y., & Li, P., (2006). Parallelization of multicategory support vector machines (PMC-SVM) for classifying microarray data. *BMC Bioinformatics*, *7*, S4–S15.

CHAPTER 14

# AUTOMATIC RECOGNITION AND CLASSIFICATION OF MEDICINAL PLANTS: A REVIEW

OLUGBENGA KAYODE OGIDAN and
ABIODUN EMMANUEL ONILE

## ABSTRACT

Some existing methods for recognizing and classifying medicinal plants are manual, cumbersome, and time-consuming. In this chapter, a comprehensive review of recognition and classification of medicinal plants using Information Communication Technologies (ICT) – Automated Techniques are presented. The study focuses on the recognition and classification of medicinal plant's leaves using image processing-based and spectroscopic identification techniques. The work reveals that the image processing-based recognition method is more predominant in literature than the spectroscopic method of recognizing medicinal plants. Analysis of previous studies reveals that image processing-based and spectroscopic recognition methods are less cumbersome, faster, and non-destructive when compared to the chemical method. The details of various implementation platforms that are required for effective recognition and classification of medicinal plants are also presented in this chapter. It is believed that with the techniques outlined in this study, more people, including non-experts using electronic devices, would be able to easily recognize and classify medicinal plants. This would offer better insights into their usefulness and conservation for the benefit of the future generation.

## 14.1 INTRODUCTION

Medicinal plants are raw materials for herbal medicines and secondary metabolites [6]. Medicinal plants (herbs) can produce biochemical compounds for biological functions. These medicinal plants have a lot of advantages. For instance, medicinal plants have less side effects when compared to Western or orthodox medicine [6, 18]. Many companies producing pharmaceutical drugs make use of medicinal plants as their raw materials [6]. Even though medicinal plants have great curative potentials, accidental drug poisoning can result if not recognized properly [1].

Despite the advantages associated with medicinal plants, its potentials are yet to be harnessed. For instance, 940 Indonesia plants are of medicinal values, but 80% of them are uncultivated [16]. Though some efforts have been made, yet Indonesia does not still have a complete medicinal plants inventory and where some information exists, it lacks a chronological order of arrangement [10]. Nigeria has a rich flora biodiversity and so many medicinal plants, but most of these medicinal plants are used by traditional herbalists, and many have not been documented [12]. China has about 30,000 plants, but many of the medicinal plants in China are yet to be discovered [8, 11].

Several challenges are associated with the development of medicinal plants such as recognition and classification of the plant species. When the plants are available, their judgment is prone to human subjectivity and errors. The botanist and other experts, who would have helped in the recognition and classification of the plants, are scarce [7, 6]. Because of urbanization, rural lands that used to be forest reserves are now being encroached through deforestation and converted to urban centers [7]. This action results in the loss of many medicinal plant species with many of them gradually going into extinction.

Because of these situations, there is a need to identify and protect the medicinal plants for the future. Protection of plants is not only the work of the government, everybody must be involved, but this will only be possible when people are able to identify and understand the plant species and its importance [2]. If this is done, the future generations will be able to take advantage of the various benefits derivable from medicinal plants [8].

With the advancement in technology, there is a shift from the traditional methods of identifying medicinal plants. The disadvantage of identification by local herbalists is that it is subjective, error-prone, and without proper documentation of the identified medicinal plants.

The rural-urban drift (migration) in which young people move from their rural communities to the urban centers in search of white-collar jobs have also made the transfer of knowledge of medicinal plants from the aged to the young people very difficult.

Another method employed in recognition of medicinal plants is the expert intervention. These experts are botanist, biochemist, and who use proven scientific methods such as chromatography techniques [13] for chemical analysis of extract recognition and classification. Although these methods give accurate results, yet the steps involved in their preparation is rigorous and time-consuming [9].

To mitigate the challenge of identifying and classification of medicinal plants for the purpose of conservation of the plants for the future generation, this chapter explores a new approach in which Information Communication Technologies (ICT) are used for the recognition and classification of the medicinal plants. This involves the use of devices such as digital camera, scanner, mobile phones, and the like to capture images of medicinal plant leaves. The captured images are then processed using image processing techniques and stored in a database (MySQL). Machine learning algorithms such as the Artificial Neural Network (ANN) or Probabilistic Neural Network (PNN) among others are then used to classify medicinal plants.

## 14.2 RECOGNITION OF MEDICINAL PLANTS

The recognition of medicinal plants in this chapter shall be grouped under two headings depending on the method of recognition: Image processing-based recognition [3, 5, 7, 15–18], and spectroscopic recognition [13].

### *14.2.1 IMAGE PROCESSING-BASED RECOGNITION*

This is a method of medicinal plants recognition in which the digital image of the plant under consideration is captured using a digital camera and preprocessed using different algorithms to extract the unique features of the plant to be used for its recognition. These features are later stored in a database [3] or used to train classifiers [6, 15. 18] for the classification of the plant.

The image of the medicinal plant is captured using a digital camera. Gray Level Co-occurrence Matrix (GLCM), which is a texture analysis method, is used for leaf feature extraction [18]. A graphical user interface (GUI) is designed using MATLAB to process the leaf by comparing the extracted features with those stored in the database and returns the one with the least amount of variance. This approach was found to be accurately giving a success rate of 94%, when 11 leaf features were used; and 91% when four features were selected and used [17].

The potentiality of plant extracts as an antibacterial agent can now be determined in a short time using image processing [15]. Formally this procedure was manually carried out in time past, but it was plagued with problems of interpretation due to human subjectivity leading to an error of judgment. With digital image processing, this procedure is carried out on plant extracts with results in 15 seconds [15]. The quantification of the luminance in this research was achieved through a computer application developed for the purpose. Neural Network was used for segmentation during the work. The quantification performance of this approach was 87.5%. Major drawback in this approach is that the quality of input image adversely affects the performance of the system.

The statistical method is used to analyze images acquired from medicinal plants using the principle that many medicinal plant images features have Gaussian distribution [7]. This method was implemented in MATLAB and tested with *Vincarosea* plant (10 species with 50 leaves). The method was found to be a fast and easy tool for medicinal plants recognition.

The flow diagram (Figure 14.1) indicates the leaf recognition and classification system [18]. These images are processed using the Prewitt's edge detection algorithm. The image processing module considers factors such as the surface orientation of the leaf samples, color properties, photo illumination, and discontinuities in depth. After this, the preprocessed images are passed to the feed-forward back-propagation neural network. The processed leaf photos are used for training the network. Testing and comparison are also performed, after which the different types of illnesses and the medicinal plants that could be used for their cure are displayed. The system has a database that provides the list of illnesses and the corresponding curative medicinal plant. The proposed system operates through a mobile application. With this system, an average accuracy of about 90% to 98% accuracy level can be achieved with 20 x 20 training set.

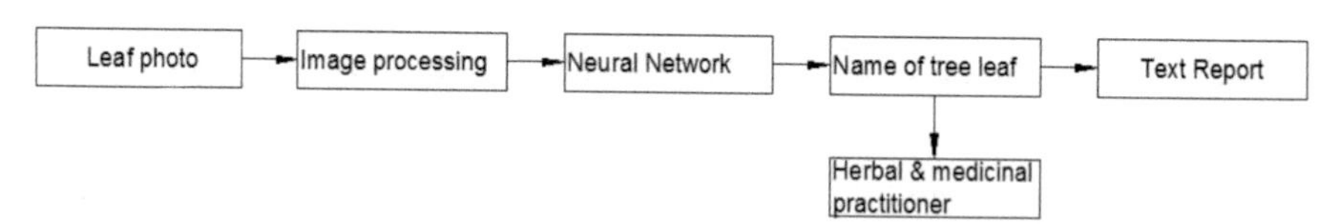


**FIGURE 14.1** Block diagram of the leaf recognition and classification system.
*Source:* Modified from Ref [18].

To achieve greater efficiency and reduction in processing time, three features (4 shapes + 3 colors + 13 texture) of medicinal plants are combined together and processed [6]. Image processing toolbox is used to process the digital image while the samples are trained with ANN classifier. The results revealed an efficiency of 94.44% as compared to 55% when three color features were used. In a similar way, Herdiyeni et al., [5] combined leaf features (such as shape, color, and texture) to identify Indonesian medicinal plants. During this procedure. Local Binary Pattern Variance (LBPV) was used to extract the leaf texture, and the shape was determined by morphological features while the color moment was used for the color. To combine these features for recognition, Product Decision Rule (PDR) was employed, and classification was carried out using PNN. To implement this work, a software (MedLeaf) was designed and deployed on an android mobile phone. The results revealed that recognition accuracy was increased to 72.16% when a combination of the three techniques (LBPV, morphology, and color moments) were used. When morphology was not used, the accuracy of the plants using LBPV and color moments was still 70.78%. This revealed that the morphology feature was not very important in the recognition process [5]. The morphology alone indicated a 17.84% accuracy (Figure 14.2).

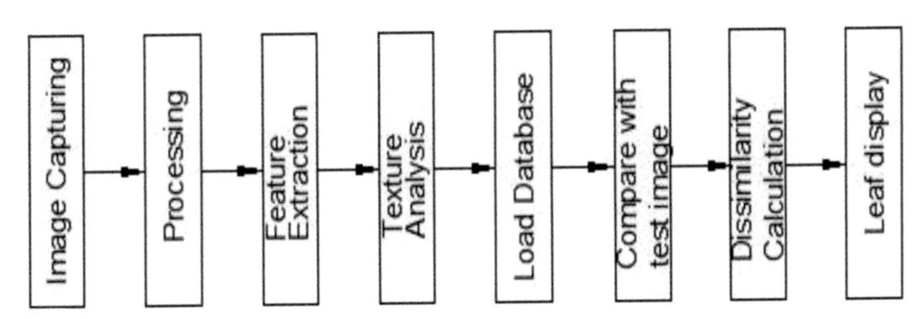


**FIGURE 14.2** Proposed flow chart for the authenticity of Basil (*Ocimum tenuiflorum*) leaf.

The authenticity of Basil (*Ocimum tenuiflorum*) leaf was carried out using a digital image processing method [3]. The leaf image was captured and preprocessed into greyscale. Features were then extracted. The extracted features included entropy, curvature, image motion, shape, aspect ratio, and compactness). From here, the extracted features were passed to the pattern recognition algorithm. Pattern recognition was performed by means of software that coded the leaf using general specification and then stored them into the MySQL database. New input images are then compared to the database content, after which similar leaves are displayed. Figure 14.2 represents the block diagram of this procedure.

One of the factors to be considered in determining the biometric characteristic of a plant is its veination. As a result, digital image processing is employed in capturing the veination of the medicinal plant [16]. Hessian matric is used for the segmentation of the medicinal plant leaf vein into primary, secondary, and tertiary veins by analyzing the Eigen values of pixels of captured images from the medicinal plant. The Hessian matric is written in $C^{++}$ language using the open Computer Vision (CV) software. The performance of this method can be improved with the number of leaf vein segmentation. For instance, the system gave 3.75% accuracy when the primary leaf vein was segmented. With the segmentation of the primary and secondary veins, the accuracy was increased to 42.5%. When the primary, secondary, and the tertiary leaf veins were segmented, the accuracy was increased to 53.75%. The uniqueness of this approach is that it is adequate for the extraction of extracting primary, secondary, and tertiary veins of medicinal plants.

Table 14.1 summarizes the methodology, level of accuracy obtained, and the implementation platform.

### 14.2.2 SPECTROSCOPIC RECOGNITION

The principle by which spectroscopy works is that it measures the molecular interactions with electromagnetic radiation. This principle can be used to identify medicinal plant leaves. An example in the Near-Infrared Spectroscopy (NIRS), which was used on five medicinal plants for the purpose of recognition and classification of these plants [13]. The NIRS was used to obtain the spectra signatures of wet and dry leaves, which shows the uniqueness of each of the leaves.

This method is less laborious when compared with chemical recognition procedures. The chemical procedures include

- Gas chromatography (GC);
- High-performance liquid chromatography (HPLC);
- Thin-layer chromatography (TLC).

It should be noted that the quality of herbal medicine and the authentication of plant variety can be determined by using as reference the unique signature obtained from NIRS. Other merits of this approach include low cost, high efficiency, good repeatability, and convenience [13]. NIRS can also be used on dry leaves with reliable results.

Spectral image analysis was performed on *Folium Uncariae* [14]. The results were consistent with the systematic botanical data, thus confirming the viability of spectral image analysis to recognize the medicinal plants. The results of species recognition also reveal a correlation of 0.0229 with sample no. of 1, 2, 5 and 3 as compared to 0.0045 and 0.0072 with a single sample no. of 2 and 5, respectively.

**TABLE 14.1** Methodology, Level of Accuracy Obtained and the Implementation Platform (In Terms of Hardware and Software [2–7, 10, 11, 13, 15–18]

| Methodology | | | |
|---|---|---|---|
| **Uniqueness** | **Accuracy** | **Implementation Platform** | **References** |
| **a. Microplate luminescence automated digital analyzer** (MILDA), which was developed to perform qualification of luminance atomically and gives a report in a spreadsheet in order to solve the problem associated with luminous image interpretation due to lack of speed and human subjectivity. **Automation** is done with algorithms including gridding algorithm, object recognition, and segmentation algorithm based on neural network learning. This is followed by the **quantization process** to summarize all object pixels intensity values in the database and represent them in a more sophisticated format. After this, the result is given in XLs format.<br>Indonesian medicinal plants under study were: fennel, temulawak, lawing, selasih, temuireng, temuputih, and temugiring. | | | |
| Potentiality of plants extracts as an antibacterial agent can be easily, quickly, and accurately determined.<br>Quality of input image is poor and brings some difficulties.<br>No real-time detection. | 87.5% quantification results from all input image samples. | Computer application (MILDA) that give the report in 15 seconds in a spreadsheet (XLS format). | [15] |
| **b.** Spectral image device was employed to measure the transmission imaging spectra of *Folium uncariae*. The transmission imaging spectra were consistent with systemic botany data for *Folium uncaria*. | | | |
| Spectral imaging analysis technique was used to construct fingerprint of medicinal plants.<br>The procedure was simple, rapid, and non-destructive. | Correlation of 0.0229 with sample number of 1, 2, 5, and 3 as compared to 0.0045 and 0.0072 with a single sample number of 2 and 5, respectively. | Implemented using a self-designed spectral imaging device | [11] |
| **c.** System was developed that uses leaves of a tree as input and provide outputs that include trees. Identify unknown plants with medicinal usage about the illness. Image processing and neural network were combined to train the model, after which pattern matching was done to recognize the leaf. To detect the shape and veins of the leaf at the image processing level, Prewitt's edge detection algorithm was used. Feedforward back propagation neural network was used to train the input images. It had a database with a list of illnesses and plants to be used to cure them. | | | |

**TABLE 14.1** *(Continued)*

| Methodology | | | |
|---|---|---|---|
| **Uniqueness** | **Accuracy** | **Implementation Platform** | **References** |
| Dynamic training: Because new leaves must be used while the program is functioning.<br>Application is connected to the mobile app and will give the name of the plant in real-time. | 90%–98% accuracy with 20x20 training sets. | Mobile set application is proposed. | [18] |
| **d.** Combination of some leaf features (texture, shape, and color) are used in the identification of medicinal plants in Indonesia. Leaf texture was extracted by Local Binary Pattern Variance (LBPV). Morphological extraction (shape) was performed. The color was extracted using color Moment features. Product Decision Rule (PDR) was used to combine leaf features while the classification was done by Probabilistic Neural Network (PNN). The 2,448 leaf images consisting of 51 species of medicinal plants were used for the experiment. | | | |
| "Morphology, LBPV, and color moments" help to provide accuracy of identification up to 72.16% as compared using one or two of the features. | 72.16% identification accuracy. | Med leaf-Software running on Android mobile OS. | [5] |
| **e.** Fusion of Local Binary Pattern (LBP) was performed for recognition of tropical medicinal plants and house plants in two steps: i) Fusion operators; ii) Classifiers combination by means of Product Decision Rule (PDR). Experiment was performed on 1440 samples consisting of 30 species of tropical medicinal plants and 30 species of House plants. | | | |
| Identification accuracy for house plants was found to be better than tropical medicinal plants.<br>Higher classification accuracy was achieved by fusion of LBP. | For tropical medicinal plants: recognition accuracy without fusion was 71%; with fusion operator is 70%; with classifier combination PDR and LBP is 77%;<br>For House plants, recognition accuracy without fusion is 79%; with fusion LBP operator is 84%; with classifier combination PDR and LBP is 86%; | Not mentioned. | [4] |

**TABLE 14.1** *(Continued)*

| Methodology | | | |
|---|---|---|---|
| **Uniqueness** | **Accuracy** | **Implementation Platform** | **References** |
| **f.** Hessian matrix is used in segmenting the veins of medicinal plants (Primary, secondary, and tertiary veins). This was done by analyzing the Eigen values of the pixels from captured images of medicinal plants. This is because; veination is considered a biometric feature (signature) when it comes to leaf classification or characterization. This approach will help botanist in the identification of medicinal plants. The 40 digital medicinal plant leaf images were used for the experiment. They were classified into four types: Pinnate, Acrododromous, Actinodromous, and Campylodromous. | | | |
| Can be used to segment tertiary veins of medicinal plants.<br>The method shows good performance in extracting primary, secondary, and tertiary features of leaves. | Accuracy of 53.75% of leaf image scored 2; and 42.5% scored 1. | C++ language in open Computer Vision (CV) library. | [16] |
| **g.** Identification was achieved by Combining plant features of shape, color, and texture.<br>Classification was done using ANN toolbox in MATLAB. The 63 images of medicinal plants leaves were used for the experiment. They include leaves of Tulsi (12), Curry leaf (12), Hibiscus (9), Neem (12), Thudhuvalai (8) and Henna (10). | | | |
| Combination of 3 features (shape, color, and texture). Reduction in computational time by 17% (from 65.36s to 54.10s) | Accuracy of 94.4% with eight input feature. | MATLAB. | [6] |
| **h.** Image and text features of medicinal plants are combined (fused) and were used for recognition purposes. Fuzzy Local Binary Patterns (FLBP) was used to extract leaf image texture. Probabilistic Neural Network (PNN) was used for image classification to improve accuracy by measuring image weighting on image similarity measurement. Text query is processed by the Sphinx search engine while BM25 is used for document similarity. PNN performance was evaluated with fusion and without fusion. The 2448 leaves images were used from 51 Indonesian medicinal plant species. | | | |

**TABLE 14.1** *(Continued)*

| Methodology | | | |
|---|---|---|---|
| **Uniqueness** | **Accuracy** | **Implementation Platform** | **References** |
| Retrieval system was improved with fusion than it is without fusion. PNN performed best with relevant image queries and irrelevant text queries. | Average Precision (AVP) of text and image information (fusion) is 0.708; while the system with information and image without text (without fusion) has AVP of 0.314;<br>PNN image classification accuracy is 59% | Web application | [10] |
| **i.** Image samples (Leaf base point, leaf tip Point) are photographed or scanned, preprocessed, and neural network is used for classification. The system is trained with known species by experts, while unknown species are inputted into the system by the user for recognition. | | | |
| Recognizes species of medicinal plants using eight features that act as inputs to the artificial neural network classifier.<br>Plays an important role in the Chinese medicinal plants' species recognition. | Not mentioned | Not mentioned | [2] |
| **j.** Near-Infrared Spectroscopy (NIRS) was used to identify and to classify five medicinal plants. Experiments were performed on fresh leaves and dried leaves of the five plants to obtain the spectral signatures of different leaves. The classification was achieved using pre-processing techniques combined with Principal Component Analysis (PCA). The leaves for the study were from Indian borage (*Kapoora Valli*), *Hibiscus rosa-sinensis* (Hibiscus), *Ocimum tenuiflorum* (Tulasi), *Solanum trilobatum* (Dhuthuvalia) and *Piper betel* (Betel). | | | |
| Unique spectral signature of dry and fresh leaves was obtained using NIRS. This can be useful in the authentication of herbal varieties as well as the quality of herbal medicine. The method is less time consuming, less laborious compared to other HPLC, GC, TLC methods. | PCA performed on fresh leaves revealed PC-1 98%; and PC-2 2% of their total variance. In the same way, PCA performed on dry leaves showed PC-1 of 70%; and PC-2 28% of their total variance. | Not mentioned | [13] |

**TABLE 14.1** *(Continued)*

| Methodology | | | |
|---|---|---|---|
| **Uniqueness** | **Accuracy** | **Implementation Platform** | **References** |
| **k.** Texture analysis method (Statistical) is used to identify and classify medicinal plants. | | | |
| Grey Level Co-occurrence Matrix (GLCM) is used to extract leaf features. | | | |
| An algorithm is written with MATLAB to compare the leaf being tested with the database and return the leaf least variance. The developed software is trained with 63 leaves in 9 categories. The algorithm was tested on different leaves, such as neem, tulsi, and hibiscus. | | | |
| The algorithm compares the texture features of input medicinal plant leaves with those in the database to automatically recognize medicinal plants.<br>A graphical user interface in MATLAB is used for the implementation. | 94% accuracy with 11 tested features; and 91% with selected four features. | Implemented in MATLAB Graphical User Interface (GUI). | [17] |
| **l.** The procedure of how to test feature is compared with the database is outlined in this chapter. Extracted from the leaf (entropy, sum, entropy variance, correlation, sum variance, etc.) are loaded into the database. Leaf under test features are used to compare database classification: Closest match is selected or recognized. Different types of tulsi leaves and other medicinal plants are used to create the database. | | | |
| ANN is not used. The only database is used for recognition/classification. | Small database affects reliability. | MYSQL is loaded into MATLAB. | [3] |
| **m.** The statistical nature of image features of medicinal plants was analyzed and computed. The simulation results confirmed that previous findings that various medicinal plant images features have Gaussian distribution. A novel classified design method in medicinal plants study is by using statistics. Experimented with *Vinca Rosea* plants (10 species of 50 leaves each = 500 leaves) | | | |
| Leaf features are Gaussian distributed. The classifier of the medicinal plant is designed using statistical tools. | *Vincarosea* White is 86% accurate; Vincarosea Pink is 52%; *Santalum album* Is 54%; and *Phyllanthus acidus* is 54%. | MATLAB | [7] |

### *14.2.3 CLASSIFICATION OF MEDICINAL PLANTS*

After recognition of medicinal plants, the next step is the classification. Several methods have been used in literature for the classification. These include the use of ANN [6, 15, 18] and PNN [4, 5]. The ANN includes training the system with inputs of medicinal plant images that are known. After this, the system is now used to classify unknown images of medicinal plants [2]. PNN is a neural network using a radial basis function, and this is a function that is shaped like a bell scaling with non-linear variable [14]. Training in PNN requires only one iteration, and it uses a Bayesian approach to obtain its general solution [14].

Another method is Hessian Matrix [16], which was used for segmentation of leaf veination into primary, secondary, and tertiary veins. Apart from these methods, researchers also employ statistical methods for classification of medicinal plants. For example. *Vincarosea* plant was classified by the statistical method, and it was also discovered that most medicinal plants have Gaussian distribution [7]. Another tool used for medicinal plant classification is the ID the database method, where MySQL database is implemented in MATLAB for medicinal plant classification [3].

## 14.3 IMPLEMENTATION PLATFORMS

Implementation of different algorithms used for recognition and classification of medicinal plants are being implemented on different platforms. Some of these include MATLAB [3, 6, 7, 17], Spreadsheet [15], $C^{++}$ language [16], and MySQL database [3], etc. Devices such as android mobile phones are used to acquire digital images of leaves and to display the output [5]. Some of these devices operate in real-time [18] so that they can continually be used to identify and classify new medicinal plants, while others operate as non-real-time devices [15].

With the variety of platforms and devices available for the recognition of medicinal plants, more people, including the non-experts, would have the opportunity to obtain necessary information about medicinal plants, their uses, and associated illness. The problem of the scarcity of experts, such as botanists who are needed to identify the plants, would be solved by the available technologies who can carry out these functions if properly utilized.

## 14.4 SUMMARY

In this chapter, the application of ICT to medicinal plants recognition and classification is considered. The chapter focuses on the recognition and classification of medicinal plants using optical and spectroscopic recognition techniques. It is found that these methods are less cumbersome, faster, and non-destructive, when compared to the chemical method.

## KEYWORDS


- **artificial neural network**
- **feature extraction**
- **image processing**
- **information communication technologies**
- **spectroscopy**


## REFERENCES


1. Fan, S., (2008). *Gelsemiumelegan* – An intangible killer. *Medpharm & Health, 4*, 36.
2. Gao, L., & Lin, X., (2012).A study on the automatic recognition system of medicinal plants. *Consumer Electronics, Communications and Networks (CECNet)*: *IEEE 2nd International Conference Proceedings, 2*, 101–103.
3. Gopal, A., (2015). Comparison procedure for the authentication of basil (*Ocimum tenui-florum*) leaf using image processing technique. *Communications and Signal Processing (ICCSP), IEEE International Conference Proceedings,* 75–78.
4. Herdiyeni, Y., & Kusmana, I., (2013). Fusion of local binary patterns features for tropical medicinal plants identification. *Advanced Computer Science and Information Systems (ICACSIS): IEEE International Conference Proceedings,* 353–357.
5. Herdiyeni, Y., & Santoni, M., (2012). Combination of morphological, local binary pattern variance and color moments features for Indonesian medicinal plants identification. *Advanced Computer Science and Information Systems (ICACSIS): IEEE International Conference Proceedings,* 255–259.
6. Janani, R., & Gopal, A., (2013). Identification of selected medicinal plant leaves using image features and ANN. *Advanced Electronic Systems (ICAES): IEEE International Conference Proceedings,* 238–242.
7. Kumar, E., & Talasila, V., (2014). Leaf features approach for automated identification of medicinal plants. *Communications and Signal Processing (ICCSP): IEEE International Conference Proceedings,* 210–214.

8. Lewis, W. H., & Elvin-Lewis, M. P., (2003). *Medical Botany: Plants Affecting Human Health* (p. 832). John Wiley & Sons, New York.
9. Mao, J., & Xu, J., (2006). Discrimination of herbal medicines by molecular spectroscopy and chemical pattern recognition. *Spectrochimica Acta Part A: Molecular and Biomolecular Spectroscopy*, *65*(2), 497–500.
10. Maulana, O., & Herdiyeni, Y., (2013). Combining image and text features for medicinal plants image retrieval. *Advanced Computer Science and Information Systems (ICACSIS): IEEE International Conference Proceedings,* 273–277.
11. Meng, Q., & Hu, C., (2011). Transmission spectrum imaging analysis of Uncaria leaves from Guangdong. *Bioinformatics and Biomedical Engineering, (iCBBE): IEEE International Conference Proceedings, 5*, 1–4.
12. Odugbemi, T. O., Akinsulire, O. R., Akinsulire, I. E., & Fabeku, P. O., (2007). Medicinal plants useful for malaria therapy in Okeigbo, Ondo State, Southwest Nigeria. *African Journal of Traditional, Complementary and Alternative Medicines*, *4(2)*, 191–198.
13. Rajesh, P. K. S, Kumaravelu, C., Gopal, A., & Suganthi, S., (2013). Studies on identification of medicinal plant variety based on NIR spectroscopy using plant leaves. *Advanced Computing Technologies (ICACT): IEEE 15th International Conference Proceedings, 15*, 1–4.
14. Ramakrishnan, S., & El-Emary, I. M., (2009). Comparative study between traditional and modified probabilistic neural networks. *Telecommunication Systems*, *40*(1/2), 67–74.
15. Rega, P. K., & Emantoko, S., (2013). Microplate luminescence automated digital analyzer for medicinal plants evaluation on quorum sensing inhibition. *QiR (Quality in Research): IEEE International Conference Proceedings,* 31–34.
16. Salima, A., Herdiyeni, Y., & Douady, S., (2015). Leaf vein segmentation of medicinal plant using the Hessian matrix. *Advanced Computer Science and Information Systems (ICACSIS): IEEE International Conference Proceedings,* 275–279.
17. Sathwik, T., Yasaswini, R., Venkatesh, R., & Gopal, A., (2013). Classification of selected medicinal plant leaves using texture analysis. *Computing, Communications, and Networking Technologies (ICCCNT): IEEE Fourth International Conference Proceedings*, 1–6.
18. Udantha, M. G., & Jayamaha, A. D., (2014). Leaf recognition and classification algorithm to be used by indigenous medicine. *Advances in ICT for Emerging Regions (ICTer): IEEE International Conference Proceedings,* 258–258.

CHAPTER 15

# NANOPARTICLE-BASED DELIVERY OF PLANT METABOLITES

SHESAN JOHN OWONUBI, EMMANUEL MUKWEVHO, and NEERISH REVAPRASADU

## ABSTRACT

The delivery of metabolites using nanoparticle-based materials has been investigated by researchers to improve their efficacy and with less toxicity. This chapter focuses on the introduction of drug delivery systems (DDS), a summary of common routes of drug administration. Then it is followed by nanotechnology; nanoparticles (NPs) culminating to nanoparticle-based DDSs. Metabolites (primary, secondary) are highlighted along with their health benefits. Finally, nanoparticle-based drug delivery of plant metabolites is introduced with reports by researchers along with examples.

## 15.1 INTRODUCTION

The ability to deliver the therapeutic agent/drug to the specific diseased site at the required dose, minimize side effects, thus getting the best therapeutic effect is a major aim of an ideal drug delivery system (DDS) [150]. A DDS is a drug carrier vehicle that aids in the conveyance of substance(s) into the body's system to aid in its efficacy and bio-distribution. The ability of a therapeutic compound to reach its intended site for action is a major challenge in the treatment of a disease and for the design of most DDS [150, 216]. Conventional DDS's overtime has challenges of being non-effective, possessing poor bio-distribution, and not being able to specifically target the site of the diseased condition [92, 146]. These challenges have led to the implementation of multidisciplinary tactic that employs nano-science and technology, which aid in development of DDS's with certain level of possible control of the delivery vehicle, which conveys these substance(s): targeted delivery, possible protection of the therapeutic agent from degradation and even in most cases leads to lower doses of the therapeutic agent required [146].

In this chapter, the focus will be on the delivery of plant metabolites using nanoparticle-based techniques. Hence, an overview of DDSs, targeted DDSs, and the ideal DDS is introduced. In addition, an introduction to nanotechnology, and a map out of what 'nanoparticles (NPs)' refer to, as well as numerous types of nanoparticle-based drug delivery types are highlighted. Finally, an introduction to metabolites, with emphasis on plant metabolites and the delivery of these metabolites using nanoparticle-based means is dwelt upon with numerable literature review.

## 15.2 DRUG DELIVERY SYSTEMS (DDSS)

DDS are therapeutic delivery agents or devices that aid the introduction of a specific substance(s) into the body with enhancement to its effectiveness and safety, by having an effect on the degree, rate, and discharge of the substance in the body. DDSs are known to be delivered via various routes of administration, dependent on the nature of the drug delivery vehicle. Irrespective of the route of administration, the therapeutic agent(s) is/are released from the drug delivery vehicle via two major forms of release systems:

- **Sustained Release Systems:** This is a therapeutic release system, which functions as a gradual discharge of the therapeutic agent over a prolonged duration, with the time being sufficient to provide the desirable effect.
- **Controlled Release Systems:** This is a therapeutic release system, which manages the conveyance of the drug at a specific rate and for a fixed duration.

Although in the literature, readers would find other forms of release systems, such as modulated release systems, variable release systems, rate-controlled release systems, etc. [77], yet these are just further variations of these two major release systems.

## 15.3 ROUTES OF ADMINISTRATION

Different routes of drug administration are utilized dependent on the drug delivery vehicle in question. Over the years, it has been confirmed that at least 90% of all drugs are delivered via oral route [77]. This has been the preferred route by patients over the years due to its ease of administration: many patients believe oral ingestion during meals is easily achieved for its convenience [154]. Although quite several limitations exist for the oral route of drug administration [150], yet it is still the most desired route because of constant advancements in the design of the DDS's using this populous route. Many other routes of administration exist because of some drugs being unsuitable for the oral route, as shown in Figure 15.1. Although, in some other literature, additional routes may be observed, yet they mostly are sub-routes under these summarized routes [150].

The race to identify an ideal drug delivery route has been on for so many years, as none of these routes, even the most popular oral route of drug delivery measures up to what researchers accept as an ideal route of drug administration. This is because quite several conditions have to be met for the route of administration to be acknowledged as ideal. The conditions that researchers highlight for an ideal route of administration include, but are not limited to [77]:

- Accessibility;
- Good contact time;
- Lack of variability;
- Large surface area;
- Low metabolic activity;
- Permeability;
- Retain adequate blood supply.

In addition to the drug(s) route of administration and the release system, the increase in the effect of the administered drug can be improved, if the DDS is a targeted one. Targeted DDS are those DDS, which target the specific diseased site; thus a stimuli trigger is usually used to initiate the release of the therapeutic agent when it comes in contact with the specific organ, cell or group of cells [1]. Targeted DDS is usually rather preferred to the conventional DDS due to some factors. Compared to conventional DDS, targeted DDS has more drug stability, higher solubility, and better absorption, all these lead to better therapeutic index, which results in the requirement of a lower dose of the drug by the patient and less toxicity to the patient [142, 171, 211]. Further classification of targeted drug delivery exists in the forms of passive, active targeting, inverse targeting, dual targeting, and double targeting [20, 22].

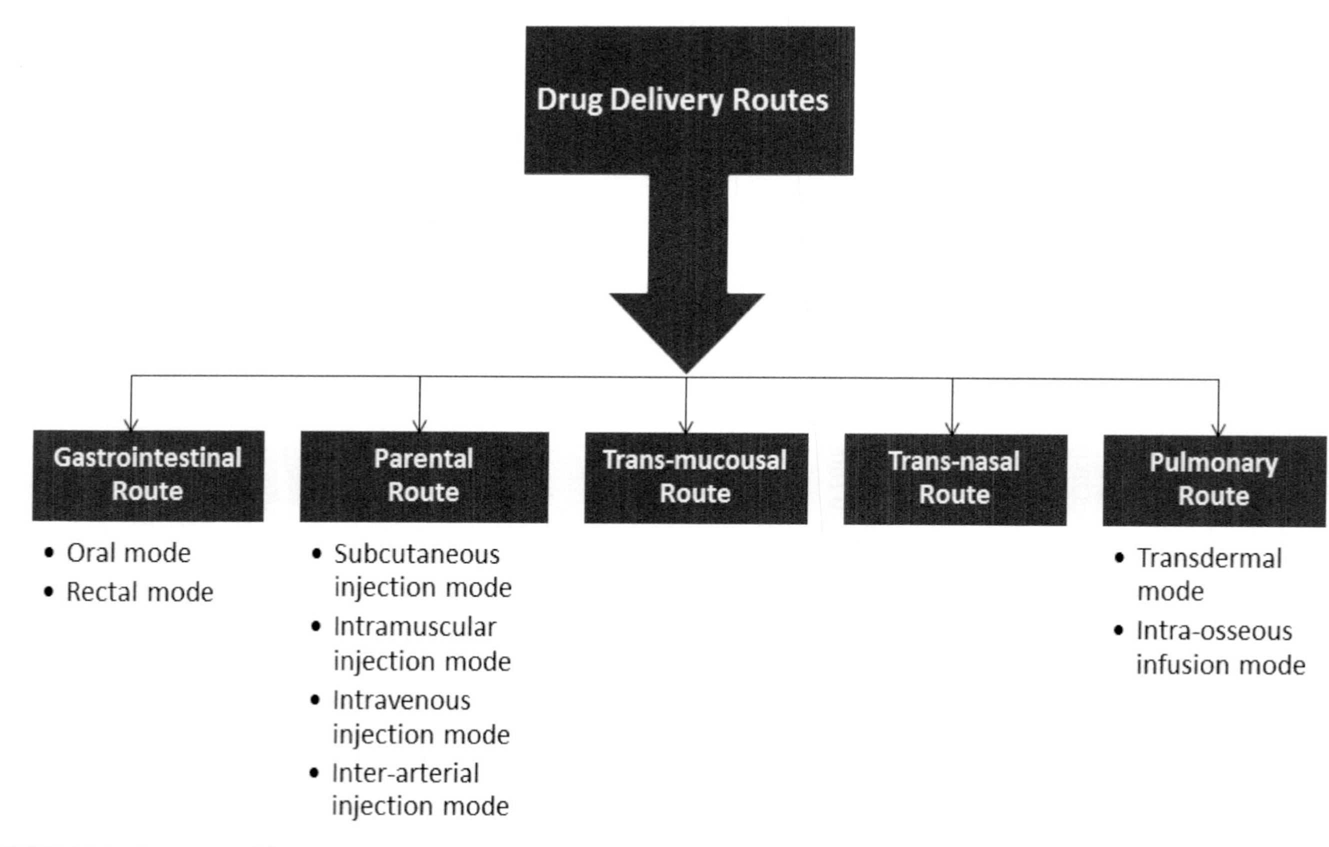


**FIGURE 15.1** Summary of drug routes of administration. Adapted from [150].

Over decades, with advances in the field of biomedical technology, the desire for ideal conditions is always increasing, and similarly, quite several factors influence the design of a DDS to perform ideally for any specific therapeutic agent(s). Depending on the therapeutic agent, conditions have been acknowledged to influence the design of the DDS [86, 164]. These factors include [150], but are not limited to:

- Barrier co-efficient;
- Dosage size;
- pH stability and aqueous stability;
- Stability.

## 15.4 NANOTECHNOLOGY

The challenges, of designing the ideal DDS and route of administration, has led to the size reduction of the therapeutic agent and establishment of the favorable route of administration, which forms the background to the use of nanotechnology by researchers today. Nanotechnology aids to manipulate individual atoms and molecules, which invariably affect the physicochemical and biological properties of the materials compared to their normal bulk forms [129]. Researchers have been exploring the use of nanotechnology for biomedical advancement; by the use of nano-scaled materials due to various benefits [23]:

- Decreased drug resistance;
- Decreased inter-patient variation;
- Decreased variability;
- Drug targeting ability [121, 126, 180];
- Enhanced therapeutic stability [26];
- Improved bio-availability [68];
- Improved dissolution rates;
- Improved patience compliance [26];
- Improved solubility of the therapeutic agent [123];
- Improved surface area;
- Improved therapeutic action;
- Lower dosage [50];
- Reduced toxicity [104].

These benefits may be possible solutions to the challenges faced with conventional DDSs and have led to the development of more efficient targeted DDS with the prospect of transforming the development of therapeutics and

DDS's, thus influencing the pharmaceutical trade [161, 212]. The applications of the use of nanotechnology are widespread as shown in Figure 15.2.

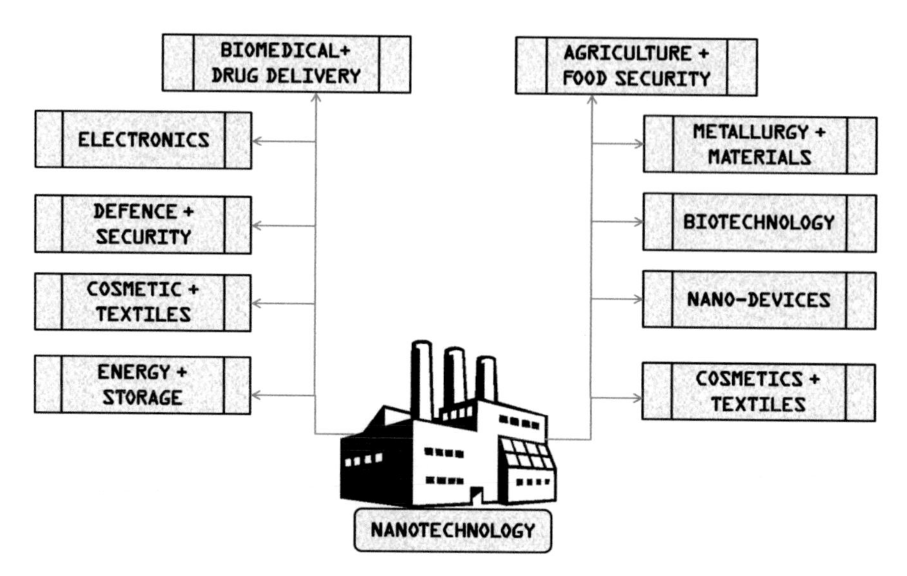


**FIGURE 15.2** Application of nanotechnology.

## 15.5 NANOPARTICLES (NPS)

A nanoparticle (NP) is simply a nano-scaled particle with at least one dimension less than 100 nm. It is the nano-scale dimension that qualifies any material as a nanoparticle. It is widely accepted that a certain bulk material will have specific physical and possibly chemical properties, but at the nano-scale, the properties of such a material are more than often not constant. Properties that are dependent on the size of the material tend to be significantly affected because of the possible change in percentage of atom at the surface of the material. Other factors are affected due to the altered surface area of the material, but not in all cases are the altered properties desirable. The essences of using NP's as DDS is to have determined the size of the therapeutic agent, hence influencing the surface properties of the agent and invariably affecting the potential release of the bio-active agents from the therapeutic and attain optimal targeted delivery and effective dose routine. However, in addition to this level of control over the size of the therapeutic agent, the benefits of existing at a nano-scaled level is also encompassing.

### 15.5.1 NANOPARTICLE CLASSIFICATION

- 1-D nanoparticles (thin films).
- 2-D nanoparticles (carbon nanotubes).
- 3-D nanoparticles (quantum dots, dendrimers, fullerenes).

### 15.5.2 NANOPARTICLE PREPARATION

NP's are usually prepared by three methods:

- Preparation from the dispersion of already formed polymers;
- Preparation from the polymerization of monomers; and
- Preparation from ionic gelation or coacervation of hydrophilic polymers.

Although additional methods such as particle replication in non-wetting templates [174] and preparation using supercritical fluid technology (SCF) [173] can be found in the literature, yet the more common methods have been highlighted above. The indications by Rolland and his colleagues may be a highlight in the future of large scale preparation of NP's as they indicated that they had total control over the prepared NP's particle size, shape, and composition [174].

Proteins, polysaccharides, and synthetic polymers have been used in NP's preparation, but the use of a specific material is dependent on the application and other factors:

- Antigenicity of the final product;
- Biocompatibility of the therapeutic agent;
- Biodegradability and toxicity of the therapeutic agent;
- Nanoparticle size required;
- Surface properties of the therapeutic agent (charge, permeability);
- Therapeutic agent's drug release profile;
- Therapeutic agent's inherent properties (stability, aqueous solubility).

## 15.6 NANOPARTICLE-BASED DRUG DELIVERY

The use of NP's for the delivery of therapeutics has been extensively researched in recently. NP's for drug delivery are often coated with materials of various kinds to aid in the controlled or targeted drug release. The drugs could be loaded via physical means as encapsulation, or chemical conjugation, the design would be dependent on the drug, the material or polymer, and the aim of the researcher. The drug release usually occurs because of diffusion or other stimuli influenced means and in the case of a polymer, possible degradation of the polymer matrix. The rate of release hence would be dependent on the solubility of the drug, the ability of the drug to diffuse through the material membrane, and the chemical interaction with the drug. This may possibly result in slow release of the drug from the polymer matrix or a burst release in some cases [134]. Several NP based DDSs are shown in Figure 15.3.

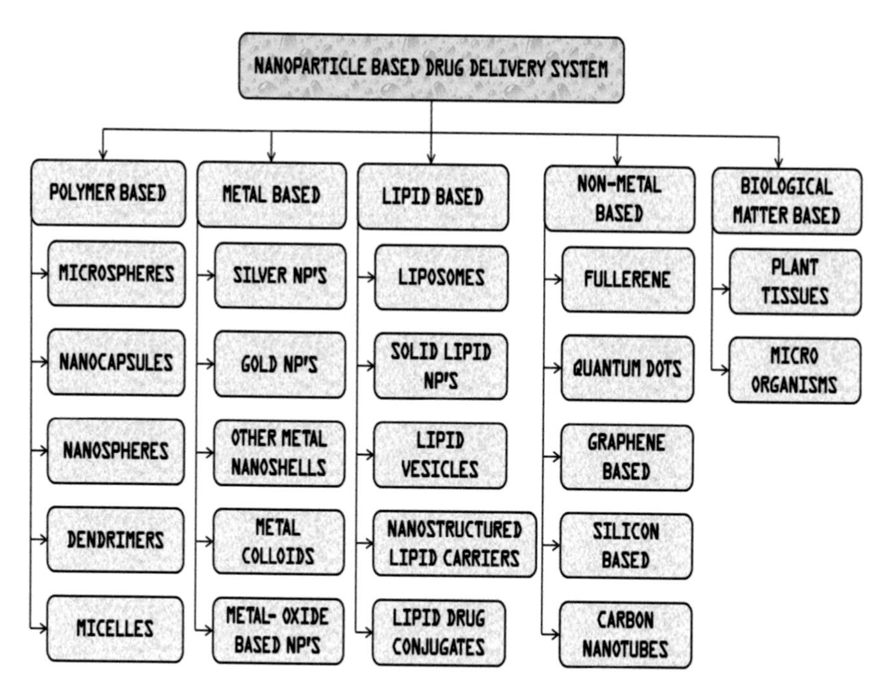


**FIGURE 15.3 (See color insert.)** Summary of nanoparticle-based drug delivery systems.

NP's more recently have been synthesized by various forms for numerous applications; ranging from antimicrobial benefits to the application as drug delivery vehicles. Several classes of NP's based DDSs exist today, from metal-based NP's to biological matter-based NP's. For example, metal-based NP's have been researched over the years and have made tremendous progress in the biomedical field. Silver NP's have been valuable for their antibacterial activity [27, 41, 59, 103, 130, 167, 168, 235]; Gold NP's with antimicrobial benefits in conjugation with antibiotics have been researched upon [25, 111, 167, 170]; their application as drug carriers have been focused on [4, 15, 36, 48, 51, 79, 108, 151, 160, 163, 198] Zinc oxide NP's for their application in water treatment and in the biomedical field [35, 53, 75, 80, 116, 120, 166, 203, 220, 225], and Titanium oxide NP's [40, 43, 98, 143, 218, 234], etc.

Some solid lipid nanoparticles (SLNs) have applications in the biomedical field as anti-inflammation, anti-cancer, and anti-microbial applications [21, 28, 84, 85, 96, 153, 155, 175, 181, 182, 185, 195, 196, 217]. Also, polymer NP's have made advances in the field of drug delivery in the form of dendrimers, microspheres, hydrogels, micelles, nanocapsules, etc. for breast cancer treatments, anti-microbial, anti-inflammatory effect, anti-tumor effects, etc. [1–3, 29, 34, 37, 38, 55, 93, 99, 107, 137, 202, 215, 228, 230, 231]. Other nonmetal - based NP's have also relevance in drug delivery, carbon-based NP's; graphene, graphene oxide, quantum dots and silicon based NP's, etc. [1, 10, 12, 13, 17, 31, 65, 67, 82, 105, 112, 122, 156, 159, 177, 178, 224, 232, 233].

Although ample success has been highlighted for the use of NP-based drug delivery providing huge benefits to the biomedical and pharmaceutical industry, yet the potential toxicological concerns need to be stated. It could be argued that the drugs themselves have been investigated pre-marketing, but the NP for use also needs to be investigated in isolation and is associated with the particular drug as well thoroughly because of the potential toxicity associated with the changes in functional properties of materials when in the nano-scale. Evidences of such toxicity have been observed by several researchers' overtime using different NP's based DDSs. The use of some carbon-based NP delivery systems has experienced toxicity in the findings by some researchers [184], e.g., Donaldson, and co-workers experienced toxicity working with carbon nanotubes [47]; and similarly, Hardman highlighted a review of toxicity and using quantum dots [72].

Many other researchers have witnessed toxic effects of using some carbon-based NP delivery system and several debates on the existence of toxicity or not have ensued [71, 78, 81, 101, 125, 135, 136, 141, 152, 165, 183, 189, 207, 214]. Also, these toxicity potentials have been identified by researchers using polymer based NP DDSs [49, 87, 97, 128, 169], metal-based NP's [5–7, 18, 19, 39, 42, 61, 70, 90, 91, 109, 119, 124, 158, 162, 179, 188] and other non-metal based NP's [56, 114, 115, 148, 157, 199, 226, 236, 237].

Due to the need for reduced toxicity, research going green, the innate capacity of biological matter to aid in the rapid synthesis and stabilizing potential of other material based NP's, several NP's have been more recently synthesized from biological matter [191]: this comes from microorganisms or plant sources. The exploitation of materials from biological matter is perceived to be more biocompatible since they are majorly naturally occurring when compared to their physiochemical counterparts [16].

Medicinal plants have become points of interest to researchers, because their pharmacological advantages can be enhanced by utilization as NP's, thus improving their efficacy [127, 140, 193]. These medicinal plants are known to be abundant in metabolites, which if enhanced can improve their therapeutic index and efficiency.

## 15.7 PLANT-BASED METABOLITES

Metabolism refers to the entire biochemical reactions carried out by a living organism. Two major pathways are encountered in the metabolism process yielding varied end products: the primary pathway and the secondary pathway. The end products of these two pathways result in what we refer to as 'metabolites.' Metabolites are simply those substances which aid in metabolism or are formed because of the process of metabolism. The primary metabolism pathway yields what are called primary metabolites that are in the minority and perform very important evident roles, as opposed to the secondary metabolites from the secondary pathway which are the majority but do not aid in any notable roles [83, 89].

### *15.7.1 PRIMARY AND SECONDARY METABOLITES*

Plants are known to produce compounds (such as alcohols, amino acids, carbohydrates, nucleotides, phytosterols, and some organic acids), which are typically key to the maintenance of normal physiology processes and are important in growth, development, and reproduction. These compounds are called primary metabolites. In contrast, plants also produce secondary metabolites (such as: sterols, terpenes, alkaloids, phenols, tannins, carotenoids, flavonoids, waxes, gums), which are not directly linked to actions that effect the growth, reproduction, and development of the living plant, but seem to have some effect on the function of ecology and the mechanism of defense by production of pigments in some cases [24]. These secondary metabolites are known to be produced from the modification of primary metabolite by enzymatic means. Figure 15.4 shows the pathways to produce some secondary metabolites.

## 15.8 HEALTH BENEFITS OF PLANT METABOLITES

The advent of green chemistry and scientific research has directed enormous focus on the use of naturally occurring products in drug discovery, for example, cyanobacterial alkaloids have been so widely researched for its cytotoxicity and protease inhibitory capacity [74]. These plant metabolites have shown potential benefits in the field of medicinal chemistry as well, with podophyllotoxin, a non-alkaloid toxin lignan leading in the development of a number of clinical drugs [197]. Plants are very diverse and huge source of metabolites that they synthesize have been established to possess therapeutic properties that help in the treatment of ailments and majority of drugs have approximately two-thirds of the bioactive content to have plant origins [131].

Several other natural products from plants have been widely researched such as:

- Curcumin;
- Artemisinin;
- Triptolide;
- Capsaicin;
- Paclitaxel;
- Breviscapine;
- Dodder;
- Naringenin;
- Genistein;
- Silymarins;
- Campotothecin;
- Berberines [8, 45, 57, 63, 64, 95, 102, 113, 147, 149, 176, 186, 187, 192, 208, 213, 219].

Though in some cases synthetic forms of the natural products are used. These synthetic forms arise from the challenge with the use of natural products in certain dosage forms to administer the natural products, the natural forms having low bioavailability, poor solubility, resulting in low therapeutic index. These challenges have led to the use of nanoparticle-based forms to retain the use of products in their natural forms but retaining the ability to be administered or targeted and possessing the benefits of nano-scaled based delivery. With advances in research, several nanoparticle-based DDSs have been reported and used for the delivery of plant metabolites [69, 204, 223].

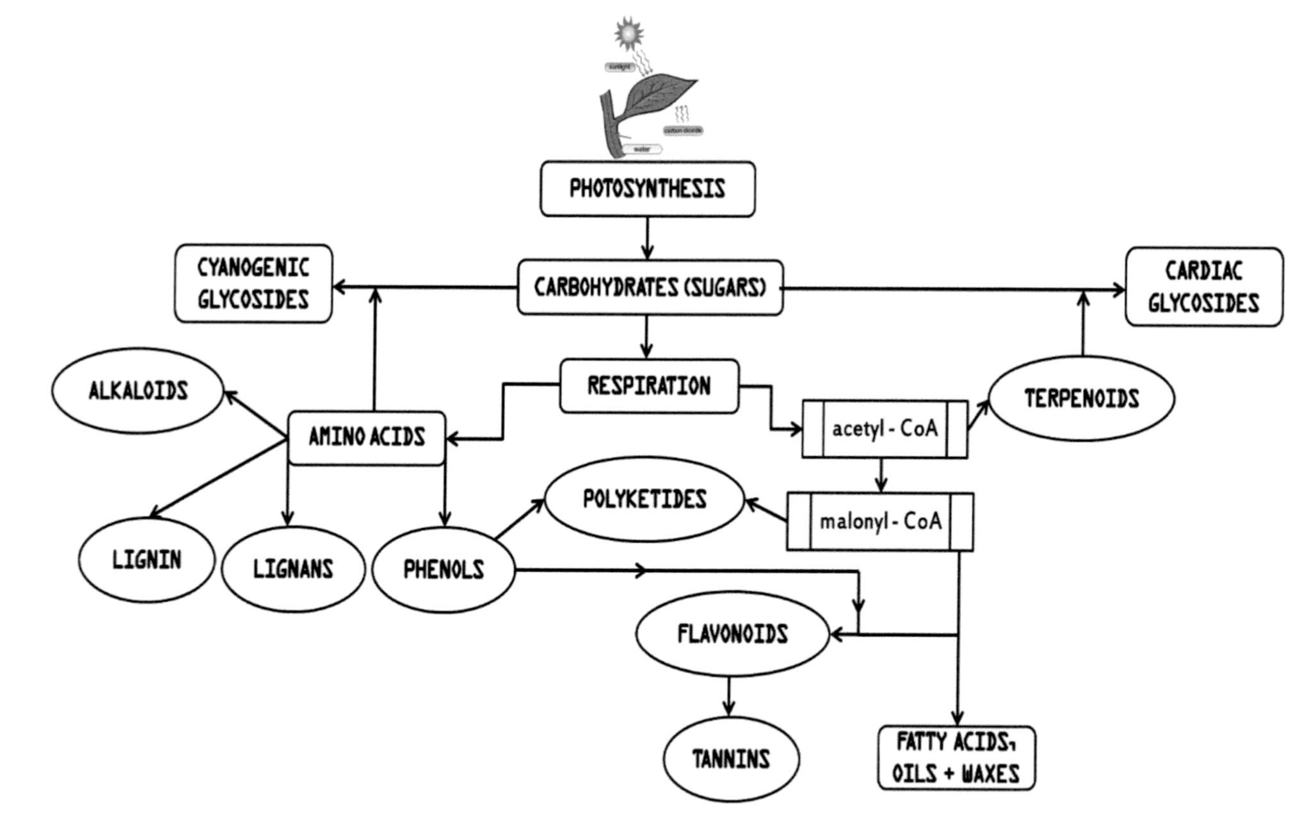


**FIGURE 15.4** **(See color insert.)** Pathway to produce some secondary plant metabolites.

## 15.9 NANOPARTICLE-BASED DRUG DELIVERY OF PLANT METABOLITES

### 15.9.1 LIPOSOMES

Liposomes, which are lipid-based drug delivery vehicles, have been utilized in the delivery of plant metabolites. Dependent on the loading capacity, scale of produce, and the nature of the plant metabolite, various method of preparation of the liposome could be used [11, 222]. There are three commonly used methods to prepare these lipid-based NPs:

- Thin film hydration method;
- Mechanical dispersion method; and
- Rotary film evaporation method.

The use of these liposomes-based DDS aids in:

- The improvement to the metabolite release;
- The ability to target the metabolites to the site required;
- The improvement to solubility and bioavailability of hydrophobic plant metabolites;
- Reduction of fluctuations in the dosage over time.

Over the years, the use of liposomes have aided in the enhancement of the bioavailability of the plant metabolites, applications; and focus of researchers has been swayed in this direction [9, 44, 144].

### 15.9.2 MICROSPHERES

Microspheres are polymer-based DDSs which are solid spherical forms with a diameter range of 1–1000 μm. They could be made of natural materials (e.g., gelatin, albumin, and chitosan) or synthetic polymers (e.g., waxes, polylactic acid), etc. They have a reputation for targeted dose-dependent effects of a pharmaceutic, because of their ability to overcome issues with reaction with plant metabolites and thus being able to maintain effective plasma concentration for long periods of time [62, 210]. They also possess the ability to enhance the therapeutic potential of plant metabolites that are poorly soluble. The microspheres usually employ varied techniques to accomplish the desirable objectives:

- Gastroretentive bloating microspheres;
- Microencapsulation;
- Emulsification;
- Calcium alginate-based beads.

### *15.9.3 SOLID LIPID NANOPARTICLES (SLNS)*

SLN's are lipid-based NPs, which are known to have their lipid excipients becoming solids at room temperature. They possess a mean diameter range of 50μm–1000μm. They are prepared by various methods [139, 209]:

- Microemulsion;
- Ultrasonification;
- Homogenization;
- Solvent diffusion method.

These SLN's have the capacity to improve the potential of targeting, prevention of degradation of the plant metabolites, reduction in toxicity, and improved encapsulation efficiency [190].

### *15.9.4 POLYMERIC NANOPARTICLES (PNPS)*

PNP's are NPs that are made from polymer materials. They may be referred to as 'nanospheres' or 'nanocapsules' by some researchers. They can be prepared by various methods [30, 194]:

- Salting out method;
- Solvent evaporation method;
- Nano-precipitation method;
- Dialysis method.

These PNP's tend to discharge the plant metabolites by the erosion of the polymer material to make the PNP's, but with regard to the plant metabolite encapsulated [30, 100, 106, 172]. The application by researchers has had widespread impact in recent times [14, 33, 46].

Table 15.1 summarizes the use of nanoparticle-based delivery systems for plant metabolites with their source.

**TABLE 15.1** Examples of Nanoparticle-based Drug Delivery of Plant Metabolites

| NP -Based | Metabolite Form | Source | References |
|---|---|---|---|
| Liposomes | Quinoline alkaloid | Camptothecin (*Camptothecin acuminata*) | [60] |
| | Polyphenolic compound | Catechin (*Camellia sinesis*) | [58] |
| | Flavonoid | Ampelopsin (*Cercidiphyllum japonicum*) | [76] |
| | Flavonoid | Silymarin (*Silymarin marianum*) | [52, 54] |
| | Bisindole alkaloid | Vincristine (*Catharanthus roseus*) | [229] |
| Polymeric Nanoparticles | Bis-isoquinoline alkaloid | Tertandrine (*Stephania tetrandra*) | [110, 221] |
| | Steroidal saponin | Glycyrrhizic acid (*Glycyrrihizaglabra*) | [205] |
| | Flavonoid | Naringenin (*Citrus paradisi*) | [200, 227] |
| | Isoquinoline compound | Berberine (*Berberis aquifolium*) | [32, 94] |
| | Diterpenoid triepoxide | Triptolide (*Tripterygium wilfordii*) | [118] |
| Microspheres | Flavonoid | Rutin (*Carpobrotus edulis*) | [88] |
| | Quinoline alkaloid | Camptothecin (*Camptothecin acuminata*) | [206] |
| | Steroidal glycoside | Ginsenosides (*Panax ginseng*) | [117] |
| | Flavonoid | Silymarin (Silymarin marianum) | [66] |
| | Pentahydroxyflavone | Quercetin (*Sophora japonica L.*) | [145] |
| Solid Lipid Nanoparticles | Diterpenoid triepoxide | Triptolide (*Tripterygium wilfordii*) | [132, 133] |

## 15.10 SUMMARY

This chapter has highlighted the use of nanotechnology in the field of biomedical technology and has revealed new perspectives in targeted drug delivery, improvements in materials using nanotechnology. Highlighting the field of nanotechnology, using materials in their nano-scaled forms, which when optimized physiochemically and biologically, possess the capacity reveal distinct characteristics opposed to when not in the nano-scale form. From the use of liposomes to the utilization of carbon NPs, the advantages of utilizing materials at a nano-scale have been showcased by researchers worldwide with applications ranging from use in the engineering field to the biomedical industry.

Although plant metabolites have been researched and effectively used to improve diseased conditions from in the early nineteenth century [73], yet in more recent time these metabolites either directly or from their derivatives have dominant roles in the manufacture of anticancer and infectious disease drugs worldwide [138, 201]. However, with advances in synthesis chemistry and combination chemistry, the interest into natural products has been on the decline more recently due to problems of dosage formulation by the pharmaceutical industries. This review has shown how the use of nanotechnology has revived the interest into the use of these natural plant metabolites, solving the problem of dosage formulation and the numerous advantages of dealing with materials at nano-scale.

## ACKNOWLEDGMENTS

This chapter write-up was motivated and mentored by Prof. Emmanuel Mukwevho, and his support is fully acknowledged.

## KEYWORDS

- **alkaloid**
- **bioavailability**
- **biocompatibility**
- **biomedical**
- **degradability**

- **flavanoid**
- **lipid-based**
- **metabolites**
- **nanoparticle**
- **nanotechnology**
- **terpenoid**


## REFERENCES


1. Abrol, S., Trehan, A., & Katare, O., (2004). Formulation, characterization, and *in vitro* evaluation of silymarin-loaded lipid microspheres. *Drug Delivery, 11*(3), 185–191.
2. Aderibigbe, B., Owonubi, S., Jayaramudu, J., Sadiku, E., & Ray, S., (2014). Targeted drug delivery potential of hydrogel biocomposites containing partially and thermally reduced graphene oxide and natural polymers prepared via green process. *Colloid and Polymer Science*, 1–12.
3. Aderibigbe, B., Sadiku, E., Jayaramudu, J., & Sinha-Ray, S., (2015). Controlled dual release study of curcumin and a 4-aminoquinoline analog from gum acacia containing hydrogels. *Journal of Applied Polymer Science, 132*, 41613.
4. Aderibigbe, B. A., & Mhlwatika, Z., (2016). Dual release kinetics of antimalarials from soy protein isolate-carbopol-polyacrylamide based hydrogels. *Journal of Applied Polymer Science, 133*(37), 43918.
5. Agasti, S. S., Chompoosor, A., You, C. C., Ghosh, P., Kim, C. K., & Rotello, V. M., (2009). Photoregulated release of caged anticancer drugs from gold nanoparticles. *Journal of the American Chemical Society, 131*(16), 5728–5729.
6. Ahamed, M., (2011). Toxic response of nickel nanoparticles in human lung epithelial A549 cells. *Toxicology In Vitro, 25*(4), 930–936.
7. Ahamed, M., Akhtar, M. J., Raja, M., Ahmad, I., Siddiqui, M. K. J., AlSalhi, M. S., & Alrokayan, S. A., (2011). ZnO nanorod-induced apoptosis in human alveolar adenocarcinoma cells via p53, survivin and bax/bcl-2 pathways: Role of oxidative stress. *Nanomedicine: Nanotechnology, Biology and Medicine, 7*(6), 904–913.
8. Ahamed, M., Akhtar, M. J., Siddiqui, M. A., Ahmad, J., Musarrat, J., Al-Khedhairy, A. A., AlSalhi, M. S., & Alrokayan, S. A., (2011). Oxidative stress mediated apoptosis induced by nickel ferrite nanoparticles in cultured A549 cells. *Toxicology, 283*(2/3), 101–108.
9. Ahmed, S., Anuntiyo, J., Malemud, C. J., & Haqqi, T. M., (2005). Biological basis for the use of botanicals in osteoarthritis and rheumatoid arthritis: A review. *Evidence-Based Complementary and Alternative Medicine, 2*(3), 301–308.
10. Ajazuddin, & Saraf, S., (2010). Applications of novel drug delivery system for herbal formulations. *Fitoterapia, 81*(7), 680–689.
11. Akerman, M. E., Chan, W. C., Laakkonen, P., Bhatia, S. N., & Ruoslahti, E., (2002). Nanocrystal targeting *in vivo. Proceedings of the National Academy of Sciences of the United States of America, 99*(20), 12617–12621.

12. Allen, T. M., & Cullis, P. R., (2013). Liposomal drug delivery systems: From concept to clinical applications. *Advanced Drug Delivery Reviews*, *65*(1), 36–48.
13. Alvarez, G. S., Helary, C., Mebert, A. M., Wang, X., Coradin, T., & Desimone, M. F., (2014). Antibiotic-loaded silica nanoparticle-collagen composite hydrogels with prolonged antimicrobial activity for wound infection prevention. *Journal of Materials Chemistry B*, *2*(29), 4660–4670.
14. Annaka, M., Mortensen, K., Matsuura, T., Ito, M., Nochioka, K., & Ogata, N., (2012). Organic-inorganic nanocomposite gels as an *in-situ* gelation biomaterial for injectable accommodative intraocular lens. *Soft Matter*, *8*(27), 7185–7196.
15. Ansari, S. H., Islam, F., & Sameem, M., (2012). Influence of nanotechnology on herbal drugs: A Review. *Journal of Advanced Pharmaceutical Technology & Research*, *3*(3), 142–146.
16. Asadishad, B., Vossoughi, M., & Alemzadeh, I., (2010). Folate-receptor-targeted delivery of doxorubicin using polyethylene glycol-functionalized gold nanoparticles. *Industrial & Engineering Chemistry Research*, *49*(4), 1958–1963.
17. Baker, S., Rakshith, D., Kavitha, K. S., Santosh, P., Kavitha, H. U., Rao, Y., & Satish, S., (2013). Plants: Emerging as nanofactories towards facile route in synthesis of nanoparticles. *Bioimpacts: BI*, *3*(3), 111–117.
18. Balunas, M. J., & Kinghorn, A. D., (2005). Drug discovery from medicinal plants. *Life Sciences*, *78*(5), 431–441.
19. Beer, C., Foldbjerg, R., Hayashi, Y., Sutherland, D. S., & Autrup, H., (2012). Toxicity of silver nanoparticles—nanoparticle or silver ion? *Toxicology Letters*, *208*(3), 286–292.
20. Bernardi, R. J., Lowery, A. R., Thompson, P. A., Blaney, S. M., & West, J. L., (2008). Immunonanoshells for targeted photothermal ablation in medulloblastoma and glioma: An *in vitro* evaluation using human cell lines. *Journal of Neuro-Oncology*, *86*(2), 165–172.
21. Bhagwat, R., & Vaidhya, I., (2013). Novel drug delivery systems: An overview. *International Journal of Pharmaceutical Sciences and Research*, *4*(3), 970.
22. Bhalekar, M. R., Pokharkar, V., Madgulkar, A., Patil, N., & Patil, N., (2009). Preparation and Evaluation of miconazole nitrate-loaded solid lipid nanoparticles for topical delivery. *AAPS Pharm. Sci. Tech.*, *10*(1), 289–296.
23. Bhargav, E., & Madhuri, N., (2013). Targeted drug delivery-a review. *World Journal of Pharmacy and Pharmaceutical Sciences*, *3*(1), 150–159.
24. Bhatia, S., (2016). Nanotechnology and its drug delivery applications. In: *Natural Polymer Drug Delivery Systems: Nanoparticles, Plants, and Algae* (pp. 1–32). Springer International Publishing, New York.
25. Boundless, (2018). *Primary and Secondary Metabolites*. https://courses.lumenlearning.com/boundless-microbiology/chapter/industrial-microbiology/ (Accessed on 30 July 2019).
26. Bresee, J., Bond, C. M., & Worthington, R. J., (2014). Nanoscale structure-activity relationships, mode of action, and biocompatibility of gold nanoparticle antibiotics. *Journal of the American Chemical Society*, *136*(14), 5295–5300.
27. Brewer, E., Coleman, J., & Lowman, A., (2011). Emerging technologies of polymeric nanoparticles in cancer drug delivery. *Journal of Nanomaterials*, 1–10.
28. Brown, A. N., Smith, K., Samuels, T. A., Lu, J., Obare, S. O., & Scott, M. E., (2012). Nanoparticles functionalized with ampicillin destroy multiple-antibiotic-resistant isolates of *pseudomonas aeruginosa* and *enterobacter aerogenes* and methicillin-resistant staphylococcus aureus. *Applied and Environmental Microbiology*, *78*(8), 2768–2774.

29. Cavalli, R., Gasco, M. R., Chetoni, P., Burgalassi, S., & Saettone, M. F., (2002). Solid lipid nanoparticles (SLN) as ocular delivery system for tobramycin. *International Journal of Pharmaceutics*, *238*(1/2), 241–245.
30. Chacko, R. T., Ventura, J., Zhuang, J., & Thayumanavan, S., (2012). Polymer nanogels: A versatile nanoscopic drug delivery platform. *Advanced Drug Delivery Reviews*, *64*(9), 836–851.
31. Chan, J. M., Valencia, P. M., Zhang, L., Langer, R., & Farokhzad, O. C., (2010). Polymeric nanoparticles for drug delivery. *Methods in Molecular Biology*, *624*, 163–175.
32. Chang, C., Peng, J., Zhang, L., & Pang, D. W., (2009). Strongly fluorescent hydrogels with quantum dots embedded in cellulose matrices. *Journal of Materials Chemistry*, *19*(41), 7771–7776.
33. Chang, C. H., Huang, W. Y., Lai, C. H., Hsu, Y. M., Yao, Y. H., Chen, T. Y., Wu, J. Y., Peng, S. F., & Lin, Y. H., (2011). Development of novel nanoparticles shelled with heparin for berberine delivery to treat Helicobacter pylori. *Acta Biomaterialia*, *7*(2), 593–603.
34. Chaturvedi, M., Kumar, M., Sinhal, A., & Saifi, A., (2011). Recent development in novel drug delivery systems of herbal drugs. *International Journal of Green Pharmacy (IJGP)*, *5*(2), 156–165.
35. Chauhan, A. S., Sridevi, S., Chalasani, K. B., Jain, A. K., Jain, S. K., Jain, N. K., & Diwan, P. V., (2003). Dendrimer-mediated transdermal delivery: Enhanced bioavailability of indomethacin. *Journal of Controlled Release*, *90*(3), 335–343.
36. Chaúque, E., Zvimba, J., Ngila, J., & Musee, N., (2016). Fate, behavior, and implications of ZnO nanoparticles in a simulated wastewater treatment plant. *Water SA*, *42*, 72–81.
37. Chen, Y. H., Tsai, C. Y., Huang, P. Y., & Chang, M. Y., (2007). Methotrexate conjugated to gold nanoparticles inhibits tumor growth in a syngeneic lung tumor model. *Molecular Pharmaceutics*, *4*(5), 713–722.
38. Cheng, Y., Man, N., Xu, T., Fu, R., Wang, X., Wang, X., & Wen, L., (2007). Transdermal delivery of nonsteroidal anti-inflammatory drugs mediated by polyamidoamine (PAMAM) dendrimers. *Journal of Pharmaceutical Sciences*, *96*(3), 595–602.
39. Cheng, Y., Qu, H., Ma, M., Xu, Z., Xu, P., Fang, Y., & Xu, T., (2007). Polyamidoamine (PAMAM) dendrimers as biocompatible carriers of quinolone antimicrobials: An *in vitro* study. *European Journal of Medicinal Chemistry*, *42*(7), 1032–1038.
40. Choi, J. E., Kim, S., Ahn, J. H., Youn, P., Kang, J. S., Park, K., Yi, J., & Ryu, D. Y., (2010). Induction of oxidative stress and apoptosis by silver nanoparticles in the liver of adult zebrafish. *Aquatic Toxicology*, *100*(2), 151–159.
41. Choi, J. Y., Kim, K. H., Choy, K. C., Oh, K. T., & Kim, K. N., (2007). Photocatalytic antibacterial effect of TiO(2) film formed on Ti and TiAg exposed to Lactobacillus acidophilus. *Journal of Biomedical Materials Research, Part B: Applied Biomaterials*, *80*(2), 353–359.
42. Chopra, I., (2007). The increasing use of silver-based products as antimicrobial agents: A useful development or a cause for concern? *Journal of Antimicrobial Chemotherapy*, *59*(4), 587–590.
43. Coradeghini, R., Gioria, S., García, C. P., Nativo, P., Franchini, F., Gilliland, D., Ponti, J., & Rossi, F., (2013). Size-dependent toxicity and cell interaction mechanisms of gold nanoparticles on mouse fibroblasts. *Toxicology Letters*, *217*(3), 205–216.
44. Cortesi, R., Esposito, E., Maietti, A., Menegatti, E., & Nastruzzi, C., (1997). Formulation study for the antitumor drug camptothecin: Liposomes, micellar solutions and a microemulsion. *International Journal of Pharmaceutics*, *159*(1), 95–103.

45. Cragg, G. M., & Newman, D. J., (2005). Plants as a source of anti-cancer agents. *Journal of Ethnopharmacology*, *100*(1/2), 72–79.
46. Crane, F. A., Mohammed, K., Blake, M. I., Uphaus, R. A., & Katz, J. J., (1970). Effects of deuteriation on alkaloid production in Atropa belladonna. *Lloydia*, *33*(1), 11–14.
47. Daoud, S. S., Fetouh, M. I., & Giovanella, B. C., (1995). Antitumor effect of liposome-incorporated camptothecin in human malignant xenografts. *Anti-Cancer Drugs*, *6*(1), 83–93.
48. Dembinski, A., Warzecha, Z., Konturek, S. J., Ceranowicz, P., Dembinski, M., Pawlik, W. W., Kusnierz-Cabala, B., & Naskalski, J. W., (2004). Extract of grapefruit-seed reduces acute pancreatitis induced by ischemia/reperfusion in rats: Possible implication of tissue antioxidants. *Journal of Physiology and Pharmacology*, *55*(4), 811–821.
49. Devi, V. K., Jain, N., & Valli, K. S., (2010). Importance of novel drug delivery systems in herbal medicines. *Pharmacognosy Reviews*, *4*(7), 27–31.
50. Donaldson, K., Aitken, R., Tran, L., Stone, V., Duffin, R., Forrest, G., & Alexander, A., (2006). Carbon nanotubes: A review of their properties in relation to pulmonary toxicology and workplace safety. *Toxicological Sciences*, *92*(1), 5–22.
51. Dreaden, E. C., Mwakwari, S. C., Sodji, Q. H., Oyelere, A. K., & El-Sayed, M. A., (2009). Tamoxifen-poly(ethylene glycol)-thiol gold nanoparticle conjugates: Enhanced potency and selective delivery for breast cancer treatment. *Bioconjugate Chemistry*, *20*(12), 2247–2253.
52. Duncan, R., & Izzo, L., (2005). Dendrimer biocompatibility and toxicity. *Advanced Drug Delivery Reviews*, *57*(15), 2215–2237.
53. Dutta, T., & Jain, N. K., (2007). Targeting potential and anti-HIV activity of lamivudine loaded mannosylated poly (propyleneimine) dendrimer. *Biochimica et Biophysica Acta*, *1770*(4), 681–686.
54. Eghtedari, M., Liopo, A. V., Copland, J. A., Oraevsky, A. A., & Motamedi, M., (2009). Engineering of hetero-functional gold nanorods for the *in vivo* molecular targeting of breast cancer cells. *Nano Letters*, *9*(1), 287–291.
55. El-Samaligy, M. S., Afifi, N. N., & Mahmoud, E. A., (2006). Increasing bioavailability of silymarin using a buccal liposomal delivery system: preparation and experimental design investigation. *International Journal of Pharmaceutics*, *308*(1/2), 140–148.
56. Elmi, F., Alinezhad, H., Moulana, Z., & Salehian, F., (2014). The use of antibacterial activity of ZnO nanoparticles in the treatment of municipal wastewater. *Water Science and Technology*, *70*(5), 763–770.
57. Elmowafy, M., Viitala, T., Ibrahim, H. M., Abu-Elyazid, S. K., & Samy, A., (2013). Silymarin loaded liposomes for hepatic targeting: *In vitro* evaluation and HepG2 drug uptake. *European Journal of Pharmaceutical Sciences*, *50*(2), 161–171.
58. Emmelius, M., Hörpel, G., Ringsdorf, H., & Schmidt, B., (1986). Polymeric micelles and liposomes as potential drug carriers. In: Chiellini, E., Giusti, P., Migliaresi, C., & Nicolais, L., (eds.), *Polymers in Medicine II: Biomedical and Pharmaceutical Applications* (pp. 313–331). Springer US: Boston, MA.
59. Eom, H. J., & Choi, J., (2009). Oxidative stress of CeO2 nanoparticles via p38-Nrf-2 signaling pathway in human bronchial epithelial cell, Beas-2B. *Toxicology Letters*, *187*(2), 77–83.
60. Erlund, I., Meririnne, E., Alfthan, G., & Aro, A., (2001). Plasma kinetics and urinary excretion of the flavanones naringenin and hesperetin in humans after ingestion of orange juice and grapefruit juice. *Journal of Nutrition*, *131*(2), 235–241.

61. Fang, J. Y., Hwang, T. L., Huang, Y. L., & Fang, C. L., (2006). Enhancement of the transdermal delivery of catechins by liposomes incorporating anionic surfactants and ethanol. *International Journal of Pharmaceutics*, *310*(1/2), 131–138.
62. Fayaz, A. M., Balaji, K., Girilal, M., Yadav, R., Kalaichelvan, P. T., & Venketesan, R., (2010). Biogenic synthesis of silver nanoparticles and their synergistic effect with antibiotics: A study against gram-positive and gram-negative bacteria. *Nanomedicine: Nanotechnology, Biology, and Medicine*, *6*(1), 103–109.
63. Flaten, G. E., Chang, T. T., Phillips, W. T., Brandl, M., Bao, A., & Goins, B., (2013). Liposomal formulations of poorly soluble camptothecin: Drug retention and biodistribution. *Journal of Liposome Research*, *23*(1), 70–81.
64. Fratoddi, I., Venditti, I., Cametti, C., & Russo, M. V., (2015). The puzzle of toxicity of gold nanoparticles. The case-study of HeLa cells. *Toxicology Research*, *4*(4), 796–800.
65. Freiberg, S., & Zhu, X. X., (2004). Polymer microspheres for controlled drug release. *International Journal of Pharmaceutics*, *282*(1/2), 1–18.
66. Fukuda, K., Hibiya, Y., Mutoh, M., Koshiji, M., Akao, S., & Fujiwara, H., (1999). Inhibition by berberine of cyclooxygenase-2 transcriptional activity in human colon cancer cells. *Journal of Ethnopharmacology*, *66*(2), 227–233.
67. Fukuda, K., Hibiya, Y., Mutoh, M., Koshiji, M., Akao, S., & Fujiwara, H., (1999). Inhibition of activator protein 1 activity by berberine in human hepatoma cells. *Planta Medica*, *65*(4), 381–383.
68. Gao, X., Cui, Y., Levenson, R. M., Chung, L. W., & Nie, S., (2004). *In vivo* cancer targeting and imaging with semiconductor quantum dots. *Nature Biotechnology*, *22*(8), 969–976.
69. Garg, R., & Gupta, G., (2010). Gastroretentive floating microspheres of silymarin: Preparation and *in vitro* evaluation. *Tropical Journal of Pharmaceutical Research*, *9*(1), 110–116.
70. Ghazani, A. A., Lee, J. A., Klostranec, J., Xiang, Q., Dacosta, R. S., Wilson, B. C., Tsao, M. S., & Chan, W. C., (2006). High throughput quantification of protein expression of cancer antigens in tissue microarray using quantum dot nanocrystals. *Nano Letters*, *6*(12), 2881–2886.
71. Goldberg, D. S., Vijayalakshmi, N., Swaan, P. W., & Ghandehari, H., (2011).G3.5 PAMAM dendrimers enhance transepithelial transport of SN38 while minimizing gastrointestinal toxicity. *Journal of Controlled Release*, *150*(3), 318–325.
72. Goyal, A., Kumar, S., Nagpal, M., Singh, I., & Arora, S., (2011). Potential of novel drug delivery systems for herbal drugs. *Indian Journal of Pharmaceutical Education and Research*, *45*(3), 225–235.
73. Hackenberg, S., Scherzed, A., Kessler, M., Hummel, S., Technau, A., Froelich, K., Ginzkey, C., Koehler, C., Hagen, R., & Kleinsasser, N., (2011). Silver nanoparticles: Evaluation of DNA damage, toxicity and functional impairment in human mesenchymal stem cells. *Toxicology Letters*, *201*(1), 27–33.
74. Haniu, H., Matsuda, Y., Usui, Y., Aoki, K., Shimizu, M., Ogihara, N., Hara, K., Okamoto, M., Takanashi, S., Ishigaki, N., Nakamura, K., Kato, H., & Saito, N., (2011). Toxicoproteomic evaluation of carbon nanomaterials in vitro. *Journal of Proteomics*, *74*(12), 2703–2712.
75. Hardman, R., (2006). A toxicologic review of quantum dots: Toxicity depends on physicochemical and environmental factors. *Environmental Health Perspectives*, *114*(2), 165–172.

76. Hartwell, J. L., (1971). Plants used against cancer: A survey. *Lloydia, 34*(2), 204–255.
77. Harvey, A. L., (2008). Natural products in drug discovery. *Drug Discovery Today, 13*(19/20), 894–901.
78. He, L., Liu, Y., Mustapha, A., & Lin, M., (2011). Antifungal activity of zinc oxide nanoparticles against *Botrytis cinerea* and *Penicillium expansum*. *Microbiological Research, 166*(3), 207–215.
79. He, Z. F., Liu, D. Y., Zeng, S., & Ye, J. T., (2008). Study on preparation of ampelopsin liposomes. *Zhongguo Zhong Yao Za Zhi. Zhongguo Zhongyao Zazhi. China Journal of Chinese Materia Medica, 33*(1), 27–30.
80. Hillery, A. M., Lloyd, A. W., & Swarbrick, J., (2008). *Drug Delivery and Targeting: For Pharmacists and Pharmaceutical Scientists* (p. 632). Taylor & Francis, Boca Raton–FL.
81. Hoshino, A., Hanada, S., & Yamamoto, K., (2011). Toxicity of nanocrystal quantum dots: The relevance of surface modifications. *Archives of Toxicology, 85*(7), 707–720.
82. Hosta, L., Pla-Roca, M., Arbiol, J., Lopez-Iglesias, C., Samitier, J., Cruz, L. J., Kogan, M. J., & Albericio, F., (2009). Conjugation of Kahalalide F with gold nanoparticles to enhance *in vitro* antitumoral activity. *Bioconjugate Chemistry, 20*(1), 138–146.
83. Hsueh, Y. H., Ke, W. J., Hsieh, C. T., Lin, K. S., Tzou, D. Y., & Chiang, C. L., (2015). ZnO nanoparticles affect bacillus subtilis cell growth and biofilm formation. *PloS One, 10*(6), e0128457.
84. Hu, X., Cook, S., Wang, P., Hwang, H. M., Liu, X., & Williams, Q. L., (2010). *In vitro* evaluation of cytotoxicity of engineered carbon nanotubes in selected human cell lines. *Science of the Total Environment, 408*(8), 1812–1817.
85. Hu, X., Hao, X., Wu, Y., Zhang, J., Zhang, X., Wang, P. C., Zou, G., & Liang, X. J., (2013). Multifunctional hybrid silica nanoparticles for controlled doxorubicin loading and release with thermal and pH dual response. *Journal of Materials Chemistry B, 1*(8), 1109–1118.
86. Hussain, M. S., Fareed, S., Ansari, S., Rahman, M. A., Ahmad, I. Z., & Saeed, M., (2012). Current approaches toward production of secondary plant metabolites. *Journal of Pharmacy & Bioallied Sciences, 4*(1), 10–20.
87. Jain, A., Jain, S. K., Ganesh, N., Barve, J., & Beg, A. M., (2010). Design and development of ligand-appended polysaccharidic nanoparticles for the delivery of oxaliplatin in colorectal cancer. *Nanomedicine: Nanotechnology, Biology, and Medicine, 6*(1), 179–190.
88. Jain, D., & Banerjee, R., (2008). Comparison of ciprofloxacin hydrochloride-loaded protein, lipid, and chitosan nanoparticles for drug delivery. *Journal of Biomedical Materials Research, Part B: Applied Biomaterials, 86*(1), 105–112.
89. Jantzen, G. M., & Robinson, J. R., (1996). Sustained and controlled release drug delivery systems. In: Banker, G. S., & Rhodes, C. T., (eds.), *Modern Pharmaceutics* (p. 575). Marcel Dekker: New York.
90. Jevprasesphant, R., Penny, J., Jalal, R., Attwood, D., McKeown, N. B., & D'Emanuele, A., (2003). The influence of surface modification on the cytotoxicity of PAMAM dendrimers. *International Journal of Pharmaceutics, 252*(1/2), 263–266.
91. Ji, P., Yu, T., Liu, Y., Jiang, J., Xu, J., Zhao, Y., Hao, Y., Qiu, Y., Zhao, W., & Wu, C., (2016). Naringenin-loaded solid lipid nanoparticles: Preparation, controlled delivery, cellular uptake, and pulmonary pharmacokinetics. *Drug Design, Development and Therapy, 10*, 911–918.

92. Jiang, H. L., Cui, Y. L., & Qi, Y., (2013). Microencapsulation of rutin in chitosan-coated alginate microspheres through internal gelation technique. *Advanced Materials Research: Trans Tech Publ.*, *716*, 455–458.
93. Jimenez-Garcia, S. N., Vazquez-Cruz, M. A., Guevara-Gonzalez, R. G., Torres-Pacheco, I., Cruz-Hernandez, A., & Feregrino-Perez, A. A., (2013). Current approaches for enhanced expression of secondary metabolites as bioactive compounds in plants for agronomic and human health purposes – a review. *Polish Journal of Food and Nutrition Sciences*, *63*(2), 67–77.
94. Jo, M. R., Bae, S. H., Go, M. R., Kim, H. J., Hwang, Y. G., & Choi, S. J., (2015). Toxicity and biokinetics of colloidal gold nanoparticles. *Nanomaterials*, *5*(2), 835–840.
95. Katsumiti, A., Gilliland, D., Arostegui, I., & Cajaraville, M. P., (2015). Mechanisms of toxicity of Ag nanoparticles in comparison to bulk and ionic Ag on mussel hemocytes and gill cells. *PloS One*, *10*(6), e0129039.
96. Kewal, K. J., (2008). *Methods in Molecular Biology* (p. 425). Humana Press, New York.
97. Khairnar, G. A., Chavan-Patil, A. B., Palve, P. R., & Bhise, S. B., (2010). Dendrimers: Potential tool for enhancement of antifungal activity. *International Journal of Pharm. Tech. Research*, *2*(1), 736–739.
98. Khemani, M., Sharon, M., & Sharon, M., (2012). Encapsulation of berberine in nano-sized PLGA synthesized by emulsification method. *ISRN Nanotechnology*, 9–18.
99. Kim, S. A., Kwon, Y., Kim, J. H., Muller, M. T., & Chung, I. K., (1998). Induction of topoisomerase II-mediated DNA cleavage by a protoberberine alkaloid, berberrubine. *Biochemistry*, *37*(46), 16316–16324.
100. Kinghorn, A. D., Chin, Y. W., & Swanson, S. M., (2009). Discovery of natural product anticancer agents from biodiverse organisms. *Current Opinion in Drug Discovery & Development*, *12*(2), 189–196.
101. Kolhatkar, R. B., Kitchens, K. M., Swaan, P. W., & Ghandehari, H., (2007). Surface acetylation of polyamidoamine (PAMAM) dendrimers decreases cytotoxicity while maintaining membrane permeability. *Bioconjugate Chemistry*, *18*(6), 2054–2060.
102. Kuhn, K. P., Chaberny, I. F., Massholder, K., Stickler, M., Benz, V. W., Sonntag, H. G., & Erdinger, L., (2003). Disinfection of surfaces by photocatalytic oxidation with titanium dioxide and UVA light. *Chemosphere*, *53*(1), 71–77.
103. Kukowska-Latallo, J. F., Candido, K. A., Cao, Z., Nigavekar, S. S., Majoros, I. J., Thomas, T. P., Balogh, L. P., Khan, M. K., & Baker, J. R., Jr., (2005). Nanoparticle targeting of anticancer drug improves therapeutic response in animal model of human epithelial cancer. *Cancer Research*, *65*(12), 5317–5324.
104. Kumari, A., Yadav, S. K., & Yadav, S. C., (2010). Biodegradable polymeric nanoparticles based drug delivery systems. *Colloids and Surfaces B: Biointerfaces*, *75*(1), 1–18.
105. Lam, C. W., James, J. T., McCluskey, R., & Hunter, R. L., (2004). Pulmonary toxicity of single-wall carbon nanotubes in mice 7 and 90 days after intratracheal instillation. *Toxicological Sciences*, *77*(1), 126–134.
106. Lamartiniere, C. A., Murrill, W. B., Manzolillo, P. A., Zhang, J. X., Barnes, S., Zhang, X., Wei, H., & Brown, N. M., (1998). Genistein alters the ontogeny of mammary gland development and protects against chemically-induced mammary cancer in rats. *Proceedings of the Society for Experimental Biology and Medicine*, *217*(3), 358–364.
107. Lara, H. H., Ayala-Núñez, N. V., Ixtepan, T. L. D. C., & Rodríguez, P. C., (2010). Bactericidal effect of silver nanoparticles against multidrug-resistant bacteria. *World Journal of Microbiology and Biotechnology*, *26*(4), 615–621.

108. Lee, C. C., Gillies, E. R., Fox, M. E., Guillaudeu, S. J., Frechet, J. M., Dy, E. E., & Szoka, F. C., (2006). A single dose of doxorubicin-functionalized bow-tie dendrimer cures mice bearing C-26 colon carcinomas. *Proceedings of the National Academy of Sciences of the United States of America*, *103*(45), 16649–16654.
109. Lee, J. H., Park, J. H., Eltohamy, M., Perez, R., Lee, E. J., & Kim, H. W., (2013). Collagen gel combined with mesoporous nanoparticles loading nerve growth factor as a feasible therapeutic three-dimensional depot for neural tissue engineering. *RSC Advances*, *3*(46), 24202–24214.
110. Leo, E., Scatturin, A., Vighi, E., & Dalpiaz, A., (2006). Polymeric nanoparticles as drug controlled release systems: A new formulation strategy for drugs with small or large molecular weight. *Journal of Nanoscience and Nanotechnology*, *6*(9/10), 3070–3079.
111. Li, G. L., Mohwald, H., & Shchukin, D. G., (2013). Precipitation polymerization for fabrication of complex core-shell hybrid particles and hollow structures. *Chemical Society Reviews*, *42*(8), 3628–3646.
112. Li, J., Wang, X., Wang, C., Chen, B., Dai, Y., Zhang, R., Song, M., Lv, G., & Fu, D., (2007). The enhancement effect of gold nanoparticles in drug delivery and as biomarkers of drug-resistant cancer cells. *Chem. Med. Chem.*, *2*(3), 374–378.
113. Li, J. J., Hartono, D., Ong, C. N., Bay, B. H., & Yung, L. Y. L., (2010). Autophagy and oxidative stress associated with gold nanoparticles. *Biomaterials*, *31*(23), 5996–6003.
114. Li, X., Lu, X., Xu, H., Zhu, Z., Yin, H., Qian, X., Li, R., Jiang, X., & Liu, B., (2012). Paclitaxel/tetrandrine coloaded nanoparticles effectively promote the apoptosis of gastric cancer cells based on "oxidation therapy."*Molecular Pharmaceutics*, *9*(2), 222–229.
115. Li, X., Robinson, S. M., Gupta, A., Saha, K., & Jiang, Z., (2014). Functional gold nanoparticles as potent antimicrobial agents against multi-drug-resistant bacteria. *ACS Nano*, *8*(10), 10682–10686.
116. Lim, Y. T., Kim, S., Nakayama, A., Stott, N. E., Bawendi, M. G., & Frangioni, J. V., (2003). Selection of quantum dot wavelengths for biomedical assays and imaging. *Molecular Imaging*, *2*(1), 50–64.
117. Lin, J. G., Chung, J. G., Wu, L. T., Chen, G. W., Chang, H. L., & Wang, T. F., (1999). Effects of berberine on arylamine N-acetyltransferase activity in human colon tumor cells. *American Journal of Chinese Medicine*, *27*(2), 265–275.
118. Lin, W., Huang, Y. W., Zhou, X. D., & Ma, Y., (2006). *In vitro* toxicity of silica nanoparticles in human lung cancer cells. *Toxicology and Applied Pharmacology*, *217*(3), 252–259.
119. Lin, W., Huang, Y. W., Zhou, X. D., & Ma, Y., (2006). Toxicity of cerium oxide nanoparticles in human lung cancer cells. *International Journal of Toxicology*, *25*(6), 451–457.
120. Lipovsky, A., Tzitrinovich, Z., Friedmann, H., Applerot, G., Gedanken, A., & Lubart, R., (2009). EPR study of visible light-induced ROS generation by nanoparticles of ZnO. *The Journal of Physical Chemistry C*, *113*(36), 15997–16001.
121. Liu, C. B., Zhang, D., Li, D. G., Jiang, D., & Xia, C., (2008). Preparation and characterization of biodegradable polylactide (PLA) microspheres encapsulating ginsenoside Rg3. *Chemical Research in Chinese Universities*, *24*(5), 588–591.
122. Liu, M. X., Dong, J., Yang, Y. J., Yang, X. L., & Xu, H. B., (2004). Preparation and toxicity of triptolide-loaded poly lactic acid nanoparticles. *Yao Xuexue Bao (Acta Pharmaceutica Sinica)*, *39*(7), 556–560.
123. Liu, S., Xu, L., Zhang, T., Ren, G., & Yang, Z., (2010). Oxidative stress and apoptosis induced by nanosized titanium dioxide in PC12 cells. *Toxicology*, *267*(1–3), 172–177.

124. Liu, Y., He, L., Mustapha, A., Li, H., Hu, Z. Q., & Lin, M., (2009). Antibacterial activities of zinc oxide nanoparticles against Escherichia coli O157:H7. *Journal of Applied Microbiology*, *107*(4), 1193–1201.
125. Liu, Z., Fan, A. C., Rakhra, K., Sherlock, S., Goodwin, A., Chen, X., Yang, Q., Felsher, D. W., & Dai, H., (2009). Supramolecular stacking of doxorubicin on carbon nanotubes for *in vivo* cancer therapy. *Angewandte Chemie. International Ed. in English*, *48*(41), 7668–7672.
126. Lo, C. W., Zhu, D., & Jiang, H., (2011). An infrared-light responsive graphene-oxide incorporated poly(N-isopropylacrylamide) hydrogel nanocomposite. *Soft Matter*, *7*(12), 5604–5609.
127. Lobenberg, R., Maas, J., & Kreuter, J., (1998). Improved body distribution of 14C-labelled AZT bound to nanoparticles in rats determined by radioluminography. *Journal of Drug Targeting*, *5*(3), 171–179.
128. Lowery, A. R., Gobin, A. M., Day, E. S., Halas, N. J., & West, J. L., (2006). Immunonanoshells for targeted photothermal ablation of tumor cells. *International Journal of Nanomedicine*, *1*(2), 149–154.
129. Luanpitpong, S., Wang, L., Davidson, D. C., Riedel, H., & Rojanasakul, Y., (2016). Carcinogenic Potential of high aspect ratio carbon nanomaterials. *Environ. Sci. Nano*, *3*(3), 483–493.
130. Mahajan, S. D., Roy, I., Xu, G., Yong, K. T., Ding, H., Aalinkeel, R., et al., (2010). Enhancing the delivery of anti retroviral drug "Saquinavir" across the blood brain barrier using nanoparticles. *Current HIV Research*, *8*(5), 396–404.
131. Makarov, V. V., Love, A. J., Sinitsyna, O. V., Makarova, S. S., Yaminsky, I. V., Taliansky, M. E., & Kalinina, N. O., (2014). "Green" nanotechnologies: Synthesis of metal nanoparticles using plants. *Acta Naturae*, *6*(1), 35–44.
132. Malik, N., Wiwattanapatapee, R., & Klopsch, R., (2000). Relationship between structure and biocompatibility *in vitro*, and preliminary studies on the biodistribution of 125I-labelled polyamidoamine dendrimers *in vivo*. *Journal of Controlled Release*, *65*(1/2), 133–148.
133. Martin, C. R., (2006). Welcome to nanomedicine. *Nanomedicine: Nanotechnology, Biology, and Medicine*, *1*(1), 5–10.
134. Martinez-Gutierrez, F., Olive, P. L., & Banuelos, A., (2010). Synthesis, characterization, and evaluation of antimicrobial and cytotoxic effect of silver and titanium nanoparticles. *Nanomedicine: Nanotechnology, Biology, and Medicine*, *6*(5), 681–688.
135. McChesney, J. D., Venkataraman, S. K., & Henri, J. T., (2007). Plant natural products: Back to the future or into extinction? *Phytochemistry*, *68*(14), 2015–2022.
136. Mei, Z., Chen, H., Weng, T., Yang, Y., & Yang, X., (2003). Solid lipid nanoparticle and microemulsion for topical delivery of triptolide. *European Journal of Pharmaceutics and Biopharmaceutics*, *56*(2), 189–196.
137. Mei, Z., Li, X., Wu, Q., Hu, S., & Yang, X., (2005). The research on the anti-inflammatory activity and hepatotoxicity of triptolide-loaded solid lipid nanoparticle. *Pharmacological Research*, *51*(4), 345–351.
138. Mohanraj, V. J., & Chen, Y., (2006). Nanoparticles – A review. *Tropical Journal of Pharmaceutical Research*, *5*(1), 3–15.
139. Monteiro-Riviere, N. A., Nemanich, R. J., Inman, A. O., Wang, Y. Y., & Riviere, J. E., (2005). Multi-walled carbon nanotube interactions with human epidermal keratinocytes. *Toxicology Letters*, *155*(3), 377–384.

140. Morimoto, Y., Horie, M., Kobayashi, N., Shinohara, N., & Shimada, M., (2013). Inhalation toxicity assessment of carbon-based nanoparticles. *Accounts of Chemical Research*, *46*(3), 770–781.
141. Moughton, A. O., Hillmyer, M. A., & Lodge, T. P., (2012). Multicompartment block polymer micelles. *Macromolecules*, *45*(1), 2–19.
142. Mukherjee, A. K., Basu, S., Sarkar, N., & Ghosh, A. C., (2001). Advances in cancer therapy with plant based natural products. *Current Medicinal Chemistry*, *8*(12), 1467–1486.
143. Mukherjee, S., Ray, S., & Thakur, R. S., (2009). Solid lipid nanoparticles: A modern formulation approach in drug delivery system. *Indian Journal of Pharmaceutical Sciences*, *71*(4), 349–358.
144. Mukherjee, S., Sushma, V., Patra, S., Barui, A. K., Bhadra, M. P., Sreedhar, B., & Patra, C. R., (2012). Green chemistry approach for the synthesis and stabilization of biocompatible gold nanoparticles and their potential applications in cancer therapy. *Nanotechnology*, *23*(45), E-article: 455103.
145. Muller, J., Huaux, F., Moreau, N., Misson, P., Heilier, J. F., Delos, M., Arras, M., Fonseca, A., Nagy, J. B., & Lison, D., (2005). Respiratory toxicity of multi-wall carbon nanotubes. *Toxicology and Applied Pharmacology*, *207*(3), 221–231.
146. Muller, R. H., & Keck, C. M., (2004). Challenges and solutions for the delivery of biotech drugs--a review of drug nanocrystal technology and lipid nanoparticles. *Journal of Biotechnology*, *113*(1–3), 151–170.
147. Muranyi, P., Schraml, C., & Wunderlich, J., (2010). Antimicrobial efficiency of titanium dioxide-coated surfaces. *Journal of Applied Microbiology*, *108*(6), 1966–1973.
148. Musthaba, S. M., Baboota, S., Ahmed, S., Ahuja, A., & Ali, J., (2009). Status of novel drug delivery technology for phytotherapeutics. *Expert Opinion on Drug Delivery*, *6*(6), 625–637.
149. Natarajan, V., Krithica, N., Madhan, B., & Sehgal, P. K., (2011). Formulation and evaluation of quercetin polycaprolactone microspheres for the treatment of rheumatoid arthritis. *Journal of Pharmaceutical Sciences*, *100*(1), 195–205.
150. Nevozhay, D., Kanska, U., Budzynska, R., & Boratynski, J., (2007). Current status of research on conjugates and related drug delivery systems in the treatment of cancer and other diseases. *Postepy. Hig. Med. Dosw. (Online)*, *61*, 350–360.
151. Nirmala, M. J., Samundeeswari, A., & Sankar, P. D., (2011). Natural plant resources in anti-cancer therapy-A review. *Research in Plant Biology*, *1*(3), 199–204.
152. Nishimori, H., Kondoh, M., Isoda, K., Tsunoda, S. I., Tsutsumi, Y., & Yagi, K., (2009). Silica nanoparticles as hepatotoxicants. *European Journal of Pharmaceutics and Biopharmaceutics*, *72*(3), 496–501.
153. Nuengchamnong, N., Hermans-Lokkerbol, A., & Ingkaninan, K., (2004). Separation and detection of the antioxidant flavonoids, rutin and quercetin, using HPLC coupled on-line with colorimetric detection of antioxidant activity. *Naresuan University Journal*, *12*(2), 25–37.
154. Owonubi, S., Aderibigbe, B., Mukwevho, E., Sadiku, E., & Ray, S., (2018). Characterization and *in vitro* release kinetics of antimalarials from whey protein-based hydrogel biocomposites. *International Journal of Industrial Chemistry*, 1–14.
155. Owonubi, S. J., Agwuncha, S. C., & Mukwevho, E., (2017). Application of hydrogel biocomposites for multiple drug delivery. In: Kumar, T., Manju, K. T., & Kessler, M. R., (eds.), *Handbook of Composites from Renewable Materials* (pp. 139–166). Wiley-Scrivener Publishing: New York.

156. Paciotti, G. F., Kingston, D. G. I., & Tamarkin, L., (2006). Colloidal gold nanoparticles: A novel nanoparticle platform for developing multifunctional tumor-targeted drug delivery vectors. *Drug Development Research*, *67*(1), 47–54.
157. Pacurari, M., Lowe, K., Tchounwou, P. B., & Kafoury, R., (2016). A review on the respiratory system toxicity of carbon nanoparticles. *International Journal of Environmental Research and Public Health*, *13*(3), 90–99.
158. Pandey, R., & Khuller, G. K., (2005). Solid lipid particle-based inhalable sustained drug delivery system against experimental tuberculosis. *Tuberculosis (Edinb)*, *85*(4), 227–234.
159. Pandey, V. P., Manavalan, R., Sundar, R. T., & Ganesh, K. S., (2003). Formulation and release characteristics of sustained release diltiazem hydrochloride tablet. *Indian Journal of Pharmaceutical Sciences*, *65*(1), 44–48.
160. Panga, X., Cui, F., Tian, J., Chen, J., Zhou, J., & Zhou, W., (2009). Preparation and characterization of magnetic solid lipid nanoparticles loaded with ibuprofen. *Asian Journal of Pharmaceutical. Science*, *4*(2), 132–137.
161. Parak, W. J., Boudreau, R., Le Gros, M., Gerion, D., Zanchet, D., Micheel, C. M., Williams, S. C., Alivisatos, A. P., & Larabell, C., (2002). Cell motility and metastatic potential studies based on quantum dot imaging of phagokinetic tracks. *Advanced Materials*, *14*(12), 882–885.
162. Park, E. J., & Park, K., (2009). Oxidative stress and pro-inflammatory responses induced by silica nanoparticles *in vivo* and *in vitro*. *Toxicology Letters*, *184*(1), 18–25.
163. Park, E. J., Yi, J., Kim, Y., Choi, K., & Park, K., (2010). Silver nanoparticles induce cytotoxicity by a Trojan-horse type mechanism. *Toxicology in Vitro*, *24*(3), 872–878.
164. Parveen, R., Baboota, S., Ali, J., Ahuja, A., Vasudev, S. S., & Ahmad, S., (2011). Oil based nanocarrier for improved oral delivery of silymarin: *In vitro* and *in vivo* studies. *International Journal of Pharmaceutics*, *413*(1/2), 245–253.
165. Pathak, S., Choi, S. K., Arnheim, N., & Thompson, M. E., (2001). Hydroxylated quantum dots as luminescent probes for in situ hybridization. *Journal of the American Chemical Society*, *123*(17), 4103–4114.
166. Patra, C. R., Bhattacharya, R., Wang, E., Katarya, A., Lau, J. S., Dutta, S., et al., (2008). Targeted delivery of gemcitabine to pancreatic adenocarcinoma using cetuximab as a targeting agent. *Cancer Research*, *68*(6), 1970–1978.
167. Petros, R. A., & DeSimone, J. M., (2010). Strategies in the design of nanoparticles for therapeutic applications. *Nature Reviews: Drug Discovery*, *9*(8), 615–627.
168. Piao, M. J., Kang, K. A., Lee, I. K., Kim, H. S., Kim, S., Choi, J. Y., Choi, J., & Hyun, J. W., (2011). Silver nanoparticles induce oxidative cell damage in human liver cells through inhibition of reduced glutathione and induction of mitochondria-involved apoptosis. *Toxicology Letters*, *201*(1), 92–100.
169. Podsiadlo, P., Sinani, V. A., Bahng, J. H., Kam, N. W., Lee, J., & Kotov, N. A., (2008). Gold nanoparticles enhance the anti-leukemia action of a 6-mercaptopurine chemotherapeutic agent. *Langmuir*, *24*(2), 568–574.
170. Popli, H., & Sharma, S. N., (1989). Trends in oral sustained release formulations. *Eastern Pharmacist*, 99–103.
171. Radomski, A., Jurasz, P., Alonso-Escolano, D., Drews, M., Morandi, M., Malinski, T., & Radomski, M. W., (2005). Nanoparticle-induced platelet aggregation and vascular thrombosis. *British Journal of Pharmacology*, *146*(6), 882–893.

172. Raghupathi, K. R., Koodali, R. T., & Manna, A. C., (2011). Size-dependent bacterial growth inhibition and mechanism of antibacterial activity of zinc oxide nanoparticles. *Langmuir, 27*(7), 4020–4028.
173. Rai, M., Yadav, A., & Gade, A., (2009). Silver nanoparticles as a new generation of antimicrobials. *Biotechnology Advances, 27*(1), 76–83.
174. Rai, M. K., Deshmukh, S. D., Ingle, A. P., & Gade, A. K., (2012). Silver nanoparticles: The powerful nanoweapon against multidrug-resistant bacteria. *Journal of Applied Microbiology, 112*(5), 841–852.
175. Rajananthanan, P., Attard, G. S., Sheikh, N. A., & Morrow, W. J. W., (1999). Evaluation of novel aggregate structures as adjuvants: Composition, toxicity studies and humoral responses. *Vaccine, 17*(7/8), 715–730.
176. Ranga, R. P., Varaprasad, K., Narayana, R. N., Mohana, R. K., & Reddy, N. S., (2012). Fabrication of Au and Ag Bi-metallic nanocomposite for antimicrobial applications. *Journal of Applied Polymer Science, 125*(2), 1357–1362.
177. Rani, K., & Paliwal, S., (2014). A review on targeted drug delivery: Its entire focus on advanced therapeutics and diagnostics. *Sch. J. App. Med. Sci., 2*(1C), 328–331.
178. Rao, J. P., & Geckeler, K. E., (2011). Polymer nanoparticles: Preparation techniques and size-control parameters. *Progress in Polymer Science, 36*(7), 887–913.
179. Reverchon, E., & Adami, R., (2006). Nanomaterials and supercritical fluids. *The Journal of Supercritical Fluids, 37*(1), 1–22.
180. Rolland, J. P., Maynor, B. W., Euliss, L. E., Exner, A. E., Denison, G. M., & DeSimone, J. M., (2005). Direct fabrication and harvesting of monodisperse, shape-specific nanobiomaterials. *Journal of the American Chemical Society, 127*(28), 10096–10100.
181. Ruckmani, K., Sivakumar, M., & Ganeshkumar, P. A., (2006). Methotrexate loaded solid lipid nanoparticles (SLN) for effective treatment of carcinoma. *Journal of Nanoscience and Nanotechnology, 6*(9/10), 2991–2995.
182. Rusin, A., Krawczyk, Z., Grynkiewicz, G., Gogler, A., Zawisza-Puchalka, J., & Szeja, W., (2010). Synthetic derivatives of genistein, their properties and possible applications. *Acta Biochimica Polonica, 57*(1), 23–34.
183. Salcher, A., Nikolic, M. S., Casado, S., Velez, M., Weller, H., & Juarez, B. H., (2010). CdSe/CdS nanoparticles immobilized on pNIPAm-based microspheres. *Journal of Materials Chemistry, 20*(7), 1367–1374.
184. Samanta, S. K., Pal, A., Bhattacharya, S., & Rao, C. N. R., (2010). Carbon nanotube reinforced supramolecular gels with electrically conducting, viscoelastic and near-infrared sensitive properties. *Journal of Materials Chemistry, 20*(33), 6881–6890.
185. Sambale, F., Wagner, S., Stahl, F., Khaydarov, R. R., Scheper, T., & Bahnemann, D., (2015). Investigations of the toxic effect of silver nanoparticles on mammalian cell lines. *Journal of Nanomaterials*, 9–17.
186. Samori, C., Ali-Boucetta, H., Sainz, R., Guo, C., Toma, F. M., Fabbro, C., Da Ros, T., Prato, M., Kostarelos, K., & Bianco, A., (2010). Enhanced anticancer activity of multi-walled carbon nanotube-methotrexate conjugates using cleavable linkers. *Chemical Communications (Cambridge, England), 46*(9), 1494–1496.
187. Sanna, V., Gavini, E., Cossu, M., Rassu, G., & Giunchedi, P., (2007). Solid lipid nanoparticles (SLN) as carriers for the topical delivery of econazole nitrate: *In-vitro* characterization, *ex-vivo* and *in-vivo* studies. *Journal of Pharmacy and Pharmacology, 59*(8), 1057–1064.
188. Sarmento, B., Martins, S., Ferreira, D., & Souto, E. B., (2007). Oral insulin delivery by means of solid lipid nanoparticles. *International Journal of Nanomedicine, 2*(4), 743–749.

189. Sayes, C. M., Gobin, A. M., Ausman, K. D., Mendez, J., West, J. L., & Colvin, V. L., (2005). Nano-C60 cytotoxicity is due to lipid peroxidation. *Biomaterials, 26*(36), 7587–7595.
190. Serpe, L., Catalano, M. G., Cavalli, R., Ugazio, E., Bosco, O., Canaparo, R., Muntoni, E., Frairia, R., Gasco, M. R., Eandi, M., & Zara, G. P., (2004). Cytotoxicity of anticancer drugs incorporated in solid lipid nanoparticles on HT-29 colorectal cancer cell line. *European Journal of Pharmaceutics and Biopharmaceutics, 58*(3), 673–680.
191. Sharma, R. A., Euden, S. A., Platton, S. L., Cooke, D. N., Shafayat, A., Hewitt, H. R., et al., (2004). Phase I clinical trial of oral curcumin: Biomarkers of systemic activity and compliance. *Clinical Cancer Research, 10*(20), 6847–6854.
192. Sharma, R. A., McLelland, H. R., Hill, K. A., Ireson, C. R., Euden, S. A., Manson, M. M., Pirmohamed, M., Marnett, L. J., Gescher, A. J., & Steward, W. P., (2001). Pharmacodynamic and pharmacokinetic study of oral Curcuma extract in patients with colorectal cancer. *Clinical Cancer Research, 7*(7), 1894–1900.
193. Sharma, V., Singh, P., Pandey, A. K., & Dhawan, A., (2012). Induction of oxidative stress, DNA damage and apoptosis in mouse liver after sub-acute oral exposure to zinc oxide nanoparticles. *Mutation Research/Genetic Toxicology and Environmental Mutagenesis, 745*(1/2), 84–91.
194. Shvedova, A. A., Castranova, V., Kisin, E. R., Schwegler-Berry, D., Murray, A. R., Gandelsman, V. Z., Maynard, A., & Baron, P., (2003). Exposure to carbon nanotube material: Assessment of nanotube cytotoxicity using human keratinocyte cells. *Journal of Toxicology and Environmental Health. Part A, 66*(20), 1909–1926.
195. Singh, D., (2015). Application of novel drug delivery system in enhancing the therapeutic potential of phytoconstituents. *Asian Journal of Pharmaceutics, 9*(4), 200–210.
196. Singh, P., Kim, Y. J., Zhang, D., & Yang, D. C., (2016). Biological synthesis of nanoparticles from plants and microorganisms. *Trends in Biotechnology, 34*(7), 588–599.
197. Singla, A. K., Garg, A., & Aggarwal, D., (2002). Paclitaxel and its formulations. *International Journal of Pharmaceutics, 235*(1/2), 179–192.
198. Sintubin, L., Verstraete, W., & Boon, N., (2012). Biologically produced nanosilver: Current state and future perspectives. *Biotechnology and Bioengineering, 109*(10), 2422–2436.
199. Soppimath, K. S., Aminabhavi, T. M., Kulkarni, A. R., & Rudzinski, W. E., (2001). Biodegradable polymeric nanoparticles as drug delivery devices. *Journal of Controlled Release, 70*(1/2), 1–20.
200. Souto, E. B., & Muller, R. H., (2005). SLN and NLC for topical delivery of ketoconazole. *Journal of Microencapsulation, 22*(5), 501–510.
201. Souto, E. B., Wissing, S. A., Barbosa, C. M., & Muller, R. H., (2004). Development of a controlled release formulation based on SLN and NLC for topical clotrimazole delivery. *International Journal of Pharmaceutics, 278*(1), 71–77.
202. Srivastava, V., Negi, A. S., Kumar, J. K., Gupta, M. M., & Khanuja, S. P., (2005). Plant-based anticancer molecules: A chemical and biological profile of some important leads. *Bioorganic and Medicinal Chemistry, 13*(21), 5892–5908.
203. Stiti, M., Cecchi, A., Rami, M., Abdaoui, M., Barragan-Montero, V., Scozzafava, A., Guari, Y., Winum, J. Y., & Supuran, C. T., (2008). Carbonic anhydrase inhibitor coated gold nanoparticles selectively inhibit the tumor-associated isoform IX over the cytosolic isozymes I and II. *Journal of the American Chemical Society, 130*(48), 16130–16138.
204. Sun, L., Li, Y., Liu, X., Jin, M., Zhang, L., Du, Z., Guo, C., Huang, P., & Sun, Z., (2011). Cytotoxicity and mitochondrial damage caused by silica nanoparticles. *Toxicology In Vitro, 25*(8), 1619–1629.

205. Suseela, P., Premkumar, K., & Saraswathy, S. D., (*2015*). Formulation, characterization and pharmacokinetic evaluation of naringenin- loaded gastroretentive mucoadhesive polymeric nanosystem for oral drug. *Journal of Drug Delivery and Therapeutics, 5*(2), 107–114.
206. Tagboto, S., & Townson, S., (2001). Antiparasitic properties of medicinal plants and other naturally occurring products. *Advances in Parasitology*, *50*, 199–295.
207. Tamber, H., Johansen, P., Merkle, H. P., & Gander, B., (2005). Formulation aspects of biodegradable polymeric microspheres for antigen delivery. *Advanced Drug Delivery Reviews*, *57*(3), 357–376.
208. Tan, M., Qiu, G., & Ting, Y. P., (2015). Effects of ZnO nanoparticles on wastewater treatment and their removal behavior in a membrane bioreactor. *Bioresource Technology*, *185*, 125–133.
209. Thakur, L., Ghodasra, U., Patel, N., & Dabhi, M., (2011). Novel approaches for stability improvement in natural medicines. *Pharmacognosy Reviews*, *5*(9), 48–54.
210. Tian, Q., Wang, W., He, X., Zhu, X., Huang, W., Zhang, C., Yuan, Z., & Chen, X., (2009). Glycyrrhetinic acid-modified nanoparticles for drug delivery: Preparation and characterization. *Chinese Science Bulletin*, *54*(18), 3121–3126.
211. Tong, W., Wang, L., & D'Souza, M. J., (2003). Evaluation of PLGA microspheres as delivery system for antitumor agent-camptothecin. *Drug Development and Industrial Pharmacy*, *29*(7), 745–756.
212. Toyokuni, S., (2013). Genotoxicity and carcinogenicity risk of carbon nanotubes. *Advanced Drug Delivery Reviews*, *65*(15), 2098–2110.
213. Umehara, K., Nemoto, K., Ohkubo, T., Miyase, T., Degawa, M., & Noguchi, H., (2004). Isolation of a new 15-membered macrocyclic glycolipid lactone, Cuscutic Resinoside a from the seeds of Cuscutachinensis: A stimulator of breast cancer cell proliferation. *Planta Medica*, *70*(4), 299–304.
214. Uner, M., & Yener, G., (2007). Importance of solid lipid nanoparticles (SLN) in various administration routes and future perspectives. *International Journal of Nanomedicine*, *2*(3), 289–300.
215. Varde, N. K., & Pack, D. W., (2004). Microspheres for controlled release drug delivery. *Expert Opinion on Biological Therapy*, *4*(1), 35–51.
216. Vyas, S. P., Sihorkar, V., & Mishra, V., (2000). Controlled and targeted drug delivery strategies towards intraperiodontal pocket diseases. *Journal of Clinical Pharmacy and Therapeutics*, *25*(1), 21–42.
217. Wagner, V., Dullaart, A., Bock, A. K., & Zweck, A., (2006). The emerging nanomedicine landscape. *Nature Biotechnology*, *24*(10), 1211–1217.
218. Wang, B., Ma, L., Tao, X., & Lipsky, P. E., (2004). Triptolide, an active component of the Chinese herbal remedy Tripterygium wilfordii Hook F, inhibits production of nitric oxide by decreasing inducible nitric oxide synthase gene transcription. *Arthritis and Rheumatism*, *50*(9), 2995–2303.
219. Warheit, D. B., Laurence, B. R., Reed, K. L., Roach, D. H., Reynolds, G. A., & Webb, T. R., (2004). Comparative pulmonary toxicity assessment of single-wall carbon nanotubes in rats. *Toxicological Sciences*, *77*(1), 117–125.
220. Wiener, E. C., Konda, S., Shadron, A., Brechbiel, M., & Gansow, O., (1997). Targeting dendrimer-chelates to tumors and tumor cells expressing the high-affinity folate receptor. *Investigative Radiology*, *32*(12), 748–754.
221. Wilczewska, A. Z., Niemirowicz, K., Markiewicz, K. H., & Car, H., (2012). Nanoparticles as drug delivery systems. *Pharmacological Reports*, *64*(5), 1020–1037.

222. Wong, H. L., Rauth, A. M., Bendayan, R., Manias, J. L., Ramaswamy, M., Liu, Z., Erhan, S. Z., & Wu, X. Y., (2006). A new polymer-lipid hybrid nanoparticle system increases cytotoxicity of doxorubicin against multidrug-resistant human breast cancer cells. *Pharmaceutical Research, 23*(7), 1574–1585.
223. Wong, M. S., Sun, D. S., & Chang, H. H., (2010). Bactericidal performance of visible-light responsive titania photocatalyst with silver nanostructures. *PloS One, 5*(4), e10394.
224. Woo, J. S., Kim, T. S., Park, J. H., & Chi, S. C., (2007). Formulation and biopharmaceutical evaluation of silymarin using SMEDDS. *Archives of Pharmacal Research, 30*(1), 82–89.
225. Xie, Y., He, Y., Irwin, P. L., Jin, T., & Shi, X., (2011). Antibacterial activity and mechanism of action of zinc oxide nanoparticles against *Campylobacter jejuni. Applied and Environmental Microbiology, 77*(7), 2325–2331.
226. Xu, H., Hou, Z., Zhang, H., Kong, H., Li, X., Wang, H., & Xie, W., (2014). An efficient Trojan delivery of tetrandrine by poly(N-vinylpyrrolidone)-block-poly(ε-caprolactone) (PVP-b-PCL) nanoparticles shows enhanced apoptotic induction of lung cancer cells and inhibition of its migration and invasion. *International Journal of Nanomedicine, 9*, 231–242.
227. Xu, X., & Burgess, D. J., (2012). Liposomes as carriers for controlled drug delivery. In: Wright, J. C., & Burgess, D. J., (ed.), *Long Acting Injections and Implants* (pp. 195–220). Springer US, Boston, MA.
228. Yadav, D., Suri, S., Choudhary, A., Sikender, M., Hemant, B. N., & Beg, N., (2011). Novel approach: Herbal remedies and natural products in pharmaceutical science as nano drug delivery systems. *International Journal of Pharm. Tech. Research, 3*(3), 3092–3116.
229. Yang, J., Han, C. R., Duan, J. F., Xu, F., & Sun, R. C., (2013). In situ grafting silica nanoparticles reinforced nanocomposite hydrogels. *Nanoscale, 5*(22), 10858–10863.
230. Yang, Q., & Ma, Y., (2014). Irradiation enhanced cytotoxicity of Zinc Oxide nanoparticles. *International Journal of Toxicology, 33*(3), 187–203.
231. Ye, Y., Liu, J., Xu, J., Sun, L., Chen, M., & Lan, M., (2010). Nano-$SiO_2$ induces apoptosis via activation of p53 and Bax mediated by oxidative stress in human hepatic cell line. *Toxicology In Vitro, 24*(3), 751–758.
232. Yen, F. L., Wu, T. H., Lin, L. T., Cham, T. M., & Lin, C. C., (2009). Naringenin-loaded nanoparticles improve the physicochemical properties and the hepatoprotective effects of naringenin in orally-administered rats with CCl(4)-induced acute liver failure. *Pharmaceutical Research, 26*(4), 893–902.
233. Zhang, H., Patel, A., Gaharwar, A. K., Mihaila, S. M., Iviglia, G., Mukundan, S., Bae, H., Yang, H., & Khademhosseini, A., (2013). Hyperbranched polyester hydrogels with controlled drug release and cell adhesion properties. *Biomacromolecules, 14*(5), 1299–1310.
234. Zhigaltsev, I. V., Maurer, N., Akhong, Q. F., Leone, R., Leng, E., Wang, J., Semple, S. C., & Cullis, P. R., (2005). Liposome-encapsulated vincristine, vinblastine and vinorelbine: A comparative study of drug loading and retention. *Journal of Controlled Release, 104*(1), 103–111.
235. Zhong, S., & Yung, L. Y. L., (2009). Enhanced biological stability of collagen with incorporation of PAMAM dendrimer. *Journal of Biomedical Materials Research Part A, 91*(1), 114–122.
236. Zhuo, R. X., Du, B., & Lu, Z. R., (1999). *In vitro* release of 5-fluorouracil with cyclic core dendritic polymer. *Journal of Controlled Release, 57*(3), 249–257.

# INDEX

**B**

## D



## E

## F

## T



## U



## V